An Introduction to Astronomical Photometry is a textbook on optical astronomical photometry intended for university students, research starters and advanced amateurs. It generally avoids delving directly into highly specialized or technical material without some preparation. Starting with an overview, the book moves on through the historical background, glossary of terms and underlying physical principles. From this base the current range of astronomical photometry is surveyed. The central two chapters deal with principles of photometer design, and data processing techniques. The remainder of the book discusses curve fitting techniques for various kinds of variable stars' photometry, including close binary stars, spotted and pulsating stars.

This book encourages interest in the practicalities of observational astronomy and the understanding of stellar astrophysics from a data-based perspective. It is primarily aimed at that gap between a popularist approach and a full-scale research level monograph.

AN INTRODUCTION TO ASTRONOMICAL PHOTOMETRY

AN INTRODUCTION TO ASTRONOMICAL PHOTOMETRY

EDWIN BUDDING
Carter Observatory, Wellington, New Zealand

Published by the Press Syndicate of the University of Cambridge
The Pitt Building, Trumpington Street, Cambridge CB2 1RP
40 West 20th Street, New York, NY 10011-4211, USA
10 Stamford Road, Oakleigh, Melbourne 3166, Australia

First published 1993

Printed in Great Britain at the University Press, Cambridge

A catalogue record for this book is available from the British Library

Library of Congress cataloguing in publication data

Budding, E., 1943–
Introduction to astronomical photometry/Edwin Budding.
p. cm.
ISBN 0 521 41867 4
1. Astronomical photometry. 2. Astrometry. I. Title
QB135.B78 1993
522'.62–dc20 92-33321 CIP

ISBN 0 521 41867 4 hardback

Contents

	Preface	*page* xi
1	**Overview**	1
1.1	Scope of the subject	1
1.2	Requirements	2
1.3	Participants	3
1.4	Targets	5
1.5	Bibliographical notes	7
2	**Introduction**	9
2.1	Optical photometry	9
2.2	Historical notes	12
2.3	Some basic terminology	22
2.4	Radiation — waves and photons	26
2.5	Bibliographical notes	27
3	**Underlying Essentials**	30
3.1	Radiation field concepts	30
3.2	Black body radiation	33
3.3	The Sun seen as a star	34
3.4	The bolometric correction	38
3.5	Absolute stellar fluxes and temperatures	40
3.6	Broadband filters — essential points	45
3.7	Surface flux and colour correlations	51
3.8	Bibliographical notes	53
4	**Themes of Astronomical Photometry**	56
4.1	Extinction	56
4.1.1	Atmospheric extinction	56
4.1.2	The atmospheric mass	58
4.1.3	Interstellar extinction	61
4.2	Broadband filters — Data and requirements	64

4.3 Photometry at intermediate bandwidths 73
4.3.1 The *uvby* system 74
4.4 Narrowband photometry 80
4.5 Photometry of extended objects 84
4.5.1 Visual photometry 85
4.5.2 Solar system photometry 87
4.5.3 Photometry of galactic nebulae 92
4.5.4 Photometry of galaxies 95
4.6 Photopolarimetry 98
4.7 Bibliographical notes 105
5 Practicalities 109
5.1 Overview of basic instrumentation 109
5.2 Detectors 116
5.2.1 Detective processes 116
5.2.2 Detector characteristics 120
5.2.3 Photomultiplier tubes 121
5.2.4 Noise in single-channel photometry 123
5.2.5 Areal detectors and enhancers 126
5.2.6 Charge coupled devices (CCDs) 129
5.3 Conventional measurement methods 132
5.3.1 The DC method 132
5.3.2 Pulse counting 135
5.4 Bibliographical notes 140
6 Procedures 142
6.1 The standard stars experiment 142
6.1.1 Parameter determinacy and set size — a simple case 158
6.2 Differential photometry 159
6.2.1 Typical photometric comparison: an example 166
6.3 Light curves of variable stars 168
6.4 Bibliographical notes 170
7 Basic Light Curve Analysis 172
7.1 Light curve analysis — general outline 172
7.2 Eclipsing binaries — basic facts 174
7.3 Hand solution of light curves 180
7.4 Computer-based light curve analysis 186
7.5 Bibliographical notes 196

8 Close Binary Systems 199
8.1 Orbital eccentricity 199
8.2 Proximity effects 201
8.3 A 16-parameter curve fitter 205
8.4 Frequency domain analysis 210
8.5 Narrowband photometry of binaries 212
8.6 Bibliographical notes 218
9 Spotted Stars 220
9.1 Introductory background 220
9.2 The photometric effects of starspots 222
9.3 Application to observations 225
9.4 Starspots in binary systems 231
9.5 Analysis of the light curves of RS CVn stars 232
9.6 Bibliographical notes 237
10 Pulsating Stars 239
10.1 Introductory background 239
10.2 The Baade–Wesselink procedure 245
10.3 Six-colour data on classical cepheids 249
10.3.1 Application to δ Cep and η Aql 251
10.4 Pulsational radii 258
10.5 Bibliographical notes 260
Appendix 262
Subject index 264
Object index 271

Preface

The book which follows has grown out of my experiences in carrying out and teaching optical astronomy. Much of the practical side of this started for me when I was working with Professor M. Kitamura at what is now the National Astronomical Observatory of Japan, Mitaka, Tokyo in the mid-seventies. Having already learned something of the theoretical side of photometric data analysis and interpretation from Professor Z. Kopal in the Astronomy Department of the University of Manchester, when I later returned to that department and was asked to help with its teaching programme I started the notes which have ultimately formed at least part of the present text. I then had the pleasure of continuing with observing at the Kottamia Observatory, beneath the beautiful desert skies of Egypt, in the days of Professor A. Asaad, together with a number of good students, many of whom have since gone on to help found or join university departments of their own in different lands of the world.

In recent years — particularly since moving to Carter Observatory — another dimension has been added to my experience through my encounters with that special feature of the astronomical world: the active amateur! In previous centuries many creative scientists were, in some sense, amateurs, but in the twentieth century the tide, for fundamental research at least, has been very much in the direction of government, or other large organization, supported professionals, no doubt with very persuasive reasons.

Nevertheless, some features of contemporary life suggest that this tide is not necessarily conclusive in its effects. If there is one feature in particular, I would cite the personal computer. The range of possibilities for participation and active investigation which are now available to individuals on their home desk-tops is already staggeringly large, and continues to increase, while the real costs of sophisticated electronics fall and demand grows as more people discover these potentialities for themselves.

A particular concept, which may become increasingly significant in the future development of astronomy, is that of the 'PC-observatory'. Much of the more routine side of observational data collection can be put under the control of a personal computer. Automatic photometric telescopes (APTs), of up to half-meter aperture class, have been developed and operated by amateurs in their backyards. Data can be gathered by the tended robot, while the human designer has the freedom to ponder and relax in the way that humans are wont. I have seen this in action right here in Wellington, but do not doubt that at least similar capabilities exist in very many other places.

In the early eighties I started a correspondance with Professor M. Zeilik of the University of New Mexico, who shared my interest in the photometry and analysis of eclipsing binary systems. This later developed into exchange visits, and in the environment of Dr Zeilik's active research and education programme, at Albuquerque and Capilla Peak, I began to appreciate more fully the momentum of the electronics revolution and its impact on optical astronomy and the propagation of information.

Enthusiasm and capabilities are thus already nascent in good measure, and against this background the appearance of a book with entry-point information, guidelines on equipment and methodology, astronomical purposes — general and specific, leading, it is hoped, towards definite new contributions in the field, seems opportune.

Introduction to Astronomical Photometry is then a textbook on astronomical photometry (essentially in the optical domain) intended for university students, research starters, advanced amateurs or others with this special interest. It avoids jumping directly into technical or formally presented information without some preparation. Each chapter is rounded off with a section of bibliographical notes. The book starts with an overview, and moves on through a historical background and glossary of terms. Then comes a chapter on the underlying physical principles of radiative flux measurement. Colour determinations, and temperature and luminosity relationships are also examined here. From this base more wide-ranging questions in current astronomical photometry are approached. The central two chapters deal with principles of photometer design, including recent advances, and some common data handling techniques for system calibration from standard star observation and the generation of light curves. The remainder of the book presents applications of photometry to selected topics of stellar astrophysics. Curve fitting techniques for various kinds of light curve from variable stars, including close binary stars, spotted and pulsating stars, are followed through. Inferences drawn from such investigations are then advanced.

There are a large number of people to whom I feel thankful for helping this book to be realized. Some of them I have mentioned already, but even if I didn't, I am sure the formative influence of Zdeněk Kopal would soon become clear to readers of the subsequent pages. Indeed, many of them were written whilst I shared his welcoming office during my sabbatical leave of 1990. Professor F. D. Kahn was principal host during my stay in Manchester, and his hospitality and that of his department helped make that year very special for me.

That period of leave, which gave me the time to collect things together, was essentially enabled through the generous support of the Carter Observatory Board, and approved by its Director, Dr R. J. Dodd, who also helped with remarks on part of the text. Useful comments were also provided by Dr J. Dyson (Manchester), Dr J. Hearnshaw (Christchurch) and Mr J. Priestley (Carter Observatory). Interest and encouragement were expressed by Dr M. Zeilik, and his colleagues and students at UNM, Albuquerque, with whom my leave started in 1990, by Drs B. Szeidl and K. Oláh during my August sojourn at the Konkoly Observatory (Budapest), and as well by Drs M. de Groot and C. J. Butler of the Armagh Observatory, where I similarly visited later that year.

Among the many others who I would like to acknowledge, though space unfortunately restricts, Mr T. Hewitt of the Computer Centre at Manchester University, who introduced me to the wonderful world of PCTEX, surely deserves mention. He helped this text materialize in a very real sense. I also thank John Rowcroft and Carolyn Hume for help with the diagrams.

Last, but not least, to my family and wife Patricia — thanks.

Wellington, New Zealand *Edwin Budding*
May, 1992.

1
Overview

1.1 Scope of the subject

In the following chapters groundwork will be laid for the methods and purposes of astronomical photometry. The outreach of the subject will be seen to be large. In the historical aspect, for example, we retain contact with the earliest known systematic cataloguer of the sky — Hipparchos — the 'father of astronomy', for his magnitude units are still in use, though admittedly in a much refined form. A special interest attaches to this very long time baseline, and a worthy challenge exists in getting a clearer view of early records and procedures.

Photometry has points of contact, or merges into, other fields of observational astronomy, though different words are used to demarcate particular specialities. Radio-astronomy, X-ray-astronomy, and so on, often concern measurement and comparison procedures which parallel the historically well-known optical domain. Spectrophotometry, as another instance, extends and particularizes information about the distribution of radiated energy with wavelength, involving additional knowledge and techniques beyond the primary concerns of this text. Astrometry and stellar photometry form limiting cases of the photometry of extended objects. Since stars are, for the most part, below instrumental resolution, a sharp separation is made between positional and radiative flux data. But this distinction can seem artifical on close examination. Thus, the most accurate positional surveys on stars refer to image dissection techniques. The blurry mobile spot of light that an Earth-based telescope forms as a star image is microscopically sampled, and its flux distribution analysed, allowing statistical procedures to fix the position of the light centroid.

But if photometry merges into more specialized fields at one side, it remains connected to simple origins at another. This has been a feature of the continuous overall growth of the subject over the last few centuries. Thus, when Fabricius noticed the variability of Mira in 1596 it was the beginning of the study of long period variable stars. Several thousand Miras are now known, each with their own peculiar vagaries of period and amplitude. The Miras are themselves just one group among a score of different kinds of variable star. Who keeps track of this vast and developing body of data?

The answer to this question reveals the special role in astronomical science for the amateur, particularly when his or her efforts are organized and collated. The human eye still plays a key part, especially in those dramatic initial moments of discovery, whether it be of a new supernova, an 'outburst' of a cataclysmic variable, or a sudden drop of a star of the R Coronæ Borealis type. An effort is made in this book to retain contact with this basic type of support; in referring quantities to eye-based units, for example, or in dealing with data that would be well within reach of small observatories, clubs or societies, or even well-endowed individuals.

On the other hand, a scientific discipline gives active motivation to serious effort, so long as frontier areas can be identified within its ambit. The final four chapters address themselves to areas of variable star research where techniques are still being developed, and answers are still unresolved. Naturally, as one approaches these areas, the field of concentration becomes more restricted. These chapters expose this process, starting from fairly mainstream topics in stellar photometry. More technical literature exists in abundance which is further on in this development.

1.2 Requirements

The remarkable spread of personal computers over recent years opens up all sorts of interesting activities, of which the control of astronomical equipment, the logging and processing of observational data, and the fitting of adequate physical model predictions are just a few — but a special few from our present point of view. High-quality optical telescopes are also increasingly available at competitive prices. Modern technology is thus placing within reach of a large number of potential enthusiasts means of dealing with observation and analysis, which would have been frontline less than a generation ago, and for the reasons indicated in the preceding section are additive to the overall course of astronomical science.

This point can be made more quantitatively. Detailed considerations will be presented in later chapters, but one of the most important specifiers is the ratio of signal to noise (S/N) — the measure of information of interest compared with irrelevant disturbances of the measurement. 'Good' measurements are associated with S/N values of 100 or over. This quality of measurement can be attained in stellar photometry for a large number of stars with relatively modest sized telescopes. Consider, for example, the few hundred thousand stars included in famous great catalogues, such as the *Henry Draper Catalogue* or the *Bonner Durchmusterung*. Basic optical photometry of such stars is possible at $S/N \gtrsim 100$, in good weather conditions at a dark sky observatory with a 'small' 10-inch aperture telescope.

Generally speaking, differential photometry of variable stars in order to produce 'light curves' which stimulate good attempts at modelling looks persuasive at $S/N \sim 100$, though this is a rather crude overall guide. There are stars whose entire variation is only of order several hundedths of a magnitude: clearly such stars require the utmost in acheivable accuracy. On the other hand eye-based estimation of stellar brightness is usually thought to be doing very well at 10% accuracy. There are many variables of large amplitude where data of this accuracy are adequate, particularly when coverage is extensive, so that observations can be averaged.

1.3 Participants

The foregoing indicates several levels of potential support to astronomical photometry. Eye-based data from skilled observers continues to have a significant place, especially with certain kinds of irregular or peculiar variable star, and appears likely to do so for the foreseeable future. Many of these observers are working with telescopes of the 10-inch class.

When a person or group with the skills and resources to combine current microcomputer capabilities with a telescope of this size, a photoelectric photometer, and sufficient awareness of photometric procedures, an order of magnitude or more of detail is added to to the information content of data obtained in a given spell of observing (cf. W. H. Allen's data on SN 1987A, Figure 1.1). New organizations have sprung up over the last few decades to support the growth in value of such work — for example, the International Amateur-Professional Photoelectric Photometry (IAPPP) association, centred at the Dyer Observatory of Vanderbilt University, Nashville, Tennessee. Relatively small and low cost, highly automated photometric telescopes (APTs), which have recently appeared in this context, offer very interesting avenues for future development of the subject.

The two components — microcomputer and telescope/photometer — can be separated. Apart from instrument and data management, a computer is also directed to data archiving and analysis. It is in this latter area where one new thrust in the present text lies. The analysis of data provides the essential

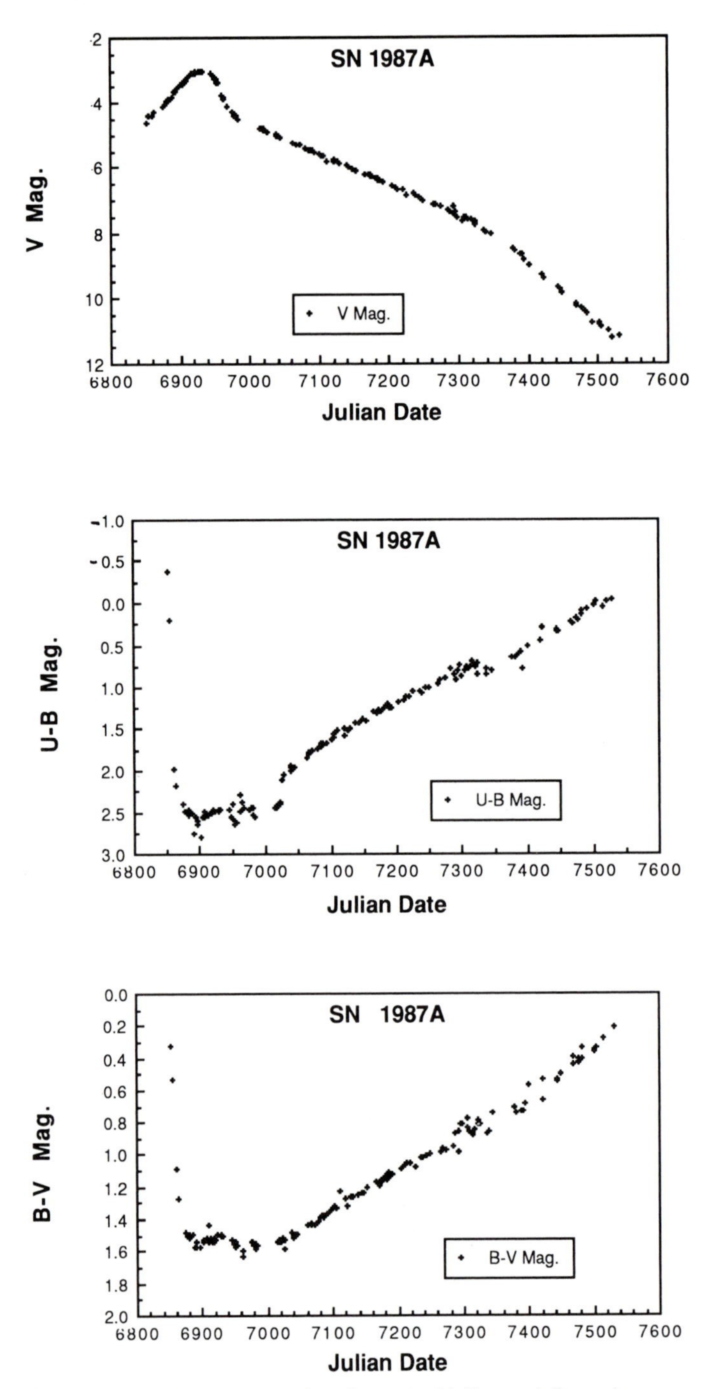

Fig. 1.1. Light (top) and colour (middle and lower) curves of supernova 1987a produced by Mr W. H. Allen of the Adams Lane Observatory, Blenheim, New Zealand. Abscissae give the Julian Date of each observation (+2440000).

link between observational production and theoretical interpretation, which appear like two halves of a net driving cycle. Of course, either side is in a continual process of growth and development, but it is hoped that this book will be helpful to students and enthusiasts of the field, interested in catching hold of procedures, somewhere near their beginning.

1.4 Targets

Astronomical photometry leads into a much wider range of topics than we have space for. There are some specific issues arising in this book, however, which can be briefly overviewed here.

Concerning broadband light curves of close binary systems, for example, — by the end of Chapter 8 we progress to a sixteen-parameter program which can describe the major features of a standard close binary model (including orbital eccentricity), where the components may be well distorted by their mutual proximity. But many light curves are more complicated than this. They show peculiar asymmetries, which may be anticipated from separate evidence, but for which the standard model is inadequate. For instance, the very close binary CQ Cep — the hot, massive Wolf–Rayet type star of the pair gives firm spectral evidence of a strong flux of matter from the surface in a very enhanced 'stellar wind'. This must entail high-energy interactions with its companion. The problems raised by its, and comparable, light curves surely call for more development of appropriate physical models.

Very close and strongly interacting binaries raise the model *adequacy* issue, which arises from time to time as the text proceeds. Unfortunately, as the physical situation becomes more complex, light curves alone do not necessarily match this. Their form may in fact become more simple — less determinate. CQ Cep (Figure 1.2) shows broadband light curves which are without the informative sharp corners of classical eclipsing binary light curves, resembling only slightly distorted sine curves. From an empirical viewpoint, such light curves are given, in principle, by only a small number of well-defined parameters. They are not very informative; alternatively, when considered in isolation, they permit a wide range of possible explanations. One way to proceed is to combine many different spectral or time-distributed data, and then seek a coherent underlying model.

Narrowband photometry of the relatively mildly interacting binaries U Cep and U Sge, presented in Chapter 8, illustrates such a process, albeit in rather a straightforward progression. From broadband light curves we infer that these stars are 'semidetached', i.e. the less massive components are filling their surrounding 'Roche' lobes of limiting dynamical stability, indeed overflowing

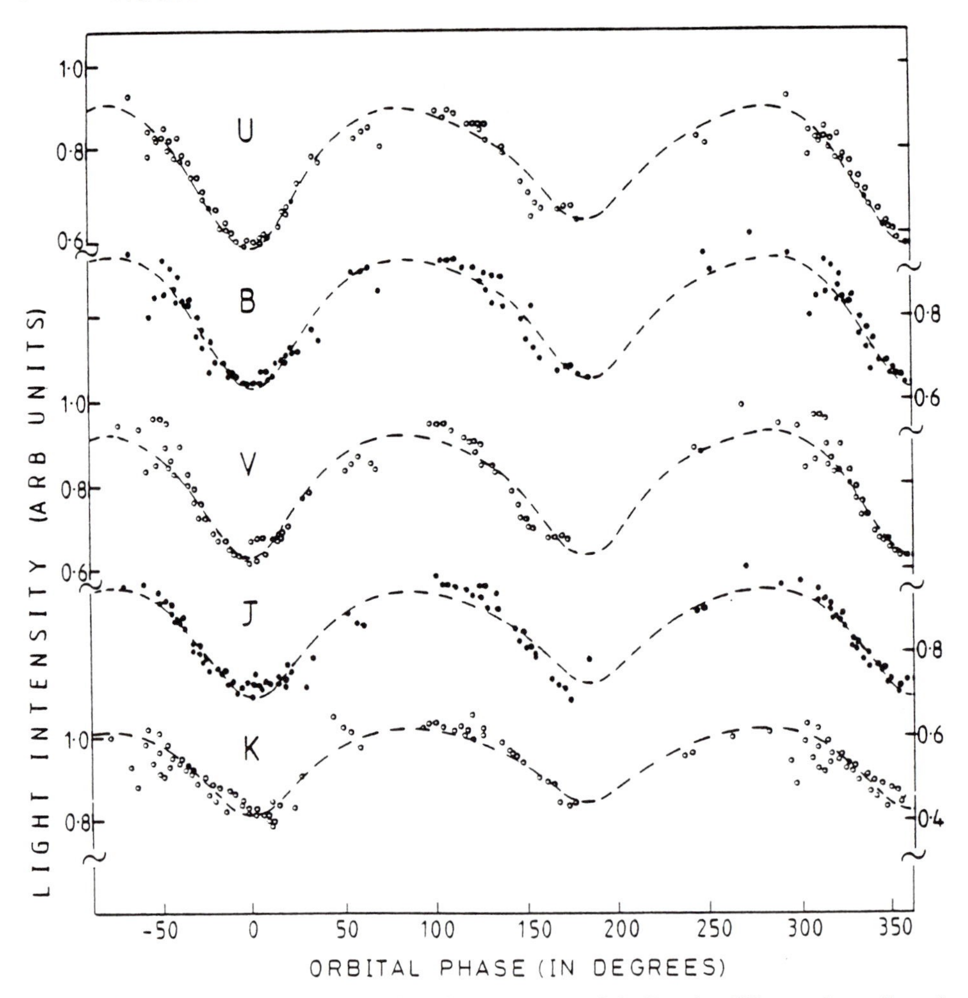

Fig. 1.2. Light curves of the binary system CQ Cep in different broadband wavelength ranges.

them, according to standard ideas on interactive binary evolution. Basic geometric parameters are derived from curve fits to such photometry. We can then approach corresponding narrowband light curves with some of the key quantities already known. In a more general situation, one seeks a *simultaneous* explanation of concurrent data sets, with information feedbacks in various directions from one curve-fitting to another.

Something like this happens in the successive approximations analysis we carry out for the spotted RS CVn type stars. In Chapter 9, great increases in observational surveillance of these "extensions to the solar laboratory" are anticipated, with the exploitation of automated photometric telescopes. But the fitting of the wave distortions in these systems is notoriously imprecise. Basically, we face a stringent information limit if we rely only on broadband photometry. Either we admit to a frustrating smallness of derivable parameter sets, or give in to the temptation to advance plausible models which can match the data well, but actually specify more information than it really contains. Again the answer will be to combine as many data as possible, spectroscopic as well as photometric, to recover a unified picture. Increased use of new techniques, such as 'Zeeman Imaging', may provide valuable steps forward in this context.

The Baade–Wesselink technique, outlined in Chapter 10, is another area where the temptation to derive and utilize numerical parameters may exceed proper caution. Even so, the suggested dangers are perhaps not that serious. Those inferences in the method which are prone to unreliability are well known, and continue to be investigated to find firmer versions. Fortunately, there are also quite independent means of testing the overall reliability of Baade–Wesselink results.

Whether or not present techniques will remain useful, we can find in them viable approaches to a fuller appreciation of the meaning of photometric data.

1.5 Bibliographical notes

Among the more general texts which have appeared recently, C. and M. Jascheks' *Classification of the Stars*, Cambridge University Press, Cambridge, 1989, reviews the basic role which photometry has played in developing our understanding for stars of all types. That book, in turn, cites M. Golay's *Introduction to Astronomical Photometry*, Reidel, 1975 and V. Straizys' *Multicolour Stellar Photometry: Photometric Systems and Methods*, Mokslas Publ., Vilnius, Latvia, 1977 (in Russian) as seminal works on astronomical photometric science. Although the present text aims at a reasonably complete introduction, occasional reference is made to such comprehensive backgrounders.

In keeping with the aim of a broad outreach indicated in Section 1.1, we would refer to the continuous communications organized by observers' societies and groups in many countries, such as the British Astronomical Association (Variable Star Section), the American Association of Variable Star Observers, the Association Française des Observateurs d'Etoiles Variables, the Variable Star Observers League in Japan, the German Arbeitskreise für veränderliche Sterne, the Variable Stars and Photometric Sections of the Royal Astronomical Society of New Zealand, and their various equivalents.

The 75th Anniversary Edition of the Journal of the AAVSO (1986) contained over 40 papers representing various fields of professional plus amateur interest, while thirty papers appeared in the Third New Zealand Conference on Photoelectric Photometry (ed. E. Budding and J. Richard), *South. Stars*, **34**, No 3, 1991.

Specific short contributions from participants in such organizations, and others, also appear in the *Information Bulletins on Variable Stars* issued by Commission 27 of the International Astronomical Union, and produced at the Konkoly Observatory of Budapest. The International Amateur-Professional Photoelectric Photometry organization addresses itself across national boundaries (T. D. Oswalt, D. S. Hall and R. C. Reisenweber, *I.A.P.P.P. Commun.* No. 42, 1, 1990), while *Peremenniye Zvezdy* records comparable activities in the Russian language. A useful set of papers relevant to this context also appeared in *The Study of Variable Stars Using Small Telescopes* ed. J. R. Percy, Cambridge University Press, Cambridge, 1986.

Quantitative data on S/N values for real photometers appear in the Optec (tradename) Manual, as well towards the end of A. A. Henden and R. H. Kaitchuck's *Astronomical Photometry*, Van Nostrand Reinhold Company Inc., New York, 1982, where a full explanation of the underlying principles is given.

Part of W. H. Allen's light curve, shown as Figure 1.1, appears at the end of the fifth edition of M. Zeilik's *Astronomy: The Evolving Universe*, John Wiley, New York, 1988, in a brief observational review of SN 1987a. Figure 1.2 appeared in D. Stickland *et al., Astron. Astrophys.,* **134**, 45, 1984.

2

Introduction

2.1 Optical photometry

Astronomical photometry is about the measurement of the brightness of radiating objects in the sky. We will deal mainly with optical photometry, which centres around a region of the electromagnetic spectrum to which the human eye (Figure 2.1) is sensitive. Indeed, photometric science, as it concerns stars, has developed out of a history of effort, the greatest proportion of which, over time at least, has amounted to direct visual scanning and comparison of the brightness of stellar images. In this context, brightness derives from an integrated product of the eye's response and the energy distribution as it arrives from the celestial source to reach the observer. Still today there is a large amount of monitoring of the many known variable stars carried out (largely by amateurs) in this way.

With the passage of time, however, there has been a general trend towards more objective methods of measurement. The use of photometers with a non-human detector element has become increasingly widespread, though the term optical remains to denote the relevant spectral range (Figure 2.2), which significantly coincides with an important atmospheric 'window' through which external radiation can easily pass. This is presumably connected with biological evolution — in fact, the maximum sensitivity of the human eye is at a wavelength close to the maximum in the energy versus wavelength distribution of the Sun's output (~5000 Å).[†] Instrumental contexts have extended the usage of optical down to ~3000 Å at the ultra-violet end of the spectrum, and ~10000 Å at the infra-red end.

Closely connected to brightness is colour. More formal statements about these terms will be made later, but colour broadly measures the difference

[†] The ångstrom unit (10^{-10} m) is frequently used in contexts where broad historical continuity is convenient.

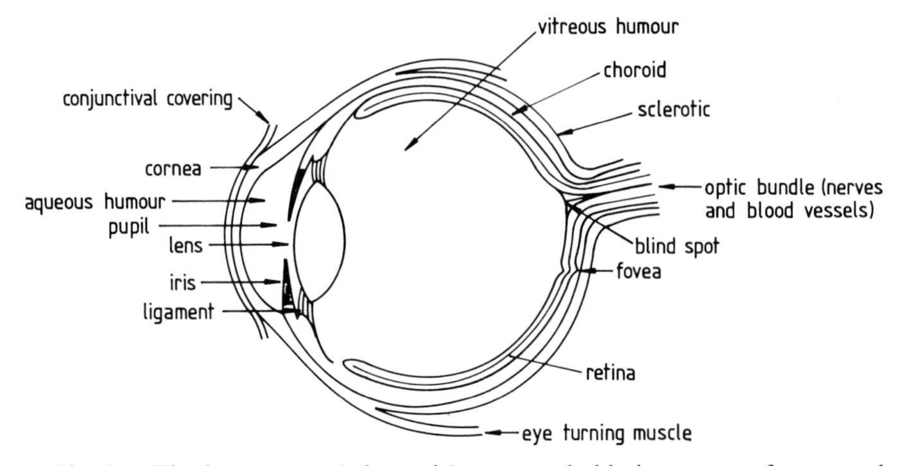

Fig. 2.1. The human eye (schematic): a remarkable instrument for general photometry with a very wide dynamic range.

in brightness of an object observed at two specified wavelength regions of observation. This is scaled in adopted units, and descriptions such as reddish, blue and so on, assigned values within the scheme. The term colour index is often used for celestial sources, and the determination of such indices forms a basic objective of the subject.

More detailed than colour determination is the specifying of relative intensity level at each wavelength over a range of the spectrum, or spectrophotometry. This would usually be distinguished from conventional photometry, because of its somewhat different background of development in instrumentation. A key point is that photometry is generally done with the simple interposition of coloured glass, or glass covered, filters in the optical path to the light detector. With a filter of varying transmission along a given direction, which can be moved and scanned during operation, it is possible to approach spectrophotometry with a similar set-up to conventional photometry, though uniform fixed-bandwidth filters are much more normal. More elaborate dispersing instruments would be employed for detailed spectrophotometry.

In pursuing the objective of fundamental determination, the framework of mean relative intensities of known light sources — stellar, i.e. point-like; and extended, as clusterings or nebulae — is evaluated and checked. Feedback from the data on such standard sources then allows calibration of any particular observer's measurement system. Alternatively, the reference framework may be extended to other objects whose brightnesses may be regarded as constant or predictable.

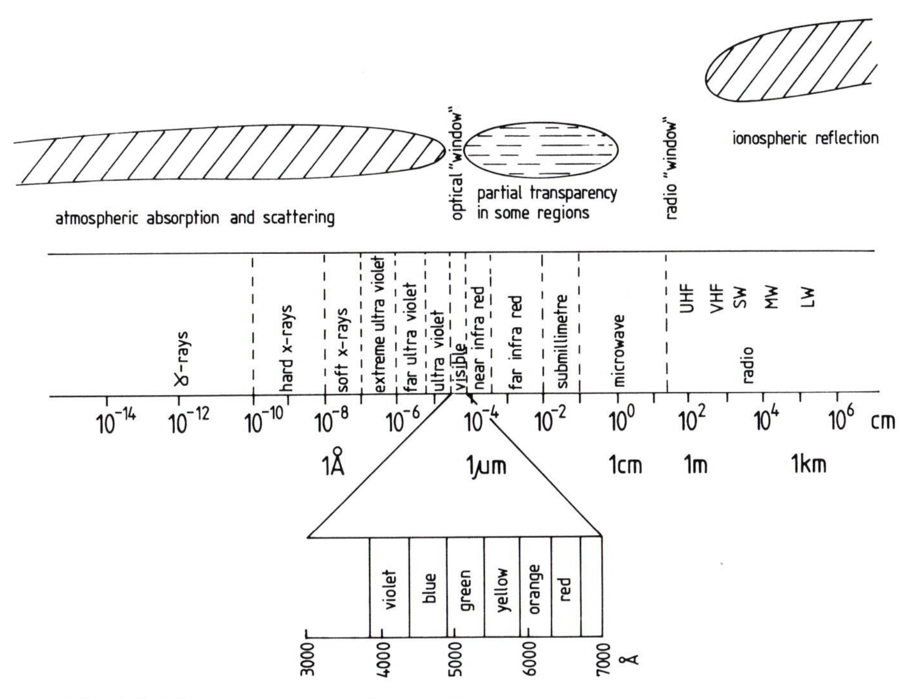

Fig. 2.2. The electromagnetic spectrum.

In comparison to most other fields of scientific measurement stellar photometry is relatively imprecise. The best that the ancients could do by naked eye methods, for instance, was to judge brightness to within about a third of the time-honoured magnitude classes: on a linear scale this means guessing the energy output of a star to within anything from about 75% to 135% of its real value. This compares with an accuracy of one part in a thousand, or better, to which they could fix position in the sky in angular coordinates. Modern photoelectric methods give individual measurements of brightness to better than 1%, but the absolute accuracy to which the entire system can be quantified is still not appreciably better than 1%. Positional determinations, meanwhile, are now attained with an accuracy of 1 in 10,000,000 or better. There are, and always have been, therefore, several powers of ten in the ratio of accuracies of specification of stellar position compared with brightness. Nevertheless, the importance of knowing the rate at which energy is radiated from the outer layers of a star for understanding stellar constitution became clear as soon as people began to develop physical ideas on this.

The calibrative side of photometry is sometimes seen as playing a supporting role to a more attention-catching activity connected with variable stars.

With variables there are again different purposes, though. Some information retrieval and processing is to check, or further refine our knowledge of the variability — such as with 'classical' variables, e.g. cepheids, or normal eclipsing binary systems. On the other hand, certain observations are made with some hope or expectation of discovery — either by finding variability in a hitherto unsuspected object, or by covering seemingly unpredictable types of irregular photometric behaviour, such as occurs, for instance, with flare stars, cataclysmic variables, BL Lacerate type objects and such-like.

Astronomical photometry was originally largely concerned with the relative brightnesses of stars, and the magnitude system in which these are expressed. Stellar surfaces are below the eye's limit of resolution, even through the world's largest telescopes, and are thus 'point-like'. Some of the most familiar astronomical objects — the Sun, Moon, the sky itself, the Milky Way — are extended, however. Their brightnesses can be expressed, in traditional units, as 'n stars of magnitude m per unit square angular measure', with the meaning that the light received from e.g. an area of surface subtending 1 arcmin by 1 arcmin (about the limit of resolution of a typical human eye) would be the same as if coming from n of mth magnitude stars (frequently $m = 10$ is used for this). There are many extended sources, but usually with relatively faint surface brightnesses. With improved linear areal detectors and more large telescopes there have been impressive developments in detailed surface photometry of faint nebulae and galaxies in recent years.

Photometry has thus a range of important roles to play in astrophysics. By providing basic reference data on stellar brightness and colour, such as via the well-known colour–magnitude diagrams, fundamental tests to ideas of stellar structure and evolution have been provided. The continual revelation of new types of photometric phenomenon in objects as varied as members of the solar system to active galactic nuclei, regions of nebulosity, supernovae, spotted stars or 'exotic' high-energy bursters ... continues to build and develop physical theory.

2.2 Historical notes

The historical basis for stellar brightness determinations centres on the magnitude system in which they are evaluated. This system goes back at least as far as the oldest known systematic efforts to compile data on stars — namely the catalogue of Hipparchos, completed at about 130 BC. Just how the system actually originated appears "lost in the mists of time", but the magnitudes of Hipparchos, as conveyed to posterity through Claudius

Ptolemy's great *Megali Syntaxis tis Astronomias*, are essentially similar to present-day values for the 1000 or so brightest stars.

During the cultural flowering of the Abbasid caliphate the astronomical works of the Greeks became known and studied in a new setting. Translated into the *Kitab al Majisti* (the Almagest) Ptolemy's treatise stimulated not only the attention of Islamic scholars but also their active experimental investigation. The enlightened Abdullah al Mamun, in the early part of the ninth century of our era, founded the renowned observatory at Baghdad, where astronomy was supported by new and improved instrumentation, as well as advances in theory and calculational methods. It was here, for example, that Al Battani (Albategnius), on the basis of new observations, substantially improved on Ptolemy's value for the precession constant.

Here also it was that Abd al Rahman Sufi decided that not only the positional determinations of the stars as given in the Almagest, but also their magnitudes, could be checked, or reassessed. Sufi published a new list of magnitudes of all the thousand or so stars of the Almagest he could actually observe. There is no doubt that Sufi's magnitudes represent an improvement in precision over the run of values attributed to Ptolemy. Magnitude values given in Ptolemy's catalogue have an average accuracy of not less than half a magnitude division, which becomes about a third of a division in Sufi's work (magnitude units were divided into three subdivisions in these ancient catalogues). The scholar of ancient astronomy, E. Knobel, pointed out that Sufi also appears to have been the first astronomer to take account of a galaxy external to our own, i.e. his was the first map to indicate the Andromeda Nebula.

The differences that existed between Sufi's and Ptolemy's magnitude values were not overemphasized. Presumably, Sufi and his followers supposed that the earlier observers had just not been assiduous enough, after all shortcomings in some of Ptolemy's positional work had already come to light. The possibility of inherent variation of starlight, while it may well have occurred to Sufi, would probably have been regarded with some demur, for the trend of opinion among the ancients, epitomized by no less an authority than Aristotle, was that the sphere of the fixed stars was something eternal and invariable ("*incorruptible*"). Sufi's improved magnitudes were therefore accepted as definitive in Ulugh Beg's famous recompilation of the classical stellar catalogue for the epoch 1437 AD — nearly five centuries after Sufi's time.

The idea of a fixed and invariable outer ætherial sphere of the stars was in harmony with prevalent philosophical concepts of the Middle Ages in

Europe; but a certain shakiness to this model was introduced by the sudden appearance in 1572 of a bright 'new star', which captured the attention of Tycho Brahe, then a 26 year old Danish nobleman looking for his calling in life. As with Hipparchos himself, who, according to Pliny, had been stimulated to compile his catalogue by just such an event some 1700 years previously, Tycho was to go on to produce his own new catalogue of the stars — though, apparently, he did not live to see the work brought to a final published form.

Another important, but rather fortuitous, event in the life of Tycho was the appearance of a bright comet in 1577, not long after the astronomer had established himself in his new observatory at Uraniborg, on a small island between Zealand and Sweden. From his series of observations of the comet Tycho was able, at last, to bring firm evidence to deny the Aristotelian proposition that nothing changes beyond the sphere of the Moon. We know also that, in his meticulous way, Tycho was marking in his catalogue manuscripts the magnitude values of some stars by dots, where he believed there was some discrepancy between previously recorded values. Tycho would not have been dismayed, therefore, by the announcement in 1596 by David Fabricius of the new appearance of what was later called Mira — the first known variable star in the normally used sense. Fabricius was in correspondance with Kepler, then about the same age as Tycho had been at the time of the new star of 1572. The following year — 1597 — both he and the now middle-aged Tycho were working together in Prague, Tycho having been forced out of his native land by political troubles.

Tycho died in 1601, a year after W. Janszoon Blaeu found another famous variable of the northern skies — P Cygni — and three years before Kepler saw his own bright new star in the form of the supernova of 1604. So it was that by the seventeenth century the concept of new and variable stars became accepted: catalogued magnitudes were checked again, more variables began to be discovered and Aristotle's immutable outer ætherial sphere began to fade into oblivion.

This development of ideas occurred side by side with the development of the telescope as a scientific instrument — championed in those early days, of course, by Galileo. It was Galileo who first made serious note of stars fainter than sixth magnitude — the faintest class of Hipparchos' subdivisions — and decided on a simple extension to seventh, eighth and so on, so as to follow in a similar pattern. The system was kept by Flamsteed and others, indeed right up to our own day when, since, in principle, the zero point of the magnitude scale still coincides with a certain average originally based on the Ptolemaic values, we still maintain a link with that first compilation of

Hipparchos (which itself may have been influenced by still earlier sources) more than 2000 years ago.

Some of the stars catalogued by Ptolemy were very far to the south of his Alexandrian sky, and, with the additional apparent displacement resulting from the precession of the equinoxes, dropped from the view of the mediæval astronomers. Some of these stars (e.g. α and β Sagittarii) attracted the notice of young Edmond Halley, who, from his observations on the island of St Helena, in 1679 provided one of the first catalogues of stars of the southern sky. Halley, still trustful of the Almagest writings, and pondering over the two magnitudes or so discrepancies, speculated on the possibility of inherent variations of this scale in the fifteen hundred years since Ptolemy had recorded their magnitudes.

Early in the eighteenth century a further important development to astronomical photometry came with the introduction of specific instrumentation by Pierre Bouguer, who might be regarded as a *patron* of much of the subject matter of this book. Bouguer was, for instance, the first to make a systematic study of the extinction of light from celestial bodies by the intervening atmosphere, and showed that the increase in magnitude (diminution in brightness) thus caused was directly proportional to the mass of intervening air. He is credited with the more basic establishing of the inverse square law diminution of light flux (*in vacuo*). He also quantified that effect, which may well have been noticed in ancient times, known as the limb darkening of the Sun, i.e. the tendency of the surface brightness to fade towards the edge of the solar disk. The effect has been a source of extensive astrophysical interest, aspects of which will be met later in this text.

Bouguer's methodology appears to have been neglected through the century that followed, though this is mollified by its ultimate dependence on that rather non-impersonal element the human eye, whatever the intervening instrumentation; up until the advent of photographic methods, at least. So even by the mid-nineteenth century when Argelander and his associates were engaged with the great undertaking of the *Durchmusterungen*, ultimately recording around half a million magnitude estimates, a very simple eye-based procedure was considered expedient; though it is also true to say that a careful, instrument-based approach, such as that followed in the more intensive work of John Herschel, had its reward in terms of a much better scale of internal accuracy.

It had been during the earlier years of the career of John Herschel's father, William, that an important class of variable star, the eclipsing binary system was discovered with the notable work of young John Goodricke. Binary systems, or double stars as they are also known, were a strong interest of W. Herschel at the time, but though he gave Algol, the binary whose eclipsing pattern was first recognized by Goodricke, close attention with the large telescopes at his disposal, he could not resolve the two components by eye. The period of the binary's orbital revolution is relatively short — less than three days — so that with a few inferences about the consequent likely angular size of the orbit at an expectable distance, Herschel's failure to resolve the components becomes not surprising. Herschel's introduction of more powerful light collectors, and his interest in the ability to resolve detail, particularly in faint and diffuse patches of light, paved the way for the photometry of extended sources; though more quantitative work on this had to await the appearance of purpose-built photometric equipment.

More advanced astronomical photometer designs, utilizing the null principle for brightness comparison, and incorporating controlled diminution of source brightnesses, e.g. by the use of polarizing agents, appeared by the middle of the last century, notably that of Zöllner, who based his photometer on a design of Arago.

It became apparent by about the middle 1800s that the traditional magnitude scale should be close to logarithmic in the received fluxes of visible starlight; a point related to the physiology of sensation as investigated by G. Fechner. The formal rule for stellar magnitudes which became generally adopted is usually associated with the name of Pogson, who, in 1856, set out a relation of the form:

$$m_1 - m_2 = -2.5 \log(f_1/f_2), \tag{2.1}$$

so a difference of 5 magnitudes (m) corresponds to a flux (f) ratio of 100.

There are some significant consequences of this logarithmic system which have favoured its retention. The first relates to the attenuation of incident radiation by the Earth's atmosphere, i.e. Bouguer's law, which can be directly validated with the logarithmic scale. Then the magnitude system adapts itself well to a differential scheme; useful, for example, if one was primarily interested in tracking the relative brightness of some particular variable star. The overall changes of brightness, associated with variations in the atmosphere's transparency from night to night or secular drifts in the response of the receiver easily drop out as zero constants on a logarithmic scale. Only when one wishes to tie in measurements with an absolute system of units (not necessarily an immediate objective) does it become required to evaluate just what (in watts per square metre, say) would correspond to the radiation from a zero magnitude star. These matters will be considered in more detail in the next chapter. Colour too, defined as a difference of magnitudes at different wavelengths, lends itself well to some quasi-empirical relationships of a simple form relating to the temperature of the radiation emitting surface.

In the last century efforts started to be made for a more systematic basis to magnitude determination through the medium of photography. For various reasons this proved not to be so straightforward a task, however, and until relatively recently photographic magnitude determinations were not greatly superior in accuracy to eye-based measures of a trained observer using specially prepared equipment — such as the meridian visual photometer, developed towards the end of the last century at Harvard College Observatory by E. Pickering and his associates. Thus, if the ancient catalogue of Sufi listed stellar magnitudes to an internal accuracy of about a third of a magnitude, the skilled observers using the Harvard meridian photometer could improve on this so that their amplitude of uncertainty was a half that of Sufi, while a tenth of a magnitude accuracy characterized photographic determinations of better quality in the first half of the twentieth century.

Despite its objective nature, there are a number of complicating and often non-linear effects which take place between the original incidence of starlight and the final forms of darkened grain stellar images in the emulsion over the exposed plate (or film). These complications make reliable extraction of a corresponding set of magnitude values a difficult exercise. A number of the early investigators, e.g. Bond, Kapteyn, Pickering, Scheiner, Bemporad (and others), looked for some empirical formula to relate magnitude with something easily measured, such as image diameter, but there was no uniformity of opinion as to what the formula should be, each worker generally preferring his own.

In the early years of the twentieth century the theory of photographic image formation was explored more fully by K. Schwarzschild, and more elaborate calibrative procedures were devised, involving things like objective partial-covering screens, plate holders capable of easy movement for repeated exposures, image plane filters, tube sensitometers and the densitometry of extrafocal images. Methods generally required the setting up of semi-empirical calibration curves, in which the measured photographic effect was plotted against a controlled variation of incident (log) flux. An interesting idea was advanced by C. Fabry for extrafocal image measurement. This made use of a small additional lens close to the main focal plane of the objective (Figure 2.3). Fabry demonstrated advantages in stability and methodology of the arrangement. The patch of light studied was always uniformly illuminated and of the same size, even from an extended object. This concept is still retained in many photometer designs.

In the early days of photography emulsions were all essentially blue-

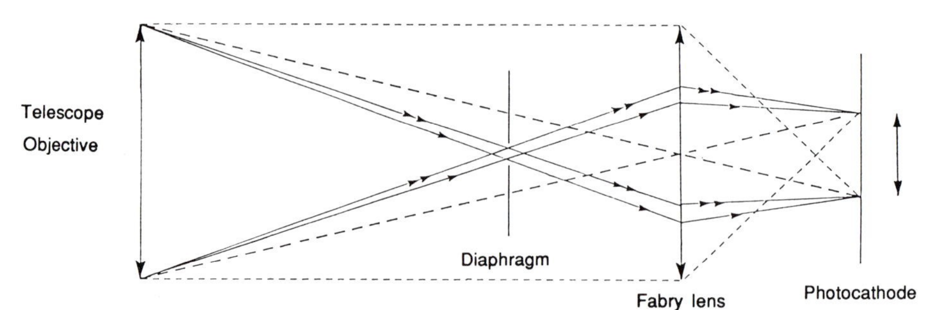

Fig. 2.3. Principle of the Fabry lens. The image on the detector is that which the small lens forms of the objective (dotted lines) as illuminated by the object of interest. Light from a stellar image (full lines), which may drift around in the focal plane, always ends up in the same place.

sensitive compared to the eye: the 'actinic' radiation which causes darkening of the silver halide grains being relatively energetic. Later 'orthochromatic', or even 'panchromatic', emulsions appeared with enhanced sensitivity at longer wavelengths. The early photographic magnitude scale did reflect a physical difference from the visual one, though, due to the difference in effective wavelengths. Hotter objects of a given visual (m_v) magnitude, which are therefore relatively brighter towards the blue region, have correspondingly reduced photographic (p_g) magnitude values. The emulsions of increased sensitivity at longer wavelengths allowed the setting up of a 'photovisual' (p_v) magnitude scale, directly comparable with the eye-based system.

Ancient catalogues tended to have a greater self-consistency in magnitude estimates for stars closer to the celestial North Pole. This is presumably related to the greater relative constancy of the intervening air mass. Pickering similarly favoured magnitude values determined by the Harvard photometry programme in the vicinity of the North Pole when it came to setting up a basic reference sequence for magnitude values. This came up with the proper establishment of the photographic magnitude scales in the 1900s. It was eventually decided, by an international congress of astronomers, that Pickering's North Polar Sequence (which originally consisted of some 47 stars) should define the basic photographic (p_g) magnitude scale. This was to be linked with the pre-existing visual scale by the requirement that the mean of all p_g magnitudes for stars of spectral type A0 in the magnitude range 5.5–6.5 be equal to the mean of the visual magnitudes for those same stars, as determined by Harvard meridian photometry. In 1912 Pickering published the photographic magnitudes of the chosen North Polar Sequence, which had by then grown in number to 96 stars.

Although difficulties with the use of the North Polar Sequence began to be found when it was used to calibrate magnitudes in other parts of the sky, the system was relatively quite accurate internally (errors generally less than five hundredths of a magnitude) and probably represented an adequate basic reference for a number of years in the photographic photometry era. A large number of secondary sequences were set up in time, with particular attention being paid to some specially selected regions, associated with the name of Kapteyn of the Groningen Observatory (Holland). The calibrations moved to the southern hemisphere, including a South Polar Sequence, and photographic photometry of the Magellanic Clouds.

Developments also occurred in the method of magnitude determination from photographic plates. Popular for a time was the Schilt type photometer, which allows a fine pencil of light from a standard lamp to be directed through the plate to be studied. The small spot of light passing through the plate could have its diameter varied, and would normally have been set as small as conveniently possible for the range of magnitudes to be measured. The beam was directed to the centre of a stellar image, where its attenuation would be maximized. The beam, thus reduced in intensity, would then be transmitted to a suitable detective device, such as a photocell and galavanometer combination.

Later a null-method arrangement was introduced by Siedentopf. Sometimes known as the iris-diaphragm type of photometer, the underlying principle is one of comparison of a beam which traverses an adjustable neutral density filter with one which passes through an iris surrounding a star of interest (Figure 2.4). The largest star image in the range to be covered would be selected first, and the iris closed down around it. Arrangements whereby the shadow of the iris and the star image field are conveniently projected onto a large screen, for easier viewing, have been produced and used (with increasing degrees of automated action) up to present times. An advantage of this method is the relatively large range of approximate linearity of the empirical calibration curves.

While work was underway to set up the North Polar Sequence reference system for photographic magnitudes, efforts were already being made in the development of a photoelectric approach. The selenium cell was introduced into astronomy by G. M. Minchin, who first performed photometry of Venus and Jupiter with it at the home observatory of W. H. S. Monck of Dublin in 1892. Among the first to obtain "well marked" effects with this device (from the Moon) was G. F. Fitzgerald, whose name is perpetuated by his original proposal of the well-known spatial contraction effect of special relativity.

The Irish pioneers of photoelectric photometry had to contend with the vagaries of their climate, relatively small telescopes and, seemingly most

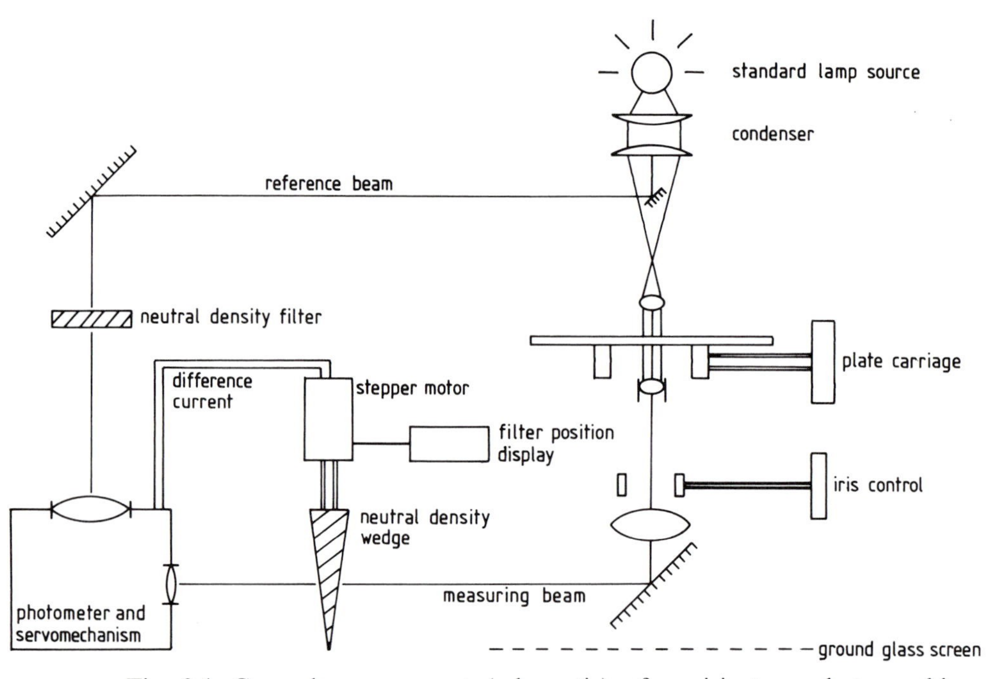

Fig. 2.4. General arrangement (schematic) of an iris type photographic photometer.

troublesome, electrometry of the signal, which depended on an older generation of quadrant electrometers of notorious instability. By opting for the alternative of measuring the photoconductive decline of resistance of an illuminated (and refrigerated) selenium cell in a Wheatstone bridge arrangement Joel Stebbins began to achieve a more continuous success in the early years of the twentieth century at Urbana, Illinois. By 1910 Stebbins had secured the first photoelectric light curve of a variable star — that of the famous eclipsing binary system Algol (Figure 2.5) — with a probable error of no greater than 0.02 magnitudes, an exceedingly accurate set of data for its time. Indeed, Stebbins was the first to establish in this way the existence of a secondary minimum to the light curve of Algol.

Despite this accuracy, photoelectric methods were still rather a long time in development; generally because of technical complications and the required brightness of the object of interest to achieve a reasonable signal to noise ratio. By 1932, however, A. E. Whitford had developed an electronic amplifier to deal with the very weak current produced by a photoemissive cell. Further improvements in photometer design and procedure were presented in G. E. Kron's PhD thesis, published in 1939. The selenium cell had by this time been replaced as a detective device by the now familiar alkaline

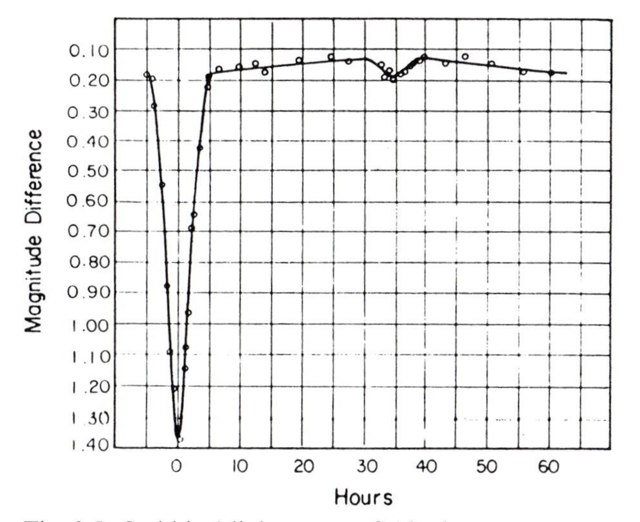

Fig. 2.5. Stebbins' light curve of Algol.

metal photocathodes, and differential measurement accuracies better than 0.01 magnitude were achievable, at least for stars brighter than about sixth magnitude. The magnitude limitation, essentially a consequence of amplification noise, was circumvented, in due course, by use of the avalanche effect of repeated secondary emissions from suitably positioned electrodes (dynodes) in the photomultiplier tube. A successful prototype photomultiplier was produced by the Radio Corporation of America by the late 1940s with the trade-name 1P21.

This type of tube was used by H. L. Johnson and W. W. Morgan in the early 1950s when they set up the three-colour *UBV* photoelectric system, which, despite certain well-known limitations, has been in widespread use right up to the present. In effect, the *UBV* magnitudes, because they have been applied to so many stars, at present probably represent the *de facto* most generally used basic reference for stellar magnitudes, though other, somewhat advantageous schemes, notably the *uvby* system, are increasingly met with. In setting up the *UBV* system, Johnson and Morgan returned to photovisual magnitudes traceable to North Polar Sequence values, and the principle of letting the relatively common A0 type stars locate, by their average, the zero point of the $B - V$ and $U - B$ colour scales. Since the system was one of essentially improved accuracy, however, ultimately it defined the magnitude scale, though in essential agreement with earlier representations. Indeed the process of defining this basic reference scale is really one of continued persistent checks and rechecks of measurements of selected stars, slowly weeding out slightly variable or peculiar stars and building up increasingly reliable averages for the remaining standards. For broadband photoelectric photometry based on that of Johnson and Morgan, this has become particularly associated with the name of Cousins, especially for stars in the equatorial regions of the sky.

The standard stars of any particular photometric system based on n filters can be usefully thought of as each occupying a precision-limited box of n-dimensional space — the 'Golay box'[†] — and progress is made by defining such boxes as carefully and acurately as possible. Such work forms a natural link between the historical and modern eras of astronomical photometry.

2.3 Some basic terminology

More specific words on some frequently encountered terms are now presented.

apparent magnitude m

The brightness of a star in the traditional system as it has evolved to its modern form. The basic formula has been given already as (2.1).

standard magnitude e.g. V, B, U

The foregoing term is still too imprecise for scientific use. The magnitude system has been more precisely defined over the last several decades, allowing different workers, using their local measurements, advantageously to convert to a generally recognized unit — for example, the standard V (for visual) magnitude.

absolute magnitude M

This takes out the distance dependence of brightness, and therefore reflects the true or intrinsic amount of light put out by the source. It is defined as the apparent magnitude which the source (in the absence of light loss in the intervening space) would have if situated at a distance of 10 parsecs.[‡] M then satisfies:

$$m - M = 5 \log \rho - 5 \tag{2.2}$$

where ρ is the distance to the source in parsecs. If interstellar extinction (see below) reduces the brightness of the source, i.e. adds A magnitudes into the apparent magnitude value, then this quantity must similarly be added to the right hand side of (2.2).

[†] This oft-cited term comes from M. Golay's informative *Introduction to Astronomical Photometry*, Reidel Publ. Co., Dordrecht, 1974.

[‡] 1 parsec (pc) is the distance at which the mean Earth–Sun distance would subtend an angle of 1 arcsecond.

monochromatic magnitude m_λ

Standard magnitudes represent the integrated product of a response function with a light flux distribution. For a monochromatic magnitude the response function is an ideal one (δ-function), being unity over a vanishingly small wavelength (or frequency) interval at the wavelength (or frequency) specified, and zero elsewhere. The concept is used in the analysis of magnitude systems.

bolometric magnitude m_{bol}

Bolometric magnitudes go beyond the restriction of magnitude at some particular wavelength, or range of wavelengths, and refer to the total power of the source integrated over all wavelengths.

bolometric correction BC

The quantity which must be added to the visual magnitude to obtain the bolometric one. In general, it is negative in value.

magnitude bandwidth

Real observations must inevitably cover some finite wavelength range. Conventionally, optical photometry is subdivided into 'narrow', 'intermediate' and 'broad' bandwidth types, depending on the order of magnitude of the ratio $\Delta\lambda/\lambda$, the effective transmission range divided by its mean wavelength. Narrowband photometry uses filters for which $\Delta\lambda/\lambda \lesssim 10^{-2}$. For intermediate-band photometry the ratio is around, or somewhat less than, 10^{-1}, while for broadband photometry we typically find $\Delta\lambda/\lambda \sim \frac{1}{4}$.

colour index C or CI

This is expressed as the difference of two magnitude values (in a given system) at different wavelengths — e.g. B (~4400 Å) – V (~5500 Å) in the UBV system.

standard star

A magnitude system is progressively defined by reference to standard stars — starting usually with relatively bright and well-known ones, such as the ten primary standards of the UBV system.

sequence

An orderly arrangement of a group of stars according to magnitudes or colours.

spectral type

Stars are classified on the basis of their observed spectral energy distribution, particularly with regard to the appearance or disappearance of certain discrete features. The types have been arranged, historically, into groups designated with capital letters — thus O, B, A, F, G, K, M represent the main types in order of decreasing temperature. These are then further subdivided into typically ten subtypes (e.g. B0, B1,...,B9). There is a strong correlation between colour index and spectral type, since surface temperature is the main underlying variable. The connection is not unequivocal, however, since both spectral type and colour index may be affected separately by other physical variables.

colour excess

Due to the strong correlation between colour index and surface temperature, if the temperature of a certain star is known *a priori* it should be possible to predict its normal colour index. Sometimes, however, stars or other astronomical sources produce anomalous effects, such as a violet or red excess, the proper interpretation of which is not always clear.

gradient Φ, G

The magnitude difference at two given wavelengths divided by the difference of the two wavelengths' reciprocals Φ gives a usually reasonably constant quantity over optical wavelength ranges, due to approximate black-body-like behaviour. Even closer to a constant is the difference between two such 'absolute' gradients for two radiation sources, – the 'relative' gradient G.

Balmer decrement measures D, c_1

The discontinuity at the end of the Balmer series of hydrogen absorption lines is a conspicuous feature in the near ultra-violet of stellar spectra. It is sensitive to surface gravity as well as temperature, consequently various photometric systems include some monitoring on either side of the discontinuity, to determine the size of the jump in energy terms, or the decrement value D.

In the intermediate bandwidth *uvby* system the colour index combination $(u - v) - (v - b)$ is found to correlate with the Balmer decrement D. This has been given the special symbol c_1.

metallicity index m_1

Stellar spectra are also dependent on the proportion of elements heavier than helium (metals) in their radiating atmospheres. Certain photometric systems allow a suitable combination of magnitude values from different filters to monitor the relative abundance of metals. In the *uvby* system this is referred to as $m_1 = (v - b) - (b - y)$.

line index e.g. α, β

The underlying idea is again that of a magnitude difference. In this case however, the context is narrowband photometry, and the narrowband filter is centred on a particular spectral line, e.g. Hα, Hβ.

2.5 log (flux) measured in this way is subtracted from a similar measure, made with an intermediate bandwidth filter of the surrounding continuum, to form the line index.

extinction k, A

Light from stars does not reach the detector directly, but passes through absorbing or scattering media — the Earth's own atmosphere, as well as the interstellar medium — whereby its intensity decreases. The extinction is related to a coefficient, which depends on the medium and wavelength of transmission. For the Earth's atmosphere k usually denotes the extinction coefficient at the observation wavelength. A is used for the interstellar medium.

interstellar reddening E

The extinction due to interstellar matter is of special interest. The reddening E measures an increase in magnitude (dimming) as one passes to shorter wavelengths, e.g. from V to B, beyond what would be appropriate for a star of the given spectral type, and after having taken account of the normal effect of the Earth's atmosphere to produce reddening.

flux $\mathscr{F}$, f

This measures the power of a radiative field passing through some unit of area, but in conventional physical units, and is thus more appropriate than the magnitude system for general physical analysis.

illuminance f

The meaning is practically the same as flux, but the usage is generally reserved for a surface illuminated by a remote source — such as a unit area above the Earth's atmosphere illuminated by starlight. The terms spectral illuminance and irradiance are also used where one wishes to specify illuminance per unit wavelength f_λ or frequency range f_ν.

luminosity L

Also a measure of the radiative power of a source, it relates to the entire output, and thus can be directly connected with absolute bolometric magnitude.

effective wavelength λ_{eff}

Observation of a source extends over a range of wavelengths centred about some nominal value. The effective wavelength is a representation of this value which allows the magnitude to be regarded as approximately monochromatic at this wavelength. There are alternative measures, such as **mean** or **isophotal**. More specific statements about these are given in Chapter 3.

2.4 Radiation — waves and photons

Though this book centres largely around photometric procedure and analysis, with certain practical applications well to the fore, the nature of light underlies these and certain basic concepts should be clear. This is not really a very straightforward matter, however. In Newton's time a 'corpuscular' or particle-like theory was considered to account satisfactorily for the evidence of experiments. On the other hand, a wave theory had also been advanced and had some proponents, so that a kind of rivalry of explanations existed for some time.

Nowadays both points of view are regarded as tenable and complementary, and either one or the other may be advantageous in different contexts. For example, interferometry of two beams of radiation emanating from a star, directed through two mirrors of known separation, but within the 'coherence length' of that radiation, in principle, enables information on the size of the star to be determined. On the other hand, the existence of such a coherence length itself testifies to some finiteness of structure within the propagating wave trains.

In the last half of the last century ideas on the wave propagation of light became clarified after a number of convincing experiments. In 1878 James Clerk Maxwell produced a set of equations describing the interaction of magnetic and electric fields, as a result of which it was shown that a wave motion, involving these fields, will sustain itself, provided this motion is characterized by a certain velocity, equal to the reciprocal of the square root of the product of the magnetic permeability and electrical conductivity of the medium in which the wave propagates. This wave motion was identified with light, and since pure space also has magnetic permeability and electrical conductivity, light propagates through it, with an 'absolute' velocity.

Maxwell himself, along with his contemporaries, appears to have conceived of some universal medium (the 'æther') permeating all space and supporting the wave motion. However, later this concept was found to be superfluous — misleading even — at any rate the notion of some entity at absolute rest and against which motions may be referred, which the existence of the supposed medium seems to imply, is not in keeping with observed results of physical experiments. It is well known that Einstein's famous theories of relativity, which revolutionized subsequent physics, grew out of this background. Many remarkable developments of modern astronomy have provided observational supports for relativity, and striking examples of progressive interaction between physical theory and observational practice.

Einstein, in another basic contribution to modern physics, gave a discussion of the photoelectric effect, whose implications are central to photoelectric photometry. This presentation, in which a given discrete energy of radiation corresponds with an energy change of a fundamental particle (e.g. an electron), leads to a view of radiation interacting with matter in a particle-like way. This view is already seen in Planck's idea that the energy emitted by a radiating body emerges in certain tiny packets or quanta. The quantum theory, which has grown out of this, associates wave properties with fundamental particles, such as electrons; as well as the reverse with photons — the modern 'corpuscles' of light.

It was known already towards the end of the nineteenth century that certain clean metal surfaces could be induced to lose negative electrical charge under the influence of ultra-violet light. Einstein's photoelectric theory neatly summarizes this effect by the equation,

$$h\nu = h\nu_0 + \frac{1}{2}mv^2, \tag{2.3}$$

where the product of Planck's constant h with the frequency of the impinging photon ν gives the input energy of the quantum. If this is greater than a particular threshold work function, associated with the frequency ν_0, and characteristic of the particular metal, or electron-yielding surface involved, then an electron can emerge with kinetic energy $\frac{1}{2}mv^2$, as specified by the last term on the right hand side.

A circuit can be set up so that electrons liberated from a suitably prepared and illuminated cathode surface are attracted to a nearby positive electrode, where their arrivals are registered. In this way, a light-responsive surface is made to form the detective element (photocathode) of a photometer.

In the basic form of the photoelectric theory, then, the energy of the emitted electron depends only on the frequency of the incident radiation and not on its intensity. However, the number of electrons emitted per second, or (negative) current, from the photoresponsive surface, does vary in direct proportion to the photon incidence rate. This proportionality is at the root of photometric methods utilizing the photoelectric effect.

2.5 Bibliographical notes

Many interesting points were made about the role of the eye in observing celestial objects in M. Minnaert's book *The Nature of Light and Colour in the Open Air*, Dover, New York, 1954. A more detailed text on the eye, exposing its optical behaviour in relation to photometry, is *Fundamentals of Visual Science*, by M. L. Rubin and G. L. Wells, C C Thomas, Springfield,

Illinois, 1972. Articles on eye-based astronomical photometry can be found among the annals of groups such as the RASNZ's Variable Stars Section, the AAVSO, and others mentioned in the previous chapter's bibliographical notes. The kind of general description of stellar colours mentioned can be found in such popular and useful primers as *Norton's Star Atlas*, A. P. Norton and J. Gall Inglis, Gall and Inglis, Edinburgh, 1910 — with re-editions every several years.

Volume 13 of *Mem. R. Astron. Soc.,* 1843, contained a compilation of the old catalogues of Ptolemy, Ulugh Beg and others, by F. Baily. E. Knobel assembled a chronological listing of information on old astronomical catalogues back to the time of Eudoxus in volume 43, 1876, of the same publication. Aspects of the tumultuous Renaissance period of astronomy are readably recounted in C. A. Ronan's *The Astronomers,* Evans Bros., London, 1964, or H. C. King's book *Exploration of the Universe*, Secker and Warburg, London, 1964. A biography of P. Bouguer may be found in *La Grande Encyclopédie*, H. Lamirault, Paris, and some of his contributions are recalled in, for example, the short biographies given by H. Ludendorff in *Populäre Astronomie*, Wilhelm Engelmann, Leipzig, 1921. The discovery of the regularity of Algol's variability and the inferences placed thereon by John Goodricke were recently reviewed by Z. Kopal in *Mercury*, **19**, 88, 1990.

A seminal article covering the whole early development of stellar photometry is that of K. Lundmark in *Handbuch der Astrophysik*, J. Springer, Berlin, **5**, 210, 1932, (see also G. Eberhard in the same series, **7**, 90, 1936), while H. F. Weaver's series of papers in *Popular Astron.* (**54**, 211, 1946 and subsequent issues) provides alternative, more descriptive, reading. An article by J. Hearnshaw on the traditional magnitude scale, and background to Pogsons's formula, recently appeared in *South. Stars*, **34**, 33, 1991. The stars of the North Polar Sequence are listed in the *Trans. Int. Astron. Union*, **1**, 71, 1922. Background on the development of photographic photometry is covered in such sources, though alternative, technically oriented material is available in reference manuals issued by some of the larger companies dealing with photographic materials. A review of the iris-diaphragm technique was given by B. E. Schaefer, *Publ. Astron. Soc. Pac.,* **93**, 253, 1981, while more information can be sought in *Modern Techniques in Astronomical Photography* (ed. R. M. West and J. L. Heudier), ESO, Geneva, Switzerland, 1978.

Minchin's early experiments in photoelectric photometry were recorded by C. J. Butler, in the *Ir. Astron. J.*, **17**, 373, 1986, and G. E. Kron recalled the pioneering work of Joel Stebbins in Chapter 2 of *Photoelectric Photometry* *of Variable Stars*, ed. D. S. Hall and R. M. Genet, Willmann-Bell Inc., Richmond Va.

The Johnson and Morgan era (cf. *Astrophys. J.* **117**, 313, 1953) was covered in a number of still useful papers in *Astronomical Techniques* (ed. W. A. Hiltner) University Press, Chicago, 1962 in articles such as those of A. Lallemand, H. L. Johnson, R. H. Hardie. The similar *Basic Astronomical Data* of the same series (ed. K. A. Strand) contains seminal papers by B. Strömgren, H. L. Johnson, S. Sharpless, W. Becker and others. The little *Astronomical Photoelectric Photometry*, ed. F. B. Wood, AAAS, Washington DC, 1953, is also a good source book on the early development of the subject.

Many of the basic terms and formulae of Section 2.3 appear in C. W. Allen's *Astrophysical Quantities*, Athlone Press, London, 1973; or H. J. Gray and A. Isaacs' *A New Dictionary of Physics*, Longman, London, 1975, provides an extensive source of related definitions.

Further general reading on modern theories of light and the photoelectric effect is to be found in, for example, *Light* (Vol. 1) by H. Haken, North Holland Publ. Co., Amsterdam and New York, 1981, while basic questions of relevance to Section 2.4 are reviewed in *What is Light?* by A. C. S. van Heel and C. H. F. Velzel, World University Library, London, Weidenfeld and Nicolson, 1968, or *The Ethereal Æther*, by L. S. Swenson Jr., Univ. Texas Press, Austin, 1972. P. Léna's *Observational Astrophysics*, Springer Verlag, Berlin and New York, 1988, includes an in-depth treatment of radiation, which it relates to general astrophysical contexts.

3

Underlying Essentials

3.1 Radiation field concepts

Consider a thermally insulated 'black' enclosure, maintained at a constant temperature, T. The enclosure has a characteristic radiation field in isolation from other sources, which depends only on T. The maintenance of this constant temperature implies an equilibrium in internal thermal emissions and absorptions. Let there be an aperture δs to the enclosure, which we take to be sufficiently small that the radiation field characterizing the interior is unaffected. Radiation emerging from the aperture is detected by a receiver presenting an area δA normal to the radius r joining the centre of the aperture to that of the detector. The external medium is regarded as transparent. The angle between the normal to δs and r is θ.

The luminous power δL picked up by the detector (Figure 3.1) is found to be proportional to $\delta A/r^2$, δs and $\cos\theta$, so that

$$\delta L \propto \delta\omega\, \delta s \cos\theta, \tag{3.1}$$

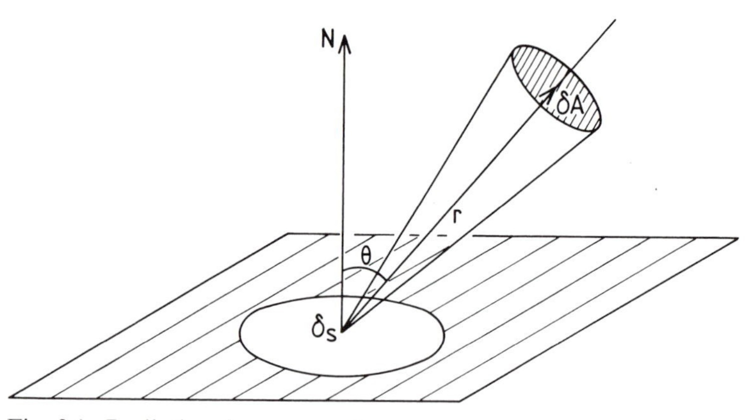

Fig. 3.1. Radiation from a small aperture.

where we have abbreviated slightly by writing $\delta\omega = \delta A/r^2$ — the 'solid angle' subtended by δA. The unit of solid angle is the steradian, defined so that an entire sphere subtends 4π steradians at an internal point, so a cone of semiangle 32.8° subtends close to 1 steradian at its apex. Sometimes the expression 'luminous flux' is met for δL, but the term power is appropriate, since δL can be measured in watts, while flux has the more restricted meaning of power passing through unit area.

The constant of proportionality in (3.1) is the intensity I — sometimes specific brightness — so that

$$\delta L = I\, \delta\omega \delta s \cos\theta. \tag{3.2}$$

Note I is an inherent characteristic of the radiation field, independent of incident solid angles, inclinations and illuminated areas. It is basically connected to the underlying scale of electric and magnetic vectors in the wavefront.

As with stellar magnitudes, the eye played an important part in the development of classical photometry, and adopted units reflect this. On the other hand, the radiation field of an ideal enclosure can be directly related to absolute quantities. A unit of luminous power, which relates well to the conceptual heated enclosure with a small aperture, but also carries something of traditional, eye-based methods, is the standard candle power or candela. This yields 1/60 of the power, per unit solid angle, emitted by a 1 cm^2 aperture of an otherwise enclosed furnace maintained at the temperature of melting platinum.† The power flowing into unit solid angle from a source of 1 candela is 1 lumen: the candela emitting uniformly over a complete sphere of 4π steradians thus produces $4\pi = 12.566\ldots$ lumens.

Consider now the flux $\mathscr{F}$ through the unit area aperture into the entire outward hemisphere. Integrating the foregoing we have,

$$\mathscr{F} = \int_{2\pi} I \cos\theta\, d\omega. \tag{3.3}$$

For isotropic radiation I is constant, and so

$$\mathscr{F} = I \int_{2\pi} \cos\theta\, d\omega = 2\pi I \int_0^1 \cos\theta\, d\cos\theta = \pi I, \tag{3.4}$$

so that the 1 cm^2 aperture of the considered isotropic furnace emits altogether 60π lumens. We will find this to be equivalent to about 0.277 watts in the visible region of the spectrum.

† The modern definition differs from this, though the unit itself is of essentially the same value.

Flux is here seen to be the cosine-weighted integral of the intensity. It is, in this way, sometimes termed the first moment of the intensity. A zeroth moment is defined by

$$4\pi J = \int_{4\pi} I \, d\omega. \tag{3.5}$$

J is easily seen to denote the mean intensity. Here we have integrated over the entire sphere. The corresponding expression for flux gives the difference between the outward hemisphere integration and the inward one (tacitly assumed to be zero in equations (3.3) and (3.4)).

A second moment, known as the 'pressure' of a radiation field, given by

$$4\pi K = \int_{4\pi} I \cos^2 \theta \, d\omega, \tag{3.6}$$

is also found with flux and mean intensity in analyses of radiation fields. In a radiating atmosphere, where there is a steady radiative flow without sources or sinks of energy, $\mathscr{F}$ can be shown to be proportional to the gradient of K with respect to the variable 'optical depth', which is the integral of local opacity over distance, in the direction of the net flow. This relationship between radiation pressure and flux is one way to express the transport of energy by radiation in steady conditions.

A variety of other units appear in photometry, associated with its visual background and the characterization of light levels or brightnesses of illuminated surfaces. Partly this stems from the replacement of CGS nomenclature by SI units — for instance, with the *phot* and the *lux*. The former indicates a luminous power passing through 1 cm^2 of 0.00147 watts (i.e. 1 lumen cm^2), the latter corresponds to the same net rate of watts, but spread over 1 m^2. The lumen and candela were conserved in value as basic units with the CGS $\rightarrow$ SI transfer. The phot, a relatively large unit, is then $10^4 \times$ the lux in value. These units can be applied to the power per unit area above the atmosphere received from a star of given magnitude (Section 3.3).†

A unit quantitatively similar to the lux for measuring the apparent brightness of an extended surface is the *apostilb*. If light illuminates a white surface so that 1 lux is perfectly diffused by each square metre it is said to have a brightness, or sometimes intensity, of 1 apostilb. Intensity is more properly regarded as power per unit area, *per unit solid angle*, however, and this is associated with the *nit* (lux sr^{-1}), which yields the same outflow as the apostilb, but apportioned into unit solid angle. The nit is thus $\pi\times$ the apostilb in value (cf. equation (3.4)). The CGS equivalents to apostilb and nit were the *lambert* and *stilb*, respectively.

† The jansky $= 10^{-26}$ W m^{-2} Hz^{-1} is a unit of irradiance frequently used in radioastronomy.

3.2 Black body radiation

The flux radiated from the black enclosure of the preceding section has a simple form of dependence on the temperature T, namely:

$$\mathscr{F} = \sigma T^4, \tag{3.7}$$

where $\sigma = 5.670 \times 10^{-8}$ W m^{-2} K. This formula is generally known as Stefan's law after the experimenter who established it; it was also verified theoretically by Boltzmann in 1884. If we substitute in the melting point of platinum for T, i.e. 2044 K, we obtain 98.97 W from the conceptualized 1 cm^2 aperture — which may be compared with the previously quoted figure of 0.277 W as visible radiation.

For the idealized enclosure, or 'black body', the distribution of radiated power as a function of wavelength takes the form

$$\mathscr{F}_\lambda = \frac{c_1}{\lambda^5[\exp(c_2/\lambda T) - 1]}, \tag{3.8}$$

where c_1 and c_2 are constants which are expressed in terms of fundamental physical quantities thus:

$$\left.\begin{array}{rcl} c_1 & = & 2\pi hc^2 \\ c_2 & = & hc/k \end{array}\right\} \tag{3.9}$$

(see Appendix for h, c and k). The constant $c_1 = 3.7418 \times 10^{-16}$ W m^2, and $c_2 = 0.014388$ mK.

The black body formula for $\mathscr{F}_\lambda$ is derived on the basis of arguments about the most probable distribution of energy levels for an assembly of equivalent oscillators. The original publication was by M. Planck in the *Annalen der Physik* in 1901, but more accessible treatments are readily found.† An important point to notice in this is that permissible energy levels are discrete. Early attempts to describe the distribution of radiation intensity using the calculus-like idea of a limiting process tending to zero, failed to give the correct result. The foregoing black body formula is a basic example of the 'quantization to h' concept of quantum physics.

Some other derivations relatable to the Planck form (3.8) are often used. The wavelength λ_{max} of the maximum $\mathscr{F}_\lambda$ satisfies a simple reciprocity with the temperature,

$$T\lambda_{max} = 0.201405c_2 = 0.0028978 \text{ mK}, \tag{3.10}$$

† See e.g. *Modern University Physics* by J.A.Richards, F.W. Sears, M.R.Wehr, and M.W. Zemansky; Addison–Wesley, Reading, Mass., 1964, p.715.

a relation known as Wien's displacement law. For wavelengths $\lambda \gg \lambda_{max}$ the classical Rayleigh–Jeans formula is approximately satisfied,

$$\mathscr{F}_\lambda = \frac{c_1 T}{c_2 \lambda^4}, \tag{3.11}$$

while for $\lambda \lesssim \lambda_{max}$ Wien's approximation

$$\mathscr{F}_\lambda = \frac{c_1}{\lambda^5 \exp(c_2/\lambda T)} \tag{3.12}$$

becomes increasingly valid as λ diminishes.

To apply the black body radiation formula in visual photometry we need an eye-sensitivity function. In Figure 3.2 a generally adopted 'photopic' (foveal vision) eye-sensitivity curve has been plotted against wavelength, together with the radiation from a black body source, i.e. the Planck function, corresponding to 2044 K. It is from integrating the product of (folding) these two functions that we derive the proportion 0.277 out of the total 98.97 W appearing in the range that the normal eye perceives. Since 0.277 W is equivalent to 60π lm, we find the often quoted result† that 680 lm = 1 W.

In conditions of very low illumination, such as occur at night in regions away from city lighting, the eye will become 'dark adapted' after a while ($\sim$ a half-hour). The rod type detectors of the retina, whose relative density increases away from the fovea, play an enhanced role in noticing very faint levels of light. The eye-sensitivity curve for these conditions is different from normal photopic vision — it is narrower, and shifted somewhat to shorter wavelengths at its peak ($\sim$ 5100 Å). Folding such a 'scotopic' sensitivity with the standard black body source yields only 0.00058 W.

3.3 The Sun seen as a star

Photometric calculations for the Sun are now appropriate.‡ There are various approaches to defining temperature for an astronomical source, but a clear physical significance attaches to the 'effective temperature' T_e of a stellar photosphere, defined as that temperature at which a black body would emit the same total power per unit area. We then utilize (3.2) and (3.7) and write

$$T_e = \left(\frac{\pi f}{5.67 \times 10^{-8}\, \delta\omega} \right)^{\frac{1}{4}}. \tag{3.13}$$

† This approach now forms the basis of the definition of the lumen, i.e. as a certain fraction ($\approx 1/680$) W.

‡ The dangers and harmful effects of direct incidence of the Sun's rays on the human eye should always be noted in practical contexts.

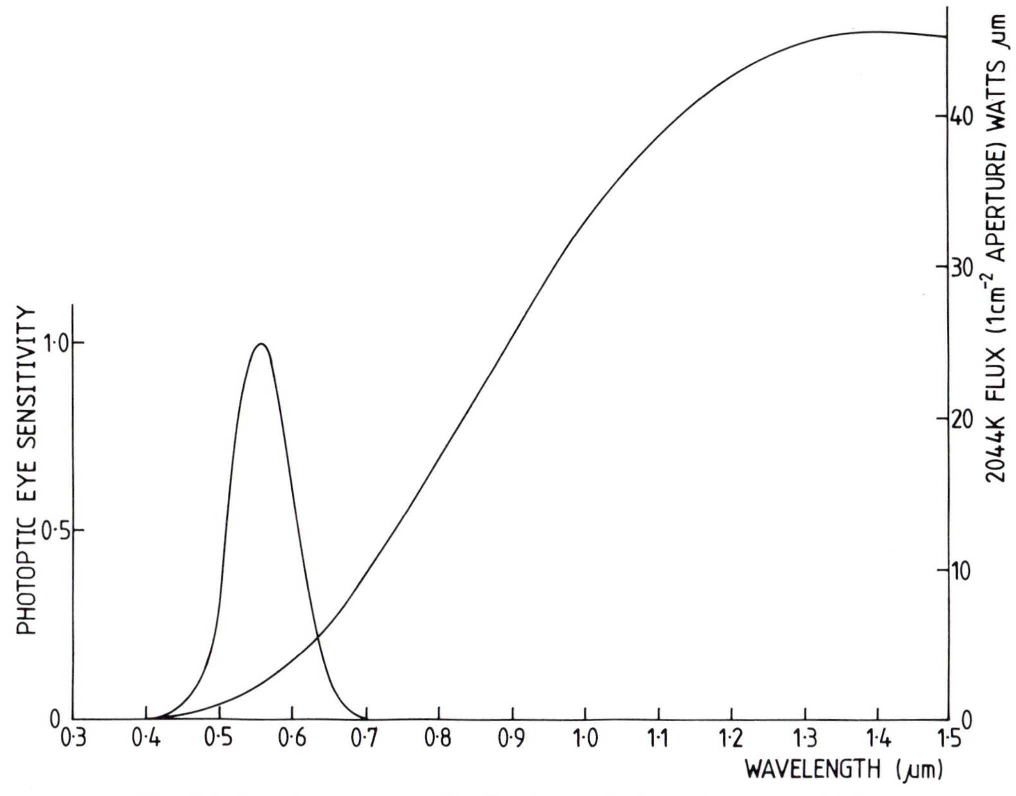

Fig. 3.2. Luminous power distribution and photopic eye sensitivity.

The Sun's mean semi-diameter is some 959.63 arcsec, i.e. its disk subtends, on average, a solid angle $\delta\omega$ of 6.800×10^{-5} sr. The quantity f denotes the mean flux, or illuminance, received by unit area normal to the Sun's rays at a point above the Earth's atmosphere, where we neglect the effects of any intervening absorption and take $f/\delta\omega$ to be the representative intensity of the whole solar surface. Values given for f (the solar constant) are about 1360 ± 10 W m^{-2}. This determination is not too direct, and involves allowances for atmospheric absorptive effects and finiteness of the bandwidth of sensitivity of the receiver — however, application of (3.13) then shows the Sun's effective temperature to be about 5770 ± 10 K. Hence, 1 m^2 of the solar photosphere emits (on average) some 63 MW. The distribution of the Sun's averaged energy spectrum with wavelength, together with that of an equivalent black body, is shown in Figure 3.3. Folded with the eye-sensitivity function the net power comes to 8.46 MW. Black body fluxes over a range of stellar surface temperatures are given in Table 3.1.

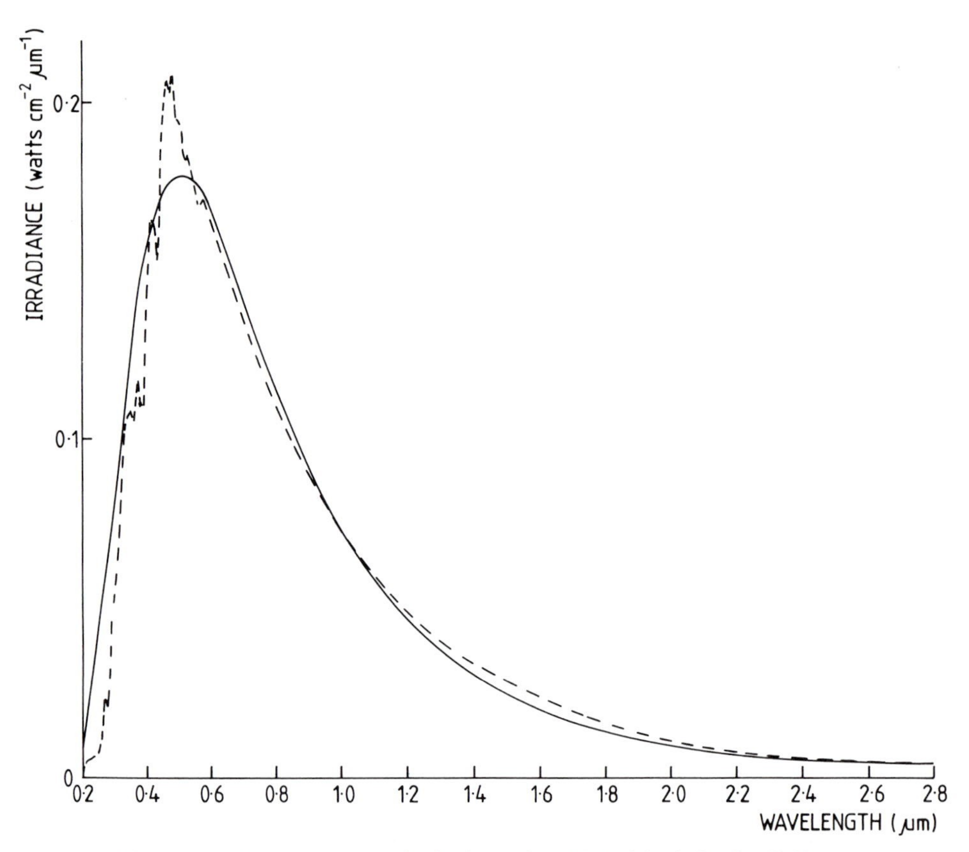

Fig. 3.3. Solar spectrum (dashed) and 5770 K black body (full).

With radius 6.96×10^8 m, and the foregoing representative temperature, the luminous output of the entire Sun amounts to 3.50×10^{28} lm, or 2.79×10^{27} cd. How much of this light is received at the Earth? In the usually adopted conditions of no temporal build-ups or losses in the energy flow, i.e. a steady state, the uniform flow of radiation entering a sphere of radius r_1 around a central source all subsequently passes through a second concentric sphere of radius r_2 ($r_1 < r_2$), so that the flux falls off as $r_1{}^2/r_2{}^2$. The flux above the atmosphere at the distance of the Earth is thus $R_{\odot}^2/\rho_{\odot}^2$ of that at the photosphere, where $R_{\odot}$ is the Sun's mean radius and $\rho_{\odot}$ is the Sun–Earth distance. (This argument simply repeats the inverse square dependence on r implicit in (3.1).) With a value of 1.496×10^{11} m for $\rho_{\odot}$, we obtain 1.25×10^5 lm m^{-2} (lux) for this photopic illuminance (= 184 W m^{-2}).

If the Sun were removed to a distance of 10 pc it would appear as a star of visual magnitude 4.82 (i.e. its absolute visual magnitude M_v). The implied increase in distance is $1/(0.1''$ in radians$) = 2062648$, so that the solid angle decreases accordingly. The corresponding increase of 31.572... magnitudes follows very precisely, but the actual value of the Sun's magnitude at its true mean distance from the Earth involves the technically awkward matter of putting the Sun on the stellar magnitude scale. The value for this — $m_v = -26.75$ — has been determined by a number of experimenters using different methods: it is probably known to an accuracy of about 0.02 mag. The v suffix (for visual) here refers to the eye's range of photopic sensitivity.

Imagine the Sun moved out further, dropping a further two magnitudes, say, to 6.82, i.e. 4.66×10^{-9} lx, which would still be just visible to a good, dark-adapted eye on a clear moonless night. We thus notice the remarkable range in sensitivity of the eye, which can accommodate incident light flux levels, from identifiable sources, over a range of 14 orders of magnitude! Light from individual visible stars varies over about three powers of ten brighter than $\sim 5 \times 10^{-9}$ lx. A dark-adapted eye can detect still lower levels of light than 10^{-9} lx — perhaps by an order of magnitude or so — but this would be from diffuse patches, noticed predominantly in the 'averted vision' of the rods, and of unclear outline.

The energy radiated by the Sun is distributed over a wide range of wavelengths, from the X-ray domain, of 1 Å or less, to detectable radio emissions at wavelengths 13 orders of magnitude or more greater; however, 96% of the energy is in the decade 0.3–3 μm, and indeed about 40% over the small photopic range. Determinations of the net output of radiation are important to a wide variety of studies, especially within the compass of earth sciences. It is notable, therefore, that until relatively recently the value of so basic a quantity as the solar constant was not known to an accuracy higher than 1%. Of course, ground-based measurements of the total radiation from astronomical sources are always influenced by the variable, energy-attenuating medium of the atmosphere. There is the powerful ultra-violet absorbing component of ozone at the short-wave end, and strong absorption bands due to water vapour in the infra-red. Dusts and aerosols introduce some extinction throughout the spectrum to an extent difficult to account for precisely, which adds to the contribution of classical Rayleigh scattering (λ^{-4} law) by atmospheric molecules.

The determination of the apparent visual magnitude of the Sun, combined with a knowledge of its spectral energy distribution and the response of the detector — the eye in the present case — allow us to put the magnitude scale on an absolute footing, or determine the zero point value in watts per square metre, in other words. Thus, if 184 W m^{-2} over the visual range corresponds to magnitude −26.75, magnitude 0.0 will correspond to

$184 \times 10^{-(0.4\times26.75)} = 3.67 \times 10^{-9}$ W m^{-2}. Similarly, the absolute visual magnitude of the Sun being 4.82, and the proportion of its total luminosity emitted in visual light taken to be $4\pi\rho_{\odot}^2 \times 184 = 5.175 \times 10^{25}$ W, we deduce that $M_v = 0.0$ would be equivalent to 4.38×10^{27} watts of isotropic visual radiation.

For bolometric magnitudes an equivalent comparison can be made. The total luminosity, using the previously quoted temperature and radius, is found to be 3.825×10^{26} W, and with a bolometric correction (Section 2.3) of –0.07 for the Sun the value of $M_{bol} = 4.75$; so that $M_{bol} = 0.0$ corresponds to an overall luminosity of 3.04×10^{28} W. The bolometric correction just cited for the Sun, i.e. –0.07, together with the value of the solar constant of 1360 W m^{-2}, allows us to calibrate the apparent bolometric magnitude scale in a similar way; thus $m_{bol} = 0.0$ corresponds to $1360 \times 10^{-(0.4\times26.82)} = 2.54 \times 10^{-8}$ W m^{-2}.

Only the very brightest few stars have apparent visual magnitudes comparable to zero. Typical measurements concern objects fainter than this by several powers of ten. With a very clear atmosphere, therefore, and a flux collector (telescope) of aperture greater than, say, 1 m^2 in area, the power received from starlight is still very small — at most, typically, 10^{-12} W, and usually much less. Even with an efficient detector the input power is considerably less than that required for its effective recording as an output accessible to human inspection. Amplification is thus necessary between input and output stages. One of the very significant technical concerns of photometry has been to minimize 'noise' in such amplification.

3.4 The bolometric correction

The calculation of the proportion of total power liberated in the visual band relates to the bolometric correction factor (BC), enabling us to transfer from visible to total components of emitted radiation and vice versa.

We have

$$\begin{aligned} BC &= m_{bol} - m_v = -2.5 \log L_{bol}/L_v + \zeta_{bol} \\ &= -2.5\{\log \int_{\infty} L_\lambda \, d\lambda - \log \int_{v} L_\lambda T_\lambda \, d\lambda\} + \zeta_{bol}, \end{aligned} \tag{3.14}$$

where ζ_{bol} is a zero constant for the bolometric magnitude scale, and the factor $T_\lambda (0 \leq T_\lambda \leq 1)$ represents the transmission of the detector (the eye, for the present example) as a function of wavelength. The right hand member in (3.14), involving the expression in curled parentheses, is always numerically less than zero; BC values are correspondingly set negative.

Table 3.1. *Temperatures, fluxes and bolometric corrections.*

T K	$\mathscr{F}$ MW m^{-2}	BC_{bb}	BC_*	T K	$\mathscr{F}$ MW m^{-2}	BC_{bb}	BC_*
1000	$0.120.10^{-6}$	–14.53		6000	99.5	–0.01	–0.08
1500	$0.387.10^{-3}$	–7.52		7000	186.7	0.0	–0.01
2000	0.0211	–4.42		8000	301.7	–0.06	–0.01
2500	0.256	–2.68		9000	441.0	–0.16	–0.08
3000	1.37	–1.65	–3.22	10000	601.3	–0.28	–0.24
3500	4.61	–1.01	–1.88	12000	972.1	–0.55	–0.67
4000	11.5	–0.59	–1.04	14000	1393	–0.83	–1.15
4500	23.5	–0.33	–0.56	16000	1849	–1.10	–1.46
5000	41.7	–0.16	–0.31	18000	2332	–1.36	–1.82
				20000	2833	–1.60	–2.03

In Table 3.1 we list BC values corresponding to $\zeta_{bol} = 2.16$ for black bodies (BC_{bb}), giving also the visual power fluxes in MW m^{-2}. Alongside the BC_{bb} values are quoted values for stars BC_*, which have come from corresponding foldings with realistic (dwarf star) photospheric irradiance models (also shown in Figure 3.4). Away from the central maximum, where the irradiance peak is close to that of eye response, the BC value becomes relatively sensitive to 'non-blackness' of the photosphere, and values given for BC_* by different authors show appreciable differences.

ζ_{bol} is chosen to make the least numerical value of the BC zero, and therefore conveniently tailors the bolometric to the apparent magnitude scale. For some practical purposes, though, the value of this constant is unimportant. Magnitude scales always carry such a zero constant question with them for basically historical reasons. Consider, though, how much the visual magnitude of a star would change by changing its effective temperature from T_1 to T_2. We can easily find

$$m_{v_1} - m_{v_2} = -10 \log (T_1/T_2) - (BC_1 - BC_2), \tag{3.15}$$

and the difference in BC values can be read from Table 3.1. In this way, stars cooler than the Sun can be seen to respond very sharply in apparent brightness to increase of temperature. We would thus anticipate a prominent role for temperature variation in accounting for the brightness changes of cool variables — for example, the long period M type variables similar to o Ceti (Mira type stars).

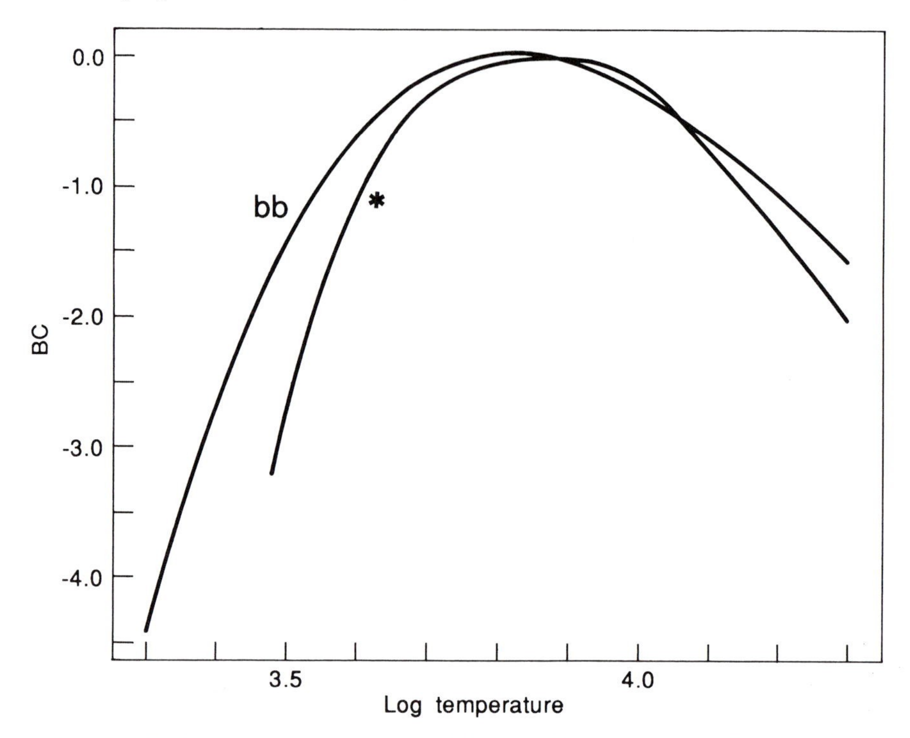

Fig. 3.4. Bolometric corrections for black bodies (bb) and stellar (*) photospheres.

3.5 Absolute stellar fluxes and temperatures

Because stars are generally spherical in shape, and uniform in their surface flux distribution, at least to a first approximation, the average intensity, for a given line of sight, is proportional to the outward flux at any surface location (Figure 3.5). The constant of proportionality comes from the integration required for hemispherical averaging. With a constant unit kernel, this is $2\pi \int_0^1 \cos\theta \, d\cos\theta = \pi$; i.e. the projected area of unit sphere. Now the outward (local) flux $\mathscr{F}$ we can write as,

$$\begin{aligned} \mathscr{F} &= 2\pi \int_0^1 I(\theta) \cos\theta \, d\cos\theta, \\ &= \pi F \text{ (say).} \end{aligned} \tag{3.16}$$

Hence, F here corresponds to the 'surface mean intensity' — the average of intensity weighted by each local unit surface area projected in the line of sight.

In the case of a non-spherical source, the corresponding averaging of intensity no longer generally involves a constant unit kernel, so the formal

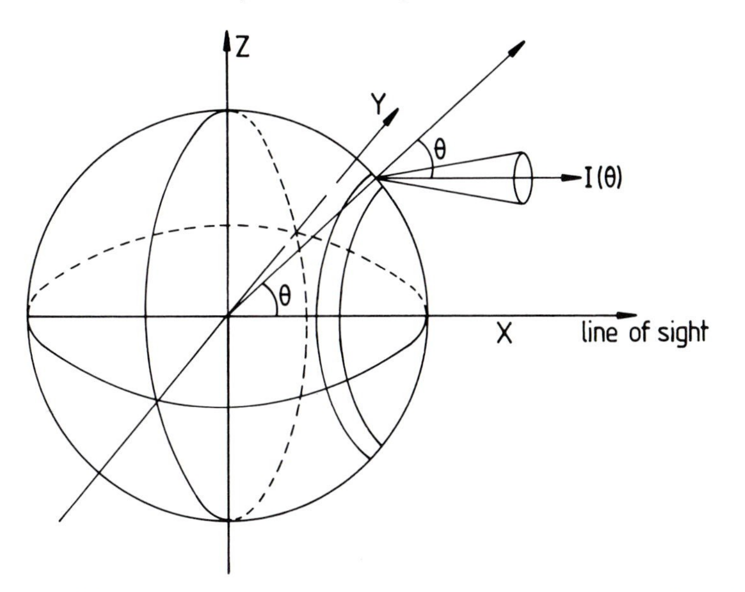

Fig. 3.5. Deriving the surface mean intensity for a hemispherical source.

resemblance to the local flux integral loses strictness, though an expression corresponding to (3.16) could still be devised, with F representing an equivalent hemisphere average of the surface intensity.

This averaging is done automatically in stellar photometry, since stellar images are, in general, much too small to be resolved even with the world's largest telescopes. A practical angular diameter $\Delta\theta$, subtending solid angle $\Delta\omega$, is then taken to be a circular mean. For the power influx over 1 m^2 at the above-atmosphere level we have an expression, corresponding to (3.13), of the form

$$f_* = F(T)\Delta\omega. \tag{3.17}$$

Here f_* can be regarded as directly determinable. In this satellite era, for instance, an appropriate bolometric detector could be located at the considered position, above the atmosphere. Practical difficulties would still be present, since real detectors always have some finite range of sensitivity, for example. The difficulties are compounded on the Earth's surface, where one has the additional problems of climatic and transparency vagaries with which to contend. Such problems, though, did not prevent Edison Pettit and Seth Nicholson from carrying out their classic series of bolometric measurements of stellar radiation in the 1920s, when from measurements of temperature increase (as little as 0.000009 K from Boss 4342) they were able to determine, in effect, quantities equivalent to the solar constant, but from stars.

The right hand side of (3.17) can be dealt with in two ways: firstly, if the angular diameter $\Delta\theta$ of the star can be determined, we can write

$$T_e = [4f_*/(\sigma\Delta\theta^2)]^{\frac{1}{4}}, \tag{3.18}$$

and thus determine the temperature of the stars in a direct way. This has been the main objective of a number of experiments to determine the scale of stellar temperature — starting with Pettit and Nicholson, and continuing by a process of steady refinement up to the present.

It is also possible to use equation (3.17) to determine stellar size, at least for stars whose temperature is relatively close to that of the Sun. Thus, if we write,

$$\frac{f_*}{f_\odot} = \left(\frac{T_*}{T_\odot}\right)^4\left(\frac{\Delta\theta_*}{\Delta\theta_\odot}\right)^2,$$

where we use * and $\odot$ to denote the star and Sun, respectively, we have

$$\frac{\Delta\theta_*}{\Delta\theta_\odot} = \frac{R_*\rho_\odot}{\rho_* R_\odot} = 10^{0.2(m_\odot - m_*)}\left(\frac{T_\odot}{T_*}\right)^2, \tag{3.19}$$

where we have converted the luminosity ratio to a bolometric magnitude difference. So now,

$$R_* = R_\odot \frac{10^{0.2(m_\odot - m_*)}}{\Pi \sin 1''}\left(\frac{T_\odot}{T_*}\right)^2, \tag{3.20}$$

where Π stands for the star's parallax in arcseconds.

In the vicinity of the Sun's temperature ($\sim$ 5000–7000 K) the bolometric correction varies relatively slowly, and since the difference in bolometric corrections is multipled by 0.2, the effect of the correction on the visual magnitude difference turns out to be much less than the accuracy with which we can specify this difference. For such stars we can use m_v values directly in (3.20).

For the temperature ratio $T_\odot/T_*$ we make use of a device, which we will call the 'linear gradient' approximation. For this we convert the Planck form (3.8) to a monochromatic magnitude scale, so that

$$\begin{aligned} m(\lambda, T) &= -2.5\log c_1 + 6.25\log\lambda + 1.086c_2/\lambda T + \\ &\quad +2.5\log[1 - \exp(-c_2/\lambda T)] + \zeta_{\lambda,T}, \end{aligned} \tag{3.21}$$

where $\zeta_{\lambda,T}$ is a constant for the particular magnitude scale chosen. The final logarithmic term in λT is small and relatively slowly varying at optical wavelengths and temperatures in the considered range; we neglect it for the present purpose. Differentiating the remaining expression, first with respect to reciprocal temperature, and then with respect to reciprocal wavelength, we find

$$\frac{\partial^2 m(\lambda, T)}{\partial(1/T)\partial(1/\lambda)} = \text{constant} = 1.086c_2. \tag{3.22}$$

This expression shows up the linear gradient idea just referred to. It has an integral of the form,

$$\frac{\partial m(\lambda, T)}{\partial(1/\lambda)} = \frac{1.086c_2}{T} + \varphi(\lambda), \tag{3.23}$$

where $\varphi(\lambda)$ is an arbitrary function of λ. The proportionality of the right hand side to reciprocal temperature directly leads to a useful interpretation for the gradient of the spectral irradiance. The difference between this function for a given star and that of a reference continuum, such as the Sun's, i.e. the relative gradient, permits photometric comparisons of temperatures.

Now integrate (3.23) over an interval corresponding to the difference $\Delta\lambda^{-1}$ in the two reciprocal wavelengths associated with the definition of a colour C, and do this at the two temperatures to be compared, $T_\odot$ and T_*, and, after a little manipulation, it emerges that

$$C_\odot - C_* = \frac{1.086c_2}{T_\odot}(1 - T_\odot/T_*)\Delta\lambda^{-1}. \tag{3.24}$$

Taking these colours to refer to conventional $B - V$ values, we rearrange (3.24) to find

$$T_\odot/T_* \simeq 1 - \Delta(B - V)/1.162, \tag{3.25}$$

which can be used in (3.20).

Consider, for an example, α^1 Cen, a star whose apparent magnitude and parallax are well known at $m_v = -0.01$ and $\Pi = 0''.745$. Its value of $B - V$ is 0.68,[†] implying a slightly cooler surface temperature than the Sun's, i.e. $T = 5620$ K. Equation (3.20) now yields $R_* = 1.3R_\odot$ — in essential agreement with cited values.

The temperature measure in this example was taken from the observed colour — the 'colour temperature'. A 'brightness temperature' may be obtained by inverting the Planck formula for the flux (3.8) at a given wavelength, and there are other approaches to temperature derivation which

[†] Such data is contained, for example, in C. W. Allen's *Astrophysical Quantities*, Athlone Press, London, 1973.

do not necessarily yield the same result for the same star. It is the *effective* temperature, however, which is the one referred to in connection with equation (3.18), dealing with integrated power and angular diameter, which has the most comprehensive role. Colour or brightness temperatures can play a supporting role, for example, over a restricted range of effective temperatures, where proportionality between the various temperature scales can be used. The establishing of dependable values of effective temperatures for stars has been a fundamental task of modern astrophysics.

Apart from the direct use of equation (3.18), or some equivalent, progress in finding effective temperatures has come from the development of sophisticated model atmospheres for stars, based on computer calculations. These predict, for a given atmospheric composition, a spectral distribution of flux which is uniquely determined at a given effective temperature (and surface gravity, on which there is a weaker dependence). Although such models can fit observed stellar irradiance data to theoretical distributions parameterized by effective temperatures, this is not the same as direct temperature determination. Thus, ultimately, such calculations have to be checked on whether their predicted effective temperatures agree with observed ones for those relatively few stars (of order a hundred) whose fluxes and angular diameters are reasonably well known.

There have been three general approaches to the determination of angular diameters: (a) direct measurement by an interferometer; (b) from the diffraction pattern observed at an occultation of a star from an intermediate 'sharp edge' — notably the Moon's limb; and (c) from eclipsing binaries which are double lined spectroscopic binaries (so that absolute parameters can then be determined), and which also have known parallax, enabling us to transfer the derived radii to angular measure. Actually, there are very few known eclipsing binaries which satisfy all these requirements. A few examples of stellar diameters, selected from a variety of sources, are presented in Table 3.2.

The bolometric correction should be a fairly smooth function of surface temperature, and the values tabulated in Table 3.2 come from interpolations on such trends. The bolometric magnitude for a given flux f is easily derived from the previously given calibration of the bolometric scale. If the measured apparent magnitude is subtracted from this value, for a given star, then the corresponding BC is obtained. Such BC values would be found to differ somewhat from the values of Table 3.2, though, in general, the discrepancies are less than a few per cent. Such differences represent the cumulative effect of errors, either in the determinations of $\Delta\theta$ or f, or in the way that the smoothed form $BC(T_e)$ has been constructed.

Table 3.2. *Angular diameters, fluxes, effective temperatures and bolometric corrections for some reference stars.*

Name	$\Delta\theta$ marcsec	f W m^{-2}	T_e	BC
β Per	0.96	5.61×10^{-9}	11600	−0.61
α Tau	24.6	3.34×10^{-8}	3590	−1.21
α Ori	54	1.04×10^{-7}	3250	−2.03
β Aur	1.05	2.45×10^{-9}	9020	−0.10
μ Gem	13.7	1.11×10^{-8}	3650	−1.1
γ Gem	1.39	4.57×10^{-9}	9180	−0.10
α CMa	6.67	1.14×10^{-7}	9350	−0.15
YY Gem	0.458	1.08×10^{-11}	3520	−1.45
α CMi	5.50	1.80×10^{-8}	6500	−0.04
α Boo	23.1	4.88×10^{-8}	4070	−0.65
α Sco	41	6.15×10^{-8}	3230	−2.06
μ_1 Sco	0.37	1.69×10^{-8}	24600	−2.4
α Her	31	2.27×10^{-8}	2900	−2.96
α Lyr	3.24	2.84×10^{-8}	9490	−0.19
β Peg	18	1.53×10^{-8}	3450	−1.96

3.6 Broadband filters — essential points

We have already met the standard *UBV* magnitude system developed notably by Howard L. Johnson. In Table 3.3 we present normalized transmission curves for the ultra-violet (U), blue (B) and visual (V) filters of this system. The mean wavelengths are approximately 3500 Å, 4400 Å and 5500 Å, and the widths at half maximum transmission ($\Delta\lambda^{1/2}$) are about 700 Å, 1000 Å and 900 Å, respectively. The normalized transmission curves are shown in Figure 3.6. Folding these transmissions with the Sun's energy distribution, we derive illuminances of $2.87\times10^{-9}, 6.42\times10^{-9}$ and 3.08×10^{-9} W m^{-2} (above the atmosphere) for magnitude zero in U, B and V, respectively. Here we have taken magnitudes −26.75 (V), −26.10 (B) and −25.97 (U) for the Sun. The standard solar irradiance, or illuminance per unit wavelength, of Thekaekara and Drummond (1971) was coupled with the filter transmissions (in place of the black body curve at 5770 K used in Section 3.3).

The luminosities in U, B and V for a star of zero absolute magnitude are obtained by multiplying these illuminances by the area of a sphere of radius 10 pc. We obtain 3.42×10^{27}, 7.68×10^{27} and 3.68×10^{27} W in U, B and V, respectively. The V data here may be compared with the previously quoted visual figures, calculated using the photopic response function.

The foregoing magnitudes for the Sun depend on where the zero point of

Table 3.3. *Transmission data for the UBV system.*

λ μm	U	B	V
0.30	0.00		
0.31	0.10		
0.32	0.61		
0.33	0.84		
0.34	0.93		
0.35	0.97		
0.36	1.00	0.00	
0.37	0.97		
0.38	0.73	0.11	
0.39	0.36		
0.40	0.05	0.92	
0.41	0.01		
0.42	0.00	1.00	
0.44		0.94	
0.46		0.79	0.00
0.48		0.58	0.02
0.50		0.36	0.38
0.52		0.15	0.91
0.54		0.04	0.98
0.56		0.00	0.72
0.58			0.62
0.60			0.40
0.62			0.20
0.64			0.08
0.66			0.02
0.68			0.01
0.70			0.01
0.72			0.01
0.74			0.00
λ_0	0.3534	0.4439	0.5538
W_0 (μm)	0.0655	0.0959	0.0868
W_2 (μm^3)	2.94×10^{-5}	1.18×10^{-4}	1.22×10^{-4}
$W_2/(\lambda_0^2 W_0)$	3.60×10^{-3}	6.24×10^{-3}	4.64×10^{-3}

the V magnitude and $B-V$ and $U-B$ colours are chosen. The zero point of the V magnitude scale was originally arranged to agree with that adopted for the North Polar Sequence visual magnitudes, and is thus in close conformity with time-honoured values. The $B-V$ and $U-B$ colours were set to be zero for the mean of a number of standard (unreddened) A0 type stars. Johnson stated that, provided certain stringent requirements on photometric

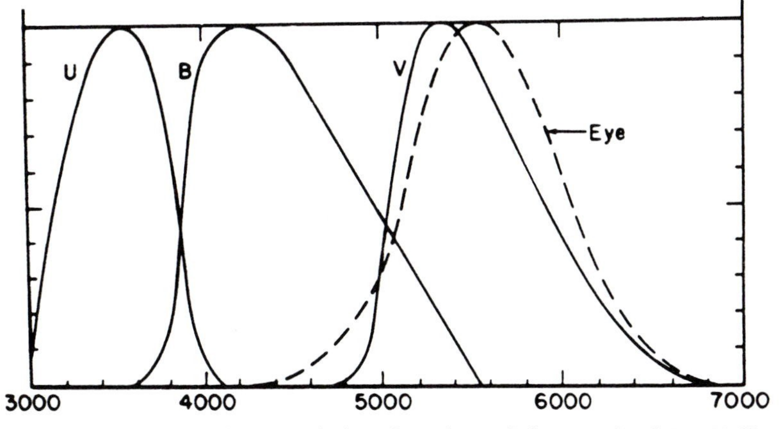

Fig. 3.6. Normalized transmission function of the standard *UBV* filters.

apparatus could be kept, a minimum of ten primary standard stars was sufficient to calibrate local photometry to the *UBV* system. These stars and their magnitudes and colours are reproduced in Table 3.4. They provide a reasonably wide baseline in colour, with types distributed towards the end points of the range. They are also all relatively bright (naked eye) stars, so that some additional procedure is implied in extrapolating the apparent magnitude scale to faint stars, e.g. precisely controllable attenuation of the illuminance from the standards, to check against any possible non-linearity of measurement.

The broadband optical filters are usually made from coloured glass, or combinations of such, as produced by manufacturers, with high and specifiably repeatable qualities. The *U* and *B* types normally use 'bandpass' filters, whose production involves the suspension of rare earth ions in the glass. Ideally, they are characterized by an inverted U-shaped transmission profile, though it seems difficult to achieve a nice symmetric form, with no 'leaks' at other wavelengths, in practice. The *V*-filter was originally a 'longpass' filter — produced from a colloid suspension of metal sulphide or selenide in the glass — it effectively shut out light of shorter wavelengths and relied on the response function of the cathode itself to cut off the longer wavelength radiation.

The bandwidths of the filters, defined as the integrals of their normalized transmissions, turn out to be 660 Å(U), 960 Å(B) and 870 Å(V), i.e. quite close to the approximate half-widths previously mentioned, and with these values the spectral irradiances for a zero magnitude star at the mean wavelength of each filter are 4.34×10^{-12} W m^{-2} Å^{-1} in U, 6.60×10^{-12} W m^{-2} Å^{-1} in B, and 3.54×10^{-12} W m^{-2} Å^{-1} in V. The use of the angstrom unit

Table 3.4. *The ten primary standard stars of the UBV system.*

Name	HD No.	V	$B-V$	$U-B$	Sp. type
α Ari	12929	2.00	1.151	1.12	K2 III
HR 875	18331	5.17	0.084	0.05	A1 V
β Cnc	69267	3.52	1.480	1.78	K4 III
η Hya	74280	4.30	–0.195	–0.74	B3 V
β Lib	135742	2.61	–0.108	–0.37	B8 V
α Ser	140573	2.65	1.168	1.24	K2 III
ϵ CrB	143107	4.15	1.230	1.28	K3 III
τ Her	147394	3.89	–0.152	–0.56	B5 IV
10 Lac	214680	4.88	–0.203	–1.04	O9 V
HR 8832	219134	5.57	1.010	0.89	K3 V

here eases direct comparisons with literature-quoted data. For example, the power emitted in discrete line features in a spectrum is often represented as of 'equivalent width' to so many angstroms of nearby continuum radiation (cf. Section 4.4).

The foregoing bandwidth values are numerically sizeable fractions of the mean wavelengths of the U, B and V filters, so that identification of U, B or V magnitudes with monochromatic ones would be a rather coarse approximation. In order to consider the effects of finite bandwidth in more detail, we expand the irradiance, here regarded in dependence only on the wavelength λ, as a Taylor series,

$$f(\lambda) = f(\lambda_0) + (\lambda - \lambda_0)f'(\lambda) + \frac{(\lambda-\lambda_0)^2}{2!}f''(\lambda_0) + \dots, \tag{3.26}$$

and reintroduce the transmission function $T(\lambda)$, so that the power which the photometric system receives, i.e. the integral of the illuminance over the spectral window of the filter f_W becomes:

$$\begin{aligned} f_W &= \textstyle\int_W f(\lambda)T(\lambda)\,d\lambda \\ &= f(\lambda_0)\textstyle\int_W T(\lambda)\,d\lambda \quad + f'(\lambda_0)\displaystyle\int_W (\lambda-\lambda_0)T(\lambda)\,d\lambda + \\ &\qquad + f''(\lambda_0)\int_W \frac{(\lambda-\lambda_0)^2}{2}T(\lambda)\,d\lambda + \dots \quad . \end{aligned} \tag{3.27}$$

Neglecting second and higher power terms in this expansion, we find simply

$$f_W = f(\lambda_0)W_0. \tag{3.28}$$

Here we have put

$$\lambda_0 = \int_W \lambda T(\lambda)\,d\lambda \Big/ \int_W T(\lambda)\,d\lambda,$$

which makes the first order term in (3.27) vanish, and thus defines the mean wavelength of the filter λ_0. We have also put

$$W_0 = \int_W T(\lambda)\, d\lambda,$$

the 'bandwidth' of the filter.

While λ_0 does not depend on the spectrum of the illuminance of the source, it is precisely representative only where the filter bandwidth is negligible in comparison to that. The transmission profile can be characterized by moments of the transmission function higher than the zeroth. The first moment W_1 is zero by definition of the mean wavelength λ_0. The second moments W_2, for the *UBV* filters, are given below Table 3.3. The square roots of the ratios W_2/W_0 are comparable to the half-bandwidths of normal U-shaped transmission profile filters.

The effect of the source power distribution over a given broadband filter is included in its 'effective' wavelength λ_{eff}, given as

$$\lambda_{eff} = \frac{\int_W \lambda f(\lambda) T(\lambda)\, d\lambda}{\int_W f(\lambda) T(\lambda)\, d\lambda}. \tag{3.29}$$

Expanding numerator and denominator in (3.29) according to (3.26) it can be shown, with a little reduction, that

$$\lambda_{eff} = \lambda_0 + \frac{f'(\lambda_0) W_2}{f(\lambda_0) W_0}, \tag{3.30}$$

where $W_2 = \int_W (\lambda - \lambda_0)^2\, T(\lambda)\, d\lambda$. The mean wavelengths, zeroth and second moments (W_0 and W_2), listed below Table 3.3, have been determined from Simpson quadrature of the data in the table.

If λ_0 is not far from the wavelength λ_{max} of the maximum of the energy distribution, $f'(\lambda_0) \simeq 0$, so that $\lambda_{eff} \simeq \lambda_0$. We have already noted that this condition is roughly satisfied for the Sun in the visual region. On the basis of a general accord between Wien's displacement law and real stellar energy distributions, stars significantly hotter than the Sun have a downward sloping continuum in the visual (negative relative gradient), while the cool 'late' type stars are still increasing their spectral irradiance through the visual region (positive relative gradient), to a maximum somewhere in the red, or infra-red. Hence, the difference $\lambda_{eff} - \lambda_0$, taking the same sign as $f'(\lambda_0)/f(\lambda_0)$, or $[d \log f(\lambda)/d\lambda]_{\lambda_0}$, can be related to equation (3.23), which we rewrite in slightly amended form as

$$\frac{d \log f(\lambda)}{-\lambda^2 d(1/\lambda)} = \frac{-(5\lambda - \Phi)}{\lambda^2}. \tag{3.31}$$

From (3.23) we deduce the gradient term Φ to be of the form $1.086c_2/T$. However, from (3.31) it can be seen that Φ may also be written as $5\lambda_{max}$.

Several approximations have been invoked here — for a start the underlying Planck function is only an approximate representation to the spectral illuminance of a real star (see Figure 3.3). The gradient could be defined to be $1.086c_2/T$, following (3.23), but such a gradient should not be exactly equal to $5\lambda_{max}$, as can be easily seen by comparison with Wien's displacement law (3.10). Finally, the effective wavelength (3.29) results from truncating a Taylor series after only its first order term. Even so, the relationship

$$\lambda_{eff} - \lambda_0 = \frac{5(\lambda_{max} - \lambda_0)W_2}{\lambda_0^2 W_0}$$

is sufficiently accurate to allow a reasonable estimation of the difference $\lambda_{eff} - \lambda_0$ in dependence on the spectral gradient.

In any case, the effective wavelength, as defined by (3.29), is not necessarily that of the best representative wavelength for the irradiance folded with the transmission function. The wavelength at which the flux multipled by the filter's bandpass equals this integrated product is the 'isophotal' wavelength λ_i, i.e.

$$f(\lambda_i) = f_W/W_0. \tag{3.32}$$

It can be shown, in a similar way to the derivation of (3.30), that

$$\lambda_i = \lambda_0 + \frac{f''(\lambda_0)W_2}{2f'(\lambda_0)W_0}, \tag{3.33}$$

which is reasonably valid so long as $f'(\lambda_0) \neq 0$, corresponding to λ_0 appreciably different from λ_{max}. In the case of $f'(\lambda_0) \sim 0$ the fluxes at all three wavelength measures, λ_0, λ_{eff} and λ_i are sufficiently close that the effect of finite bandwidth loses significance.

Since we can write,

$$\frac{f''}{f'} = \frac{f'}{f} + \frac{f d^2 \log f}{f' d\lambda^2},$$

we find that

$$\left(\frac{f''}{f'}\right)_{\lambda_0} = \frac{5(\lambda_{max} - \lambda_0)}{\lambda_0^2} + \frac{(2\lambda_{max} - \lambda_0)}{\lambda_0(\lambda_0 - \lambda_{max})}. \tag{3.34}$$

Hence, for hot stars, for which $\lambda_0 > \lambda_{max}$, we deduce that the second term on the right hand side in (3.34) will be rather smaller, in absolute value, than the first, and so $(\lambda_i - \lambda_0) \rightarrow \frac{3}{5}(\lambda_{eff} - \lambda_0) < 0$. For cool stars for which $\lambda_0 < \lambda_{max}$ the second term is of low influence ($\sim -2/\lambda_0$) compared to the

Table 3.5. *Typical photon count rates for a bright star and medium size telescope.*

	U	B	V
mag	5.0	5.0	5.0
k loss	0.5	0.4	0.25
$\bar{\nu}$	8.33×10^{14}	6.82×10^{14}	5.45×10^{14}
η	0.07	0.1	0.05
A (m^2)	1	1	1
N	1.8×10^{6}	8.5×10^{6}	3.2×10^{6}

first $(5\lambda_{max}/\lambda_0^2)$, so that $(\lambda_i - \lambda_0) \rightarrow \frac{1}{2}(\lambda_{eff} - \lambda_0) > 0$. These differences appear to become large for large λ_{max}, but, in practice, λ_{max} would seldom be greater than $\sim 3\lambda_0$ for optical filters and visible stars. Stars for which $(\lambda_{max} - \lambda_0) \sim \lambda_0$ can be seen to introduce differences $(\lambda_{eff} - \lambda_0) \sim 0.01$ μm, using the data given in Table 3.3.

The similar trend of the more accessible λ_{eff} toward λ_i throughout the range of observed spectral types at optical wavelengths is thus evident. Numerical quadratures for real stars and broadband filters have shown that the discrepancies between these various wavelength representations are such that a difference of $\gtrsim$0.01 magnitudes could occur if the monochromatic magnitude at the mean wavelength λ_0 were simply used regardless of the combined effect of finite bandwidth and spectral gradient. If λ_{eff} is used as the equivalent monochromatic wavelength reductional errors drop to only around a tenth of this.

In Table 3.5 the previously derived irradiance data are applied to a telescope of aperture $A = 1$ m^2, directed to a star of magnitude 5.0. Allowance is made for some typical atmospheric extinction (k loss). The number of photons of energy $h\bar{\nu}$ required to account for the influx is then reduced, by a typical detector + filter efficiency factor η, to yield the representative count rate per second N.

3.7 Surface flux and colour correlations

Equation (3.18) can be rewritten in the form:

$$\log T_e = 0.25 \log f_* - 0.5 \log \Delta\theta + 7.119, \qquad (3.35)$$

with f_* in W m^{-2}, $\Delta\theta$ in marcsec, and T_e in K. But, from rearranging (3.14), we can also write

$$0.25 \log f_* = \text{const.} - 0.1(m_v + BC). \tag{3.36}$$

Now, from the solar calibration of Section 3.3, we find the logarithm of the illuminance of a star of zero bolometric magnitude is –11.595. Substituting (3.36) into (3.35), therefore, we derive

$$\log T_e + 0.1BC = 4.220 - 0.1m_v - 0.5 \log \Delta\theta\,. \tag{3.37}$$

The usefulness of this relationship has been stressed in the work of T.G. Barnes, D.S. Evans and others. The right hand side is a potentially measurable quantity — at least for nearer or larger stars, or certain eclipsing binary components. From (3.17), and the fundamental relationship between illuminance and the magnitude scale (1.1), we deduce that this right hand side, F_V say, is equal to 0.25× the logarithm of the surface flux in V light, together with some additive constant. The left hand side is dependent on essentially the temperature only, and should show a tight relationship to an appropriately chosen colour; such as equation (3.25), for instance, suggests for $B - V$. In fact, Barnes and Evans found a particularly well-correlated and stable relationship of the form:

$$F_V = a + b(V - R)\,, \tag{3.38}$$

where R denotes the magnitude corresponding to a red filter, which is one of a group extending the UBV system to longer wavelengths (R, I, J, K, L, M, N) introduced by Johnson in 1965. a and b are constants, having values as follows:

$$\begin{aligned}
-0.17 < (V - R) < 0.00 \quad & a = 3.977, \quad b = -1.390;\\
0.00 < (V - R) < 1.26 \quad & a = 3.977, \quad b = -0.429;\\
1.26 < (V - R) < 4.20 \quad & a = 3.837, \quad b = -0.320.
\end{aligned}$$

With these calibrations in place, the value of F_V can be derived for a star from the relatively easily measured value of $V - R$. This is now, in turn, applied to (3.37) allowing the angular diameter to be evaluated. This would clearly be useful when either a star's distance, or its size is known — application of the angular diameter value allows the determination of the other quantity — size or distance.

In an early application of the foregoing to the Mira type variable R Tri, for example, Barnes and Evans, using a distance derived from an adopted period–absolute magnitude relationship, determined the actual radius values at several points through the light cycle from corresponding angular diameter values. In Chapter 10 a similar idea is pursued for a classical cepheid type variable star.

Table 3.6. *Empirical Main Sequence.*

Spectral type	log (Mass) $M_\odot$	log (Rad) $R_\odot$	log (T_e)	M_V
B0	1.185	0.760	4.475	−3.3
B5	0.720	0.505	4.190	−0.88
A0	0.324	0.262	3.974	0.96
A5	0.215	0.185	3.911	1.87
F0	0.126	0.117	3.844	2.72
F5	0.070	0.072	3.810	3.38
G0	0.006	0.018	3.772	4.25
G5	−0.044	−0.026	3.762	5.01
K0	−0.106	−0.083	3.715	5.93
K5	−0.259	−0.227	3.633	8.20
M0	−0.375	−0.339	3.604	9.90
M5	−0.58	−0.54	3.513	13.0
M8	−1.0	−0.96	3.418	19.5

The correlation (3.38) has an essentially empirical nature. As it happens, Barnes and Evans found the $V - R$ index to be a reasonably stable colour monitor of surface flux in the visual over a wide range of stellar types and luminosity classes. Generally speaking, however, there is no strong *a priori* reason why a connection between colour and effective temperature, which may be valid over a limited domain, should extend itself over a wide range of spectral classes or surface gravities without losing significant accuracy.

Eclipsing binaries which have well-determined pairs of radial velocity curves, though without established parallaxes, directly furnish only masses and radii of the component stars. The absolute luminosities are of equal physical significance as the masses, but in order to determine these for stars of known radii we must also know their temperatures. These can be derived from colour–temperature relations, such as the foregoing. In Table 3.6 an empirical Main Sequence has been provided from running smooth curves through the double-lined eclipsing binary solutions for radii, masses and M_V values of normal unevolved stars quoted by D. M. Popper (1980). Popper's M_V values were determined from using a Barnes–Evans type relation to find the temperatures, and then applying his own bolometric correction scale.

3.8 Bibliographical notes

Fundamentals of photometry related to the vision based units are covered in J. W. T. Walsh's *Photometry*, Constable and Co., London, 1953. (The

SI system of units is expounded in S. Dresner's *Units of Measurement*, Harvey, Miller and Medcalf, Aylesbury, 1971, where the modern definitions of the candela and lumen are given, and it is pointed out that these were not changed by the CGS → SI transition.) The moment integrals of the intensity are also covered in numerous astrophysical texts which discuss radiative transfer — e.g. V. Kourganoff's *Basic Methods in Transfer Problems*, Clarendon Press, Oxford, 1952.

The original references on black body radiation laws are: A. J. Stefan, *Wien Ber.*, **79**, 397, 1879; L. Boltzmann, *Ann. Physik (ser. 3)*, **31**, 291, 1884; M. Planck, *Ann. Physik (ser. 4)*, **4**, 553, 1901; W. Wien, *Phil. Mag. (ser. 5)*, **43**, 214, 1894; Lord Rayleigh, *Phil. Mag.*, and **49**, 539, 1900; J. H. Jeans, *Phil. Mag. (ser. 6)*, **17**, 229, 1905. A variety of useful alternatives are given in Allen's *Astrophysical Quantities*, 1973, Athlone Press, London, (pp104–107).

For a recent review of measurements of the total solar luminosity, see H. S. Hudson, *Annu. Rev. Astron. Astrophys.*, **26**, 473, 1988. The solar irradiance distribution used in Figure 3.3 was from M. P. Thekaekara, and A. J. Drummond, in *Nature Phys. Sci.*, **229**, 6, 1971. An updated version of this was given by H. Neckel and D. Labs, in *Sol. Phys.*, **90**, 205, 1984 (see also G. Thuillier, *et al.*, *Science*, **225**, 182, 1987.) This was discussed recently by E. A. Makarova, *et al.* in *Astron. Zh.*, **66**, 583, 1989, who provided an alternative reference distribution. More discussion on the absolute calibration of the magnitude scale can be found in *IAU Symp. 54: Problems of Calibration of Absolute Magnitudes and Temperaturess of the Stars*, (ed. B. Hauck and B. E. Westerlund) Reidel, Dordrecht, 1973 (e.g. A. D. Code's article, p131).

The classic papers on stellar bolometry and angular diameters of E. Pettit and S. Nicolson *Astrophys. J.*, **68**, 279, 1928, and G. P. Kuiper *Astrophys. J.*, **88**, 429, 1938, were updated in D. L. Harris' review in *Basic Astronomical Data* (ed. K. A. Strand) University Press, Chicago, 1963, p273. Further revisions have been made with the aid of checks on effective temperature determinations via colour temperature calibrations, particularly since the publication of the 'Barnes–Evans relation' by T. G. Barnes and D. S. Evans, *Mon. Not. R. Astron. Soc.*, **174**, 489, 1976, and T. G. Barnes, D. S. Evans and T. J. Moffett, *Mon. Not. R. Astron. Soc.*, **183**, 285, 1978. An increased supply of data from above the absorbing layers of the atmosphere, or through windows in the infra-red region has also enabled improved net flux determinations, leading to calibrations such as that of A. D. Code *et al.*, *Astrophys. J.*, **203**, 417, 1976, and D. S. Hayes, *IAU Symp. 80: The H–R Diagram*, 1978, p65. D. M. Popper, *Annu. Rev. Astron. Astrophys.* **18**, 115, 1980, provided empirical data from analysis of binary stars to compare with such scales. A review of stellar temperatures was given by E. Böhm-Vitense,

Annu. Rev. Astron. Astrophys. **19**, 295, 1981, and it may be worthwhile to repeat here her apology about numerous other contributions to which space limitation prevents reference. The data of Tables 3.1 (BC_*) and 3.2 have come from combining the information in the foregoing sources, and taking appropriate averages to smooth trends.

The ten primary standards of Table 3.4 come from H. L. Johnson's article in *Basic Astronomical Data* (ed. K. A. Strand) University Press, Chicago, 1963, p204. The *UBV* filter transmission coefficients whose normalized forms have been presented in Table 3.3, originally appeared in *Astrophys. J.*, **114**, 522, 1951. More detailed discussions of these filters can be found in e.g. M. Golay's *Introduction to Astronomical Photometry*, Reidel, Dordrecht, 1974, and V. Straizys' *Multicolour Stellar Photometry*, Mokslas Publ., Vilnius, 1977. Much of Section 3.6 has been adapted from Chapter 2 of Golay's text with some differences of treatment and application.

The Barnes–Evans relation was introduced in 1976, and many papers have followed in its wake. Increased detective capability in the infra-red has allowed fuller integration of the flux distribution, giving rise to the 'infra-red flux method' of D. E. Blackwell *et al., Mon. Not. R. Astron. Soc.,* **221**, 427, 1986, so that the process of stellar temperature determination continues to be checked and refined (cf. e.g. S. Arribas and C. Martinez Roger, *Astron. Astrophys.,* **215**, 305, 1989).

Table 3.6 has been adapted from Budding's (1982) article in *Investigating the Universe* (ed. F. D. Kahn) Reidel, Dordrecht, p271.

4

Themes of Astronomical Photometry

4.1 Extinction

Extinction, or the attenuation of light as it passes through a medium, can be explained on the basis of the equation for the transport of radiation, written as

$$\mu \frac{dI_\lambda}{ds} = -\kappa_\lambda \rho I_\lambda + j_\lambda, \tag{4.1}$$

where I_λ represents the intensity (see (3.2)) at wavelength λ and in the direction $\arccos \mu$ to the outward direction of spatial coordinate s, while κ_λ denotes a general extinction coefficient per unit mass of the medium, of density ρ, through which the radiation is passing. Emission of intensity j_λ adds into the beam in this formulation, but in the usual situation for optical transmission from source to detector j_λ is effectively zero. The intensity then has the simple form:

$$I_\lambda = I_{\lambda_0} \exp(-\int_0^s \kappa_\lambda \rho \, ds/\mu). \tag{4.2}$$

4.1.1 Atmospheric extinction

We can apply (4.2) to the illuminance f above and below the Earth's atmosphere to find the radiation intercepted by a real detector at an Earth-based observatory. The illuminance will be the star's surface mean intensity multiplied by the solid angle which its disk subtends, (3.17), and we write

$$f(\lambda, \zeta) = f_0(\lambda) \exp[-k(\lambda) X(\zeta)], \tag{4.3}$$

or in magnitudes

$$m(\lambda, \zeta) = m_0(\lambda) + 2.5 \log_{10} e \, k(\lambda) X(\zeta), \tag{4.4}$$

where the extinction coefficient $k(\lambda)$ is regarded as a function of wavelength (taken to be uniform through the atmosphere), and $X(\zeta)$ represents the mass of air through which the radiation penetrates from the source at zenith distance ζ; $m_0(\lambda)$ denotes the apparent magnitude above the atmosphere. Equation (4.4) is sometimes called Bouguer's law — its convenient linear form in the magnitude scale has already been mentioned: though, strictly speaking, this linearity only applies for the monochromatic magnitude $m(\lambda, \zeta)$. In applying (4.4) to broadband, e.g. UBV, photometry, account has to be be taken of finiteness of the passband of the filters. Departures from a strict linearity will rise with increasing bandwidth of the filter, with a significance which depends on the accuracy standards set or required. The human eye, for example, although essentially a quite broadband detector, has accuracy limitations which would allow the preceding linear form to be quite adequate.

In order to take account of the finite passband effects at a given wavelength λ_0, we rewrite equation (3.27) as

$$f_W(\zeta) = f(\lambda_0, \zeta)W_0 + \frac{1}{2}f''(\lambda_0, \zeta)W_2.$$

We substitute from (4.3) for $f(\lambda_0, \zeta)$ in this, and then convert from the linear scale of fluxes to the logarithmic scale of magnitudes, using natural logarithms multiplied by -1.086 ($= -2.5\log_{10}e$). This allows the convenient approximation $\ln(1 + x) \simeq x$, valid for small x, and enables the exponential terms to be directly simplified. After a little manipulation we find

$$\begin{aligned} m_{\lambda_0}(\zeta) &= m_{\lambda_0}(0) + 1.086X(\zeta)\Big\{k(\lambda) + \\ &+ \frac{W_2\, k'(\lambda) d\log f_0(\lambda)}{W_0\, d\lambda} + \frac{W_2 k''(\lambda)}{2W_0}\Big\}_{\lambda=\lambda_0} - \\ &- \frac{1.086k'(\lambda_0)^2 X(\zeta)^2 W_2}{2W_0}, \end{aligned} \tag{4.5}$$

where $m_{\lambda_0}(0)$ is not quite the same as $m_0(\lambda_0)$ would be in (4.4), since it refers to a finite bandwidth filter centred at λ_0, and so has absorbed a small term (of order W_2/W_0) in $f_0''(\lambda)$, which does not depend on the air mass.

We have seen in (3.31) that $d\log f/d\lambda$ can be approximated as $5(\lambda_{max} - \lambda_0)/\lambda_0^2$, which, using (3.23) and (3.24), we are able to rewrite using a colour index (say, $B - V$). A form, ideal or observationally obtained, for the dependence of extinction coefficient $k(\lambda)$ on wavelength is also useful. A power-law is often adopted, e.g.

$$k(\lambda) = \text{const.}\lambda^{-n}, \tag{4.6}$$

and for the Rayleigh law of scattering on atmospheric molecules, $n = 4$. Such a form may be more or less valid in practice, depending on observing and climatic conditions during the observations, but we can thereby analyse (4.5); $k'(\lambda)$ can be replaced by $-nk(\lambda)/\lambda$, and $k''(\lambda)$ by $n(n+1)k(\lambda)/\lambda^2$.

We may now examine the various contributions on the right hand side in (4.5):

(i) The principle term $\sim k(\lambda)X(\zeta)$, which shows the approximate validity of Bouguer's law, as expected.

(ii) The colour-dependent term $\sim 5n(\lambda_{max} - \lambda_0)W_2/\lambda_0^3 W_0$. Over the range of encountered surface temperatures of stars $(\lambda_{max} - \lambda_0)/\lambda_0$ is, in general, of order unity. $W_2/W_0\lambda_0^2$ is typically of order 5×10^{-3} for filters like the UBV ones (Table 3.3). Hence, this term is of order a few hundredths of a magnitude, and not negligible for detailed work.

(iii) The term in k'' works out to be of order $n(n+1)W_2/2\lambda_0^2 W_0$ which, though not negligible, only appears as a constant factor multiplying the air mass, and thus may be absorbed into the main extinction term.

(iv) Finally, the second order term in the air mass, though amplified by the coefficient n from the power-law $k \sim \lambda^{-n}$, nevertheless is also multiplied by k^2, in addition to the small term $W_2/\lambda_0^2 W_0$. Since k is of order 10^{-1}, this term will be of order 10^{-3} for broadband filters, and so probably negligible in general, though not if the most precise reductions are pursued.

An empirical test on real data is likely to show the extent of non-linearity of Bouguer's law. What is often found, though, are irregularities of a greater scale than the minor contributions (ii)–(iv), due to an atmosphere which has more complicated (and time-dependent) properties than those of our simplified model.

4.1.2 The atmospheric mass

The mass of air through which starlight passes should, at least for lower values of the zenith angle ζ, be proportional to $\sec\zeta$. This is based on regarding the atmosphere as made up of plane stratified layers. The approximation is tolerable for zenith angles up to 60°. To account for the effects of curvature of the atmospheric layers use is frequently made of the formula

$$\begin{aligned} X(\zeta) = \sec\zeta & -0.0018167(\sec\zeta - 1) - \\ & -0.002875(\sec\zeta - 1)^2 - \\ & -0.000808(\sec\zeta - 1)^3 - \ldots, \end{aligned} \tag{4.7}$$

which is correct (at sea level) to within 0.001 for $X(\zeta)$ up to about 6.8, and has at least 1% accuracy for $X(\zeta) < 10$.

How do we determine ζ? Here we outline some general procedures for coordinate transformations, which will be used in this and other sections of the text. Readers already familiar with such details may proceed directly to the result (4.9).

Three rotation matrices are introduced which take the form:

$$\mathbf{R}_x(\theta) = \begin{vmatrix} 1 & 0 & 0 \\ 0 & \cos\theta & \sin\theta \\ 0 & -\sin\theta & \cos\theta \end{vmatrix},$$

$$\mathbf{R}_y(\theta) = \begin{vmatrix} \cos\theta & 0 & -\sin\theta \\ 0 & 1 & 0 \\ \sin\theta & 0 & \cos\theta \end{vmatrix},$$

$$\mathbf{R}_z(\theta) = \begin{vmatrix} \cos\theta & \sin\theta & 0 \\ -\sin\theta & \cos\theta & 0 \\ 0 & 0 & 1 \end{vmatrix},$$

where θ represents a positive (anticlockwise) rotation about the positive axis given as the suffix to $\mathbf{R}$. These matrices operate on a three coordinate set (x, y, z), representing some fixed point, arranged as a vector $\mathbf{x}$. The result of the matrix vector product is a new three coordinate set corresponding to new axes, transformed as a result of the operation $\mathbf{R}$ about whichever of the three axes was chosen to rotate about. A succession of n operations corresponds to repeated multiplications of the type $\mathbf{R}_n \cdot \mathbf{R}_{n-1} \cdots \mathbf{R}_1 \cdot \mathbf{x}$.

We carry out a sequence of rotations which takes us from the west-line to the line of sight to the star, via the celestial equator, as an example (Figure 4.1). We start with an alt-azimuth system of axes and the origin is identified with the observer's position. The z-axis points to the zenith, the x-axis points horizontally to the west, and the y-axis, by the requirement of a three-dimensional right handed orthogonal set, points south. We rotate first about the x-axis by the angle $90 - \phi$, where ϕ is the observer's latitude. This will bring the z-axis into alignment with the polar axis (OP). Next we rotate about the newly directed z-axis by $90 - H$, where H is the star's hour angle. A third rotation about the new y-axis by $-\delta$, where δ is the star's declination, brings the x-axis into alignment with the line of sight to the star. We identify this orientation by setting $x_1 = 1, y_1 = 0, z_1 = 0$, i.e. the direction cosines for the star in the new system. Symbolically, for the transformation to the unit vector following from this sequence of rotations, we have:

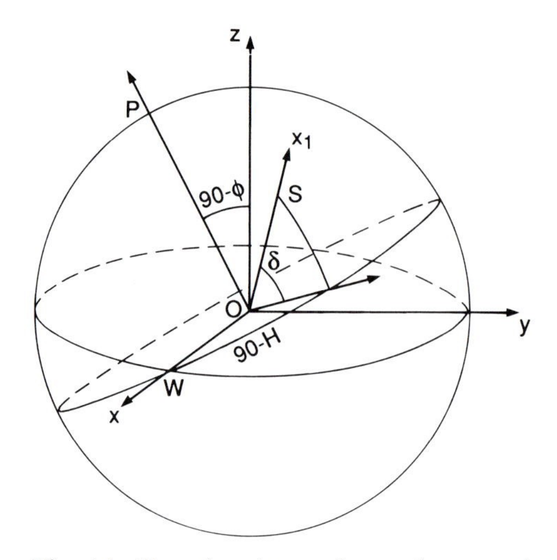

Fig. 4.1. Rotational transformations on the celestial sphere.

$$\mathbf{x}_1 = \mathbf{R}_y(-\delta)\cdot\mathbf{R}_z(90 - H)\cdot\mathbf{R}_x(90 - \phi)\cdot\mathbf{x}, \tag{4.8}$$

or putting in the full components

$$\begin{vmatrix} 1 \\ 0 \\ 0 \end{vmatrix} = \begin{vmatrix} \cos\delta & 0 & \sin\delta \\ 0 & 1 & 0 \\ -\sin\delta & 0 & \cos\delta \end{vmatrix} \begin{vmatrix} \sin H & \cos H & 0 \\ -\cos H & \sin H & 0 \\ 0 & 0 & 1 \end{vmatrix} \begin{vmatrix} 1 & 0 & 0 \\ 0 & \sin\phi & \cos\phi \\ 0 & -\cos\phi & \sin\phi \end{vmatrix} \begin{vmatrix} \lambda \\ \mu \\ \nu \end{vmatrix},$$

where λ, μ, ν are the direction cosines of the star in the original (alt-azimuth) system. An important point in the whole procedure is its reversibility; hence we may write for the sought components of the vector to the star in the alt-azimuth system:

$$\begin{vmatrix} \lambda \\ \mu \\ \nu \end{vmatrix} = \begin{vmatrix} 1 & 0 & 0 \\ 0 & \sin\phi & -\cos\phi \\ 0 & \cos\phi & \sin\phi \end{vmatrix} \begin{vmatrix} \sin H & -\cos H & 0 \\ \cos H & \sin H & 0 \\ 0 & 0 & 1 \end{vmatrix} \begin{vmatrix} \cos\delta & 0 & -\sin\delta \\ 0 & 1 & 0 \\ \sin\delta & 0 & \cos\delta \end{vmatrix} \begin{vmatrix} 1 \\ 0 \\ 0 \end{vmatrix},$$

where, since the rotations now have been reversed, the off-centre-diagonal elements change sign. It is the third component ($\nu = \cos\zeta$) of the vector on the left hand side of the equation whose evaluation prompted the present discussion. This can be determined in terms of the information we should have on the star (i.e. δ and H) by progressively multiplying out the matrix-vector combination from the right hand side in the foregoing. The bottom line of the vector, thus multiplied out, gives $\cos\zeta = \cos\phi\cos H\cos\delta + \sin\phi\sin\delta$, or, what we actually required,

$$\sec\zeta = (\cos\phi\cos H\cos\delta + \sin\phi\sin\delta)^{-1}. \tag{4.9}$$

An alternative derivation using well-known formulae of spherical trigonometry is also possible, of course. The foregoing matrix and vector approach, however, with its systematic procedure for the signs of angular arguments and functions, has some advantages in avoiding possible sign ambiguities.

4.1.3 Interstellar extinction

Unlike atmospheric extinction, which generally impedes astronomical photometry (though its effects can usually be accounted for), interstellar extinction is itself a source of inherent interest, yielding insights into the nature of the interstellar medium. Before its effects were properly determined there was a mistaken impression of the scale of distances of galactic proportions. We now know that it can reduce starlight, at visual wavelengths, by as much as 2 magnitudes over a thousand parsecs of space, with even heavier extinction in some localized regions of smaller size. The effect is sensitive to the choice of direction: away from the galactic plane the absorption declines quite steeply, within the plane there is considerable clumpiness and irregularity to the light reduction.

The reasoning which took us from (4.1) to (4.5) also applies to the interstellar extinction integrated over broadband filters. The inverse λ-dependence is not so steep as for atmospheric molecules, however, except perhaps in the infra-red, where, longward of about 0.8 μm, the decline becomes relatively swift. Over most of the optical spectrum the extinction law has a reasonably uniform λ^{-1} form (Figure 4.2). There have been differences between authorities on the range of validity of this uniform law, and there are technical difficulties hindering accurate calibrations of relevant quantities, e.g. independently found distances of highly reddened stars.

From about 4000 Å shortwards the variation of magnitude of light loss with reciprocal wavelength becomes less steep, but more dependent on the choice of galactic direction — which and how many galactic 'arms' the light passes through. The form of the inverse power-law dependence on wavelength suggests a distribution of scattering particles generally larger than atmospheric molecules is responsible for the extinction — dust or grains are usually thought of, having typical sizes comparable to the wavelength of optical radiation.

The units were changed in passing from (4.2) to (4.3). In the expression (4.2) κ has the dimensions of area ('cross-section') per unit mass, the coefficient $k(\lambda)$ in (4.3), however, is normalized to the units of (atmospheric mass)$^{-1}$, and the areal aspect is lost sight of. $k(\lambda)$ can be converted back to a cross-section per unit mass if its usually quoted value is divided by

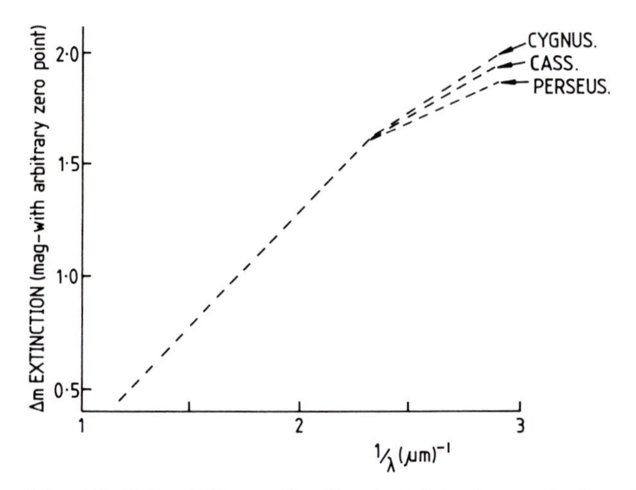

Fig. 4.2. Interstellar extinction trend in the optical region.

the 'column density' of the atmosphere (mass of a column of air with unit cross-sectional area and length equal to the mean height of the atmosphere $\approx$ 10350 kg m^{-2}). In the interstellar context there is no real equivalent to atmospheric mass — the corresponding part of the extinction exponent reverts to a column density. A possible unit for this is solar masses per square parsec along the direction to a given star, of determinable interstellar extinction. The interstellar equivalent to $k(\lambda)$ (extinction per unit column density, thus expressed) would then typically turn out to be of order unity at optical wavelengths. In any case, interstellar extinction follows the form (4.5) to increase the apparent magnitude of the source. It is made up of a principal term — a product of cross-section per unit mass, column density of the interstellar matter and the scaling constant 1.086 — together with small contributions (of order W_2/W_0) proportional to the colour and the square of the column density.

Photometric information is often sought for a group of cosmogonically related objects, e.g. a cluster of stars, in the form of a colour–magnitude (or two-colour) diagram. For practical purposes, the individual cluster members are regarded as all at essentially the same distance from Earth, with some intervening interstellar matter. If there were a uniform extinction law, then every star would be displaced from a position corresponding to its intrinsic magnitude m_{λ_1}, and colour C_{λ_1,λ_2}, by an amount depending on the extinction A_{λ_1}, and the excess E_{λ_1,λ_2} of extinction at the shorter wavelength over that at the longer. This displacement, the same (to first order) for all the stars in the cluster, fixes the cluster's 'reddening line'.

This suggests a method of probing the interstellar medium to obtain column densities to particular clusters. Turning to the two-colour diagram, the additive term $5\log\rho$ to apparent magnitude m_{λ_1}, (equation (2.2)), is subtracted out; colour–colour diagrams of various clusters can thus be directly compared. Given the direction of the reddening line, which, in the first approximation characterizes all interstellar space, the amount of movement required to superpose one cluster's distribution of points on another of similar type fixes the relative amount of interstellar matter. If $k(\lambda)$ for the interstellar medium were simply proportional to λ^{-1} then the slope of the reddening line would be $(\lambda_1^{-1} - \lambda_2^{-1})/(\lambda_3^{-1} - \lambda_4^{-1})$, where λ_1 and λ_2 are the wavelengths characterizing the magnitudes whose difference defines one of the two colours; λ_3 and λ_4 are similarly associated with the other colour. For the frequently used *UBV* system this approximation for the ratio E_{UB}/E_{BV} is about 1.3. This slope is appreciably greater than typical values found in practice ($\sim$ 0.8). In actuality, though, extinction effects applying to real *UBV* observations of clusters are more complicated than would be apparent from such a *prima facie* approach.

In the first place, the form of the extinction law shortward of about 4000 Å no longer varies simply like λ^{-1}, and there is a clear dependence on direction. Such effects are better interpreted against a three-dimensional map of the galactic environment through which we are observing. Looking along the galactic arm in which the Sun is located, for example, close to the disk in the direction of the constellation Cygnus, the relative increase of extinction with decreasing wavelength maintains itself rather well compared with other directions. In a roughly opposite galactic longitude, towards Orion, distinct fall-offs from the law beyond 4000 Å are observed, particularly near regions of nebulosity. Furthermore, second order effects in the extinction, which affect wide bandwidth filters, can be seen from (4.5) to introduce terms which would alter the linear prescription by a few per cent.

As a practical approach, relationships of the type:

$$\frac{A_{\lambda_1}}{E_{\lambda_1\lambda_2}} = R + a_1 C_{\lambda_1\lambda_2} + a_2 E_{\lambda_1\lambda_2},$$

or

$$\frac{E_{\lambda_3\lambda_4}}{E_{\lambda_1\lambda_2}} = S + b_1 C_{\lambda_1\lambda_2} + b_2 E_{\lambda_1\lambda_2}, \tag{4.10}$$

are sought, where R, S, a_1, a_2, b_1 and b_2 are constants. A frequently quoted version of (4.10), for instance, is

$$E_{U-B}/E_{B-V} = 0.72 + 0.05E_{B-V}$$

(b_1, in this case, being small compared with S and b_2). Analysis of the equations (4.10), on the basis of general expressions for extinction of the type (4.5), leads to a quantitative understanding of the properties of the interstellar medium.

4.2 Broadband filters — Data and requirements

The most well known of the broadband photometric systems, i.e. *UBV*, evolved, with the addition of *U*, from one in which eye responses to starlight were compared with the more blueward actinic reaction of a photographic emulsion. From the start it was realized that the difference in magnitudes of stars in these two regions — what has come down to us as $B - V$ — should correlate with some physical variable: surface temperature being the most likely.

It was also realized early on that stars of a given colour and located at a similar distance, e.g. by being components of gravitationally bound systems, could have quite a large difference in intrinsic brightness, and so stars must come in inherently different sizes regardless of their surface temperature. The most basic subdivision is between 'giants' and 'dwarfs'. Dwarfs are much more commonly occurring stars spatially; it is dwarfs which make up the well-known Main Sequence on colour magnitude diagrams. For stars of a surface temperature similar to the Sun, and cooler, however, there are far too many relatively bright examples in the galactic field than can be accounted for by the known proportion of cool dwarfs. The giants, indeed, on a sequence of their own, become very conspicuous in the galactic field for the later spectral types G to M.

Giants and dwarfs occupy categories in another scheme based on size rather than surface temperature — the 'luminosity' classification. This classification reflects much more directly on stellar structure than inherent mass: both giants and dwarfs may exist for stars of a given mass. The scheme developed from the spectroscopic work of W.W. Morgan, P.C. Keenan and E. Kellman in the 1940s. Giants are assigned luminosity class III; dwarfs are class V. Dwarfs and giants are not the only size groups. There are super-giant stars (classes I and II), subgiants (class IV), as well as subdwarfs and occasional 'stragglers' that don't seem to fit easily into any clear luminosity class type.

One of the more prominent issues in the development of stellar astrophysics has concerned that very noticeable ultra-violet feature of spectra — the limit of the Balmer series of hydrogen lines at 3760 Å and the formation of the Balmer continuum shortward of this limit. The spectrum in this region is sensitive to the surface gravity of the source. It is not difficult to visualize, qualitatively, why this should be — it is a consequence of a reduced pressure, or more specifically electron pressure, in the atmospheres of giants relative to dwarfs. Simple arguments can be advanced as to why the pressure near the visible surface of a star should decline rather swiftly with increasing stellar radius at a given overall mass (e.g. a dimensional argument suggests pressure $\propto$ radius^{-4}).

The populations of the various levels of excitation, and the proportions of ionized atoms in the photospheric sources of the radiation will be influenced in their probable arrangement by the ambient electron concentration. These gravity-dependent populations determine the relative strengths of spectral features. This gives a reason to monitor photometric behaviour at the Balmer series limit; to measure the Balmer 'jump' or 'decrement' which occurs there. The U filter, whose maximum transparency lies on the other side of the jump to that of the B filter, thus made a natural entrance to a developing observational programme for rapid retrieval of useful information from a limited input of photon flux.

This leads on to questions about what information can be delivered from a given photometric system, what can be sought, and how these questions relate to each other. Temperature and size (luminosity) have been seen to be among the more basic targets of a three-filter system — the interstellar extinction is also a potential derivable, as well as information on the density and composition of the interstellar medium. Stellar population type, metallicity or element abundance ratios reflect more or less detailed information on the chemical composition of a given star. Quantities which measure these can be sought from a suitably chosen filter set, but more output derivables imply more input specifiers. Other facts of interest, e.g. masses, structure or evolution may be not so immediately derived from photometry alone, but inferred when other kinds of data are added.

If the UBV system arose in a natural way from preceding methods and available resources, that, unfortunately, is no guarantee of its optimal suiting to physical requirements. Drawbacks which are often commented on include the fact that it is the Earth's atmosphere, and thus something outside the desirable laboratory control concept, which defines the short-wave cutoff of the U filter. The B and U filters overlap each other in their response curves, and indeed both overlap the Balmer jump itself, thus a main underlying target of the $U - B$ colour measure appears not so efficiently achieved. In the original setting up of the UBV system it was the photoelectric cathode response function which determined the long wavelength cutoff, and

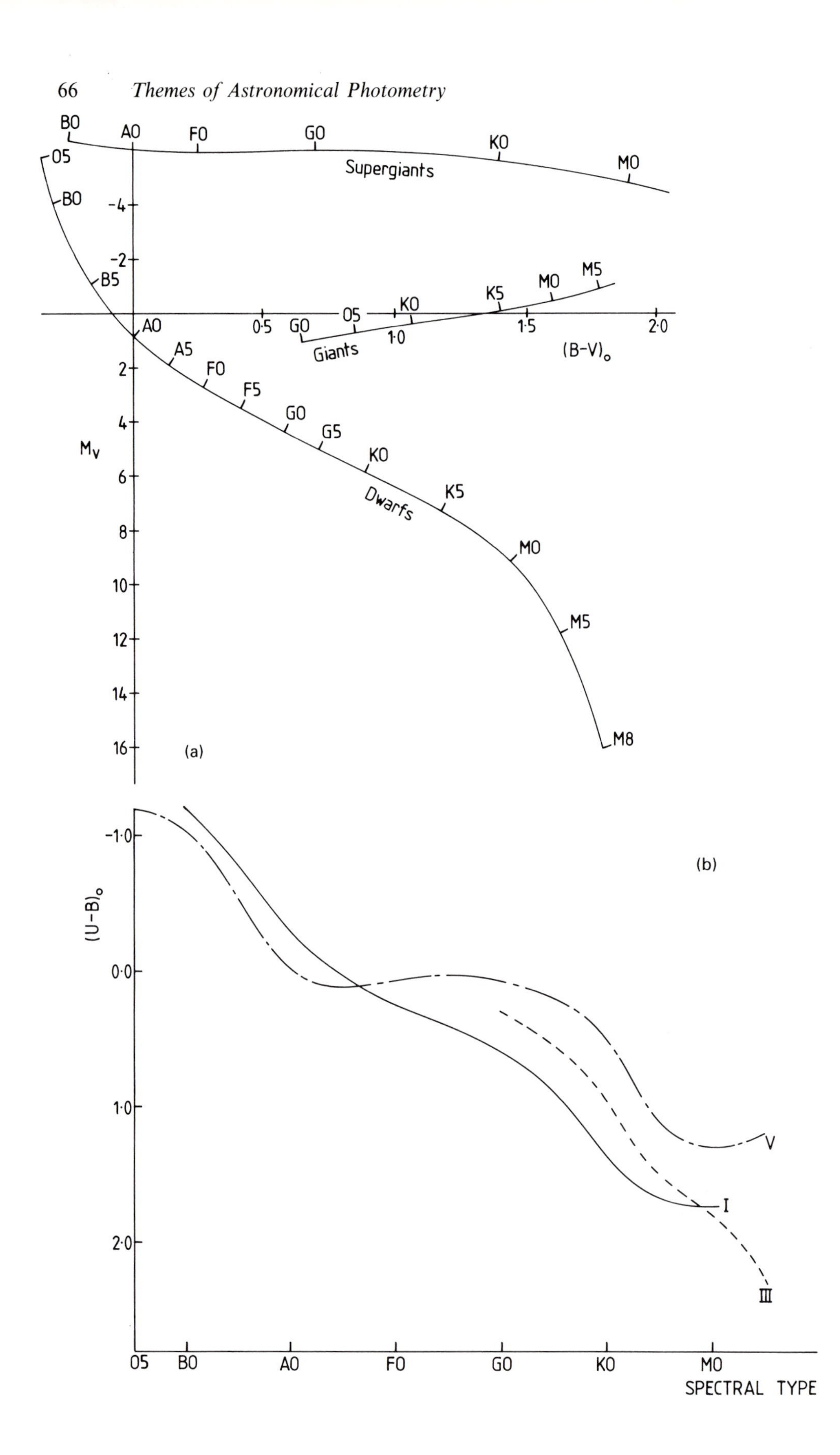

B0
A0
F0
G0
K0
M0
O5
Supergiants
B0
-4
-2
B5
M5
M0
K5
K0
A0
0·5
G0
O5
1·0
1·5
2·0
Giants
(B−V)o
A5
2
F0
F5
4
G0
G5
Mv
K0
6
K5
Dwarfs
8
M0
10
M5
12
14
16
M8
(a)
−1·0
(b)
(U−B)o
0·0
1·0
V
I
2·0
III
O5
B0
A0
F0
G0
K0
M0
SPECTRAL TYPE

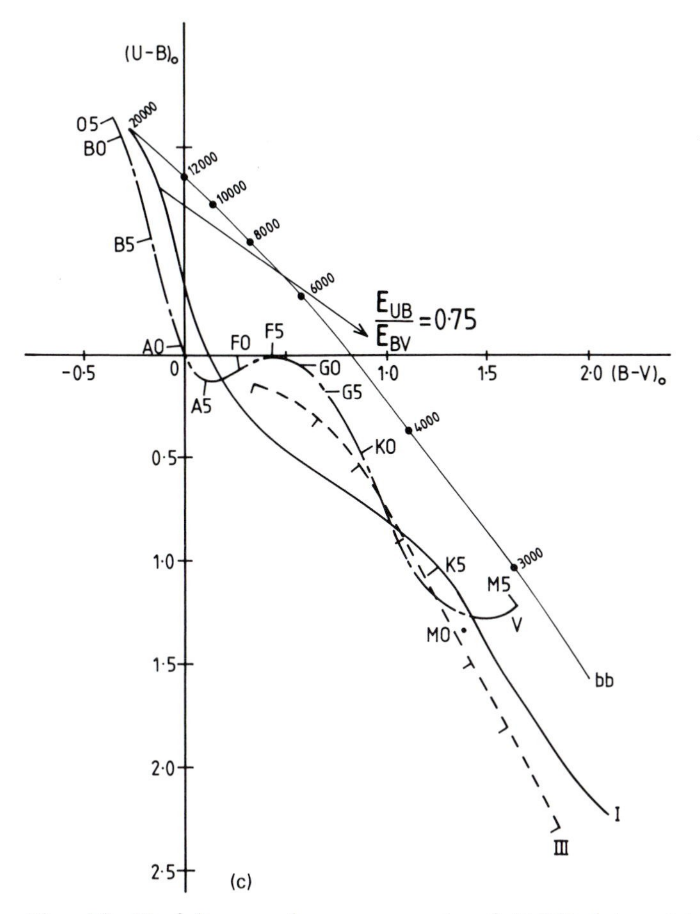

Fig. 4.3. Useful general arrangements of UBV data: (a) in the classic Hertzsprung–Russell form of colour versus absolute magnitude, (b) with $U - B$ against spectral type, and (c) as the two-colour diagram, $B - V$ versus $U - B$.

the astronomer was then subject to uncontrolled and somewhat unknown consequences of manufacturing.

But to see what the UBV system can achieve in practice consider the three diagrams which make up Figure 4.3. These diagrams have been constructed from data accumulated over many different investigations of field stars of known parallax, eclipsing binaries and clusters. They represent the general trends for stars of similar composition to the Sun, though they are not exact for particular cases.

Figure 4.3(a) is the basic schematic colour $(B - V)$ versus magnitude (absolute) M_V diagram. This diagram, or a theoretical version of it, is often called a Hertzsprung–Russell, or simply H–R, diagram after the two astronomers who first brought it to light. It has been found to be an extremely useful way of presenting astrophysical information on stars. Dwarfs, giants and supergiants separate out into their own regions of the diagram. Note the conventional sense of colour index increasing to the right implies that surface temperature increases to the left. Interstellar extinction moves stars downward and to the right, along the reddening lines appropriate for the stellar type and intervening medium's constitution.

Figure 4.3(b), which plots $U-B$ against spectral type, shows the effects of the Balmer decrement variation. Moving down from the top left hand corner of the diagram we find increasing absorption shortward of the Balmer limit. This causes $U-B$ to increase much more rapidly than would follow from the decline of photospheric temperature. In the temperature range of late B spectral types the Balmer features grow strong, and reach a maximum round about type A0 for the dwarf sequence (V). Thereafter, the Balmer decrement decreases, so that some of the excess $U-B$ it caused is made up, and the sequence bends back up again. The Balmer decrement remains relatively strong in giants (III) and supergiants (I), compared with dwarfs, at later, cooler spectral types. Reduced electron pressure at a given temperature in a higher luminosity star assists ionization, which implies a more positive $U-B$. This separation between the sequences is called the luminosity effect in $U-B$. The wobbles which occur lower down these curves are not related to the Balmer decrement, which becomes progressively weaker, but are due to the increasing net influence of absorptions by lines and bands in the transmission windows of the filters.

Figure 4.3(c) shows the two-colour $B-V$ versus $U-B$ diagram, sometimes called the *UBV* diagram. Since $U-B$ responds to the luminsity class of the star, this diagram can be regarded as a mapping of the $B-V$: M_V diagram, with the advantage of immediacy of the ordinate variable. The diagram also shows the trend of black bodies (bb) at different temperatures. The slope of the reddening line shown is 0.75. Somewhat different slopes could apply for different inherent colours and interstellar environments, as discussed previously. We see from this diagram that a unique unreddened starting position can only be deduced for an individual star which is either very early (O down to late B types) or very cool. There is intrinsic ambiguity for single intermediate type stars since more than one intersection of the reddening line can occur with the loci of the various stellar sequences, though it is sometimes possible to use *a priori* knowledge about the interstellar medium, or stellar properties, to allow a selection between alternative interpretations of results. For stars in a cluster, however, continuity of the sequence would remove such ambiguities.

The idea of automatic subtraction of the reddening displacement of an observed $U-B$ value for a given star is contained in the Q parameter of the UBV system, which was introduced with the system in 1953. It is defined thus:

$$Q=U-B-E_{UB}(B-V)/E_{BV}=(U-B)_0-E_{UB}(B-V)_0/E_{BV}. \quad (4.11)$$

Since E_{UB}/E_{BV} is a known constant, at least to a first approximation, (4.11) indicates a way of deriving the unreddened indices, because the $(U-B)_0$ index can be regarded as a known function of $(B-V)_0$ for certain types of stars. For B type dwarfs, for example, a reduced form of (4.11),

$$(B-V)_0 = 0.332Q, \quad (4.12)$$

has sometimes been quoted.

Although the linear relationship between $U-B$ and $B-V$ indices, implied by (4.12), is no longer so good outside of a rather restricted range of conditions, the idea of a self-compensating parameter has been maintained with other schemes involving different filters than UBV. Indeed, the RGU combination is rather well suited to a reddening-free parameter. Its characteristic wavelengths were especially chosen so that the ratio E_{UG}/E_{GR} would approximate unity. While still regarded as a broadband system, the bandwidths of the R, G and U filters are only about a half those of the UBV. The system was introduced by W. Becker in the 1940s, and its applications have been mainly in photographic statistical surveys of the galactic distributions of various stellar types. A more direct separation of disk and halo populations can often be made using RGU rather than UBV photometry.

It is not only interstellar reddening which causes the received continuum to deviate from that produced by light scattering, free-free or free-bound recombinations of atoms in the source. The formation of absorption lines at discrete wavelengths has a cumulative feedback on the continuum. The net effect over a broadband region may be either flux reduction — 'blocking' by the sum of individual line absorptions; or the compensatory squeezing out of more flux in some other region — 'backwarming'. These blocking or backwarming effects, collectively referred to as 'blanketing', depend on the relative amounts of different absorbing atoms or ions in the source which are contributing at the given wavelength. The effects become relatively enhanced for cooler stars (Figure 4.3(b)). In the differential photometric comparison of stars, at given surface temperature and gravity, measures of blanketing thus trace composition, particularly with regard to the relative abundances of heavier elements — 'metals' in astrophysical parlance.

Just as it was possible to use reddening lines in the two-colour diagram to study the nature and amount of interstellar extinction, so 'blanketing lines' can be drawn, which relate to the relative metallicity, or stellar population type, of the source object. Unlike reddening lines, however, which have a roughly constant slope at all positions in the two-colour diagram, blanketing lines are dependent on local source conditions. Empirical formulae for such lines in given sequences have been produced, and, while of an exploratory character, they are of interest in matching theory to observation for stellar atmospheres. More purpose-oriented filter systems than *UBV* are preferred for such research.

Before the introduction of *RGU*, Stebbins and Whitford had compiled interesting information on stars of different types, including effects of the interstellar medium, using a set of six broadband filters. Two of these, *U* and *G* are not too dissimilar from the *U* and *V* of Johnson's *UBV*. In place of *B*, however, are two filters: *V* (for violet) whose transmission peak is located at around 4000 Å, and a *B* which peaks at around 4500 Å, but, unlike the Johnson *B* filter, extends its transmission on the long wavelength side to around 6000 Å. There are two filters transmitting at longer wavelengths than visual — *R*, peaking at about 6800 Å, and *I*, with peak at about 10 000 Å. The system was originally used with a photoconductive cæsium oxide cell, and, using a 60-inch telescope, signal to noise ratios were acceptable down to ninth magnitude.

The basic information on these filters is summarized in Table 4.1, which may be compared with that at the bottom of Table 3.3. The ordinate scale for the quadratures was taken directly from the original publication, where the transmission data were not normalized, but included the cell's response function, which has therefore been folded in with the W_0 and W_2 values of Table 4.1. In this way there is an indication of relative net response over the wavelengths.

Figure 4.4 shows some of the results of observations made with this six-colour system and comparisons with corresponding differential colour curves of black bodies. Straight lines give reasonable first order approximations to these curves, so that surface temperature is relatively well determined. The existence of inherently more information in a six, over a three, data point specification for a star is noteworthy, however. It allows an extra specification on accuracy for a basic parameter like temperature. The prolongation of the Balmer decrement through the supergiant colour curves (i.e. the relative drop-off in *U*) is also easily noticed in Figure 4.4.

Early recognition of the general applicability of a $1/\lambda$ law to interstellar extinction was substantially reinforced by Stebbins and Whitford using

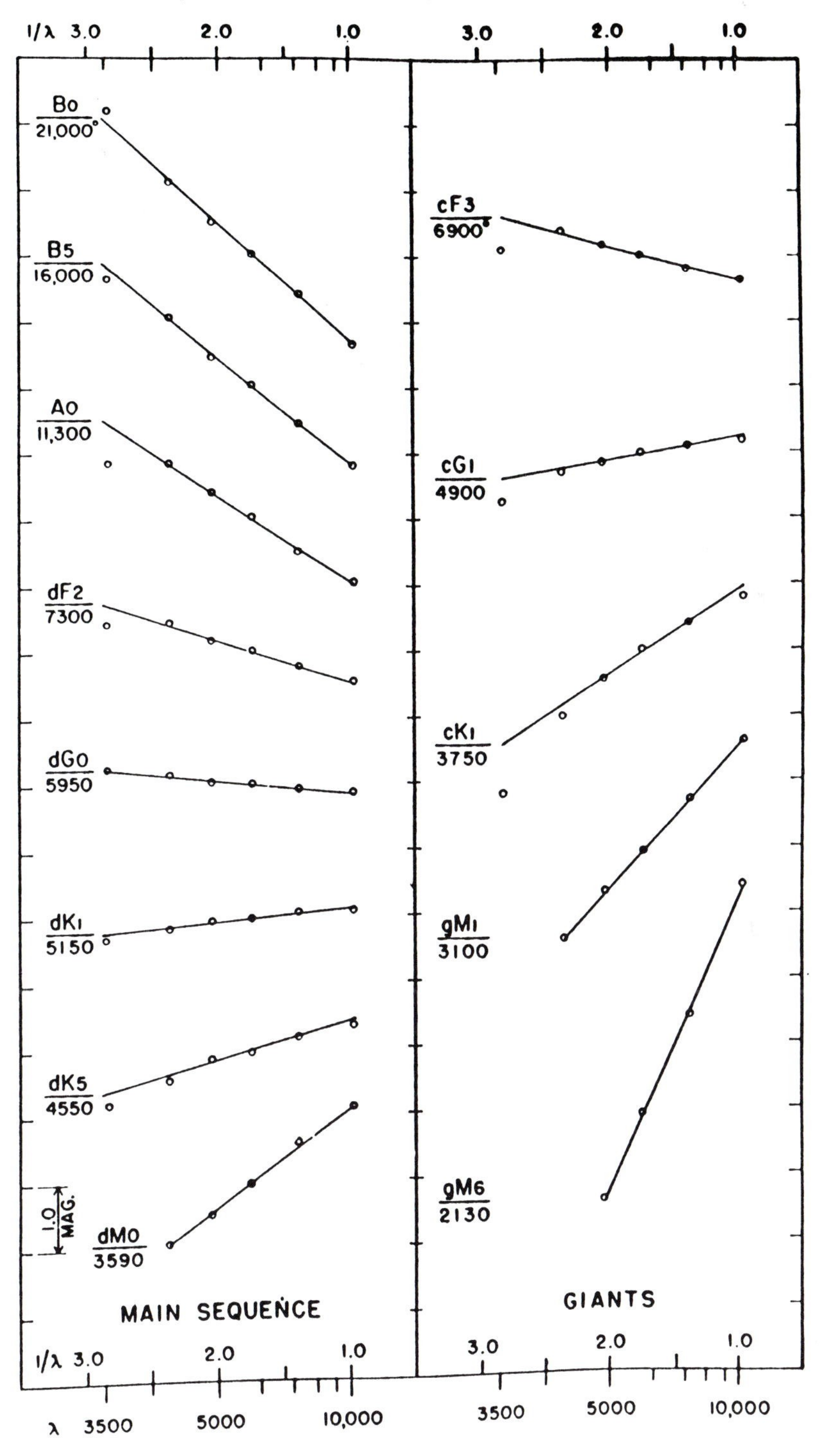

Fig. 4.4. Stellar photometry in the six-colour system of Stebbins and Whitford.

Table 4.1. *The six-colour system of Stebbins and Whitford.*

	U	V	B
λ_0 (μm)	0.352	0.422	0.489
W_0	2.42	3.78	4.36
W_2	8.76×10^{-4}	3.19×10^{-3}	8.15×10^{-3}
$W_2/(\lambda_0^2 W_0)$	2.92×10^{-3}	4.74×10^{-3}	7.81×10^{-3}
	G	R	I
	0.570	0.718	1.026
	4.53	6.91	8.50
	1.08×10^{-2}	3.36×10^{-2}	6.49×10^{-2}
	7.32×10^{-3}	9.45×10^{-3}	7.25×10^{-3}

their six-colour system. In Chapter 10 we will encounter six-colour observations of classical cepheid variables, where the advantage of the broad wavelength base index $G - I$ to effective temperature determination is used. The development of efficient and commercially available photomultiplier cells, at relatively reasonable costs, with, however, emissive surfaces responsive mainly in the visual to ultra-violet region (and not in the infra-red), may have prevented a more rapid and widespread adoption of the six-colour system. The more recent appearance of low-noise solid-state detectors of wider spectral range has allowed resumption of this, or comparable, work.

The advantages of a greater number of data points in a photometric specification, especially with regard to a suitable combination of magnitude differences enabling the separation of, for example, interstellar reddening and gravity, have led to numerous efforts to extend, subdivide, or modify the basic UBV system. Golay supplemented it with a $B_1B_2V_1G$ filter set, in which the original B band is subdivided into a B_1 and B_2 pair (approximately). He also arranged the V_1 and G combination to cover the V range. These filters are narrower than the original UBV, but they still somewhat overlap each other's response curves (B_1 and B_2; V_1 and G). $B_2 - V_1$ provides the basic temperature-related colour index of the system; other parameters are formed by appropriately scaled differences of colours. Information supplied by the various indices has been related to luminosity class and metallicity, taking into account auxilliary (computed) information to evaluate interstellar extinction.

It is perhaps more easy to expect interesting new information to come from extending the wavelength coverage of broadband photometry, though, rather than subdividing its range. Johnson did this about a decade after the introduction of the original UBV with an extension well into the infra-red:

Table 4.2. *The (UBV)RIJKLMN system: filter summary.*

	R	I	J	K
λ_0 (μm)	0.694	0.878	1.25	2.20
W_0	0.207	0.232	0.296	0.578
W_2	1.17×10^{-3}	1.75×10^{-3}	4.25×10^{-3}	1.82×10^{-2}
$W_2/(\lambda_0^2 W_0)$	1.18×10^{-2}	9.82×10^{-3}	9.19×10^{-3}	6.53×10^{-3}

	L_1	L_2	M	N
	3.54	3.44	5.03	10.3
	0.885	0.706	1.13	4.31
	8.00×10^{-2}	3.93×10^{-2}	0.221	13.7
	7.27×10^{-3}	4.72×10^{-3}	7.72×10^{-3}	2.97×10^{-3}

the so-called *UBVRIJKLMN* system. Summarizing data on these filters are given in Table 4.2 in the same way as below Table 3.3. The data in this table have been derived from Johnson's original paper, where alternative detectors were used for the L filter, resulting in the L_1 and L_2 columns. In practice, the lead sulphide cell, corresponding to L_2, has been adopted as L in subsequent observations.†

One of the main early applications of the system was in the study of interstellar extinction at wavelengths beyond the optical region. It was found that the approximate $1/\lambda$ dependence breaks down into others, generally of higher power in $1/\lambda$, but quite dependent on the choice of direction. There is thus evidence of different distributions of grain sizes in different cosmic locations. The question of a low-level neutral (i.e. independent of wavelength) contribution to interstellar extinction has provoked some attention, but efforts to answer this from photometric monitoring over extended wavelength intervals have not been conclusive so far. Such studies progress in wavelength regions on either side of the optical in the present age of space-based astronomy.

4.3 Photometry at intermediate bandwidths

The broadband systems discussed hitherto are characterized by several key points. Firstly, they have had a fairly clear development from historic photometry performed by such basic devices as photographic plate or eye. Secondly, the filters associated with these broad transmissions are usually, at least in the optical domain, made from coloured glass, or coloured glass

† An H-filter was later added to the system. Its characteristics are: $\lambda_0 = 1.65\,\mu$m, $W_0 = 0.281$, $W_2 = 2.11\times10^{-3}$, $W_2/(\lambda_0^2 W_0) = 2.78\times10^{-3}$

combinations, with broad, but not necessarily symmetric, transmission functions. Thirdly, such broad transmission characteristics allow the integrated Taylor expansion (3.26) to have reasonable validity for foldings with a real stellar irradiance function. At high resolution this irradiance has numerous discrete irregularities, which make the Taylor expansion derivatives awkward to define. In any case, very narrow wavelength sampling is likely to cause a particular brightness temperature, in the case of one filter, or colour temperature, in the case of two, to deviate from the closer approximation to source effective temperature which such measures tend to at broader passbands.

The special peculiarities of high-resolution sampling can be avoided, whilst also escaping difficulties coming from the second order terms in the Taylor development (3.26), with an appropriate choice of *intermediate* bandwidth filters. Such methods touch on the foregoing points in the following respects. Firstly, they have been led by astrophysical requirements, rather than historical continuity. Secondly, while it is still possible to devise an intermediate bandpass filter from a suitable overlap of shortpass and longpass edges in a combination of coloured glasses, it is generally much more convenient to employ filters of the interference type. Modern manufacturing techniques allow passbands with stringent requirements on transmissions and mean wavelengths to be effectively realizable over a wide optical range. Thirdly, part of this stringency in requirements is concerned with the positioning of the bands in relation to strong spectral features. These should be absent from the transmission range of any intermediate-band filter intended for determining a temperature scale; or, at least, such effects, if present in one filter, should be removed in an appropriate difference of magnitudes from two filters.

4.3.1 The uvby *system*

The most well-known intermediate-band system is the *uvby* four-filter combination, introduced following the published ideas of B. Strömgren, and the observational work of D. Crawford starting in the late 1950s. Essentially all the previous criticisms of the *UBV* system are now overcome. Magnitudes are effectively monochromatic, and defined by filters whose transmission functions do not overlap. Details on the filters are listed in Table 4.3 and the transmission curves are shown in Figure 4.5.

The basic visual magnitude parameter, generally regarded as V in the *UBV* system, is often taken to scale directly with the y-magnitude (y for yellow), since the mean wavelengths of both filters is about the same (5500 Å). This is not exactly true, however, since the heterochromatic V-magnitude includes

Table 4.3. *The* uvby *filter characteristics.*

	u	v	b	y
λ_0 (μm)	0.347	0.411	0.467	0.546
$\Delta\lambda^{(1/2)}$	0.038	0.02	0.01	0.02
$W_2/\lambda_0^2 W_0$ ($\times 10^4$)	1.50	0.80	0.75	0.96

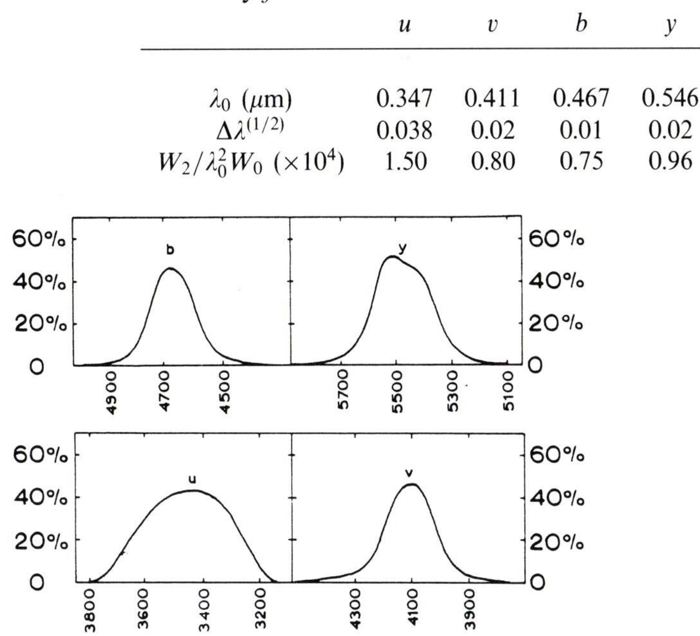

Fig. 4.5. Transmission curves of the *uvby* filters.

a small colour-dependent term, when compared with a monochromatic one, as can be seen from (3.27). If we convert that relation to the magnitude scale we obtain

$$\begin{aligned} -2.5 \log_{10} f_W &= -1.086 \ln f_W \\ &= -1.086 \ln \left\{ f(\lambda_0) W_0 \left[1 + \frac{f''(\lambda_0) W_2}{2 f(\lambda_0) W_0} \right] \right\}, \end{aligned} \tag{4.13}$$

to terms in the second order of small quantities.

Applying this to the V-filter, we have, to a sufficient accuracy,

$$V = \zeta_v - 1.086 \ln[f(\lambda_0) W_0] - 0.543 \frac{f''(\lambda_0)\, W_2}{f(\lambda_0)\, W_0}, \tag{4.14}$$

where ζ_v denotes an appropriate zero constant to tie in the scale with an adopted calibration (Section 3.3).

For the y-filter we can simply write

$$y = \zeta_y - 1.086 \ln[f(\lambda_0) W_y], \tag{4.15}$$

where λ_0 is the same (0.55μm) in both cases, and W_y represents the integral of the flux multiplied by the corresponding transmission of the y-filter. (Since only the zeroth moment need be considered for this filter for the present purpose, there is no loss of generality in dropping the zero suffix.)

Combining the foregoing, we have

$$V = y - 1.086 \ln \frac{W_0}{W_y} - 0.543 \frac{W_2 f''(\lambda_0)}{W_0 f(\lambda_0)} + \zeta_v - \zeta_y. \tag{4.16}$$

Using the Planck formula, and from the discussion which follows (3.31), with a little manipulation we find $f'(\lambda_0)/f(\lambda_0) \simeq 5(\lambda_{max} - \lambda_0)/\lambda_0^2$, and $f''(\lambda_0)/f(\lambda_0) \simeq 30(\lambda_{max} - \lambda_0)^2/\lambda_0^4$. Combining the first of these with the definition of the magnitude scale, we find,

$$\frac{5(\lambda_{max} - \lambda_0)}{\lambda_0^2} \simeq \frac{0.921(b - y)}{\Delta\lambda} + \zeta_c, \tag{4.17}$$

where the b to y wavelength difference $\Delta\lambda = 0.08$ μm. The quantity ζ_c arises from the somewhat arbitrary fixing of $b - y$ to be zero at a certain spectral type of dwarf star, or, in other words, a certain value for λ_{max}. Since $\lambda_{max} \simeq \lambda_0 \simeq 5500$ Å for stars whose $b - y$ is around 0.55, it follows that $\zeta_c \simeq -0.921 \times 0.55/0.08 = -6.33$. Therefore, removing the various zero terms in (4.16) by taking the V and y scales to be coincident for stars of colour $b - y = 0.55$, we can find,

$$V = y - 0.12[(b - y) - 0.55]^2. \tag{4.18}$$

This formula shows the expected scale of departure of y from strict V over a range of $b - y$ values on the basis of a Planck-like approximation to the radiative power distribution. In a real situation, however, such departures would be verified, as with all individual systems, from a programme of standard star observations.

The colour $b - y$ can be expected to parallel $B - V$ in its function of temperature measurer — though it is not necessarily the best suited magnitude difference combination for this purpose over the whole range of stellar types. The blocking by weak lines (the *uvby* filters tend to avoid strong lines for the most part) in b and y filters is, however, about the same for cooler stars, enhancing the effectiveness of $b-y$ as a temperature indicator in this range. To a first order approximation, where the linear gradient concept would apply, $b - y \simeq 0.68(B - V)$.

In the *UBV* system the third parameter $U - B$ was one which reflected intrinsic luminosity, via the Balmer decrement variation. With the four magnitudes *uvby*, we can regard B as having split into the v (for violet) and b (blue) pair, offering different possibilities concerning, not only the Balmer absorption edge, but also more general line absorption effects. The

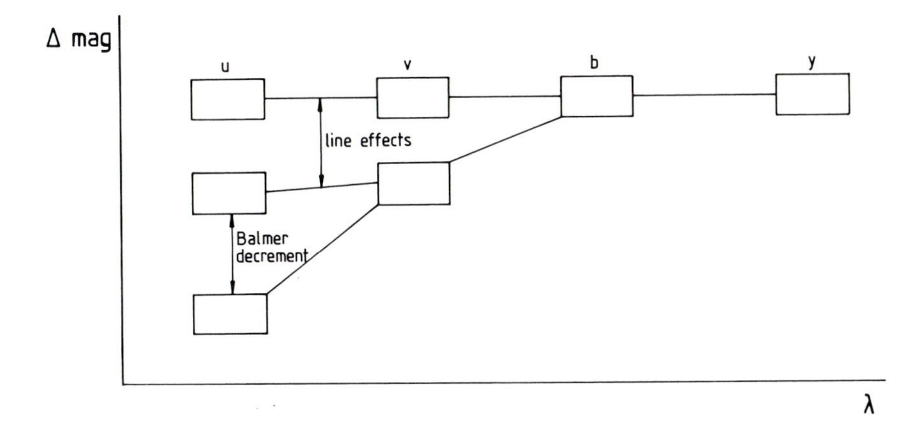

Fig. 4.6. The roles of the Balmer jump and line blocking in the four-colour system.

situation is well illustrated by Figure 4.6, which is one of Crawford's earlier explanatory schematics. The line blocking, roughly constant for b and y filters, rises more rapidly towards v, but then an additional effect adds in at u due to the Balmer discontinuity. The strategy to deal with this was the introduction of the parameters c_1 and m_1, to measure Balmer decrement and line blocking, respectively. They are defined thus:

$$\left.\begin{aligned} c_1 &= (u-v)-(v-b), \\ m_1 &= (v-b)-(b-y). \end{aligned}\right\} \qquad (4.19)$$

The c_1 index goes through a maximum (about 1.2 in value) for stars of early A, late B spectral type (dwarfs), in accordance with the known rise of the Balmer decrement for such stars. For giants this maximum occurs at a later spectral type, or larger (intrinsic) $b-y$ value. A few such examples can be noticed on Figure 4.7. For unreddened stars away from the maximum c_1 settles at a value of about 0.4.

Round about spectral type A0 the m_1 index indicates more particularly relative iron abundance. For F and G type stars it has the role of a general tracer of composition. It increases in value from about zero for some early type stars to 0.6 for late type giants. A deviation Δm_1 is defined for a star by subtracting from its m_1 value that corresponding to a member of the Hyades cluster at the same $b-y$ colour. It has been shown that Δm_1 scales directly with the relative iron abundance [Fe/H], with a proportionality constant of about 12 multiplying Δm_1.

Hence the four basic parameters of the four colour system: y ($\simeq V$), $b-y$, c_1 and m_1, measure, respectively, apparent magnitude, colour temperature,

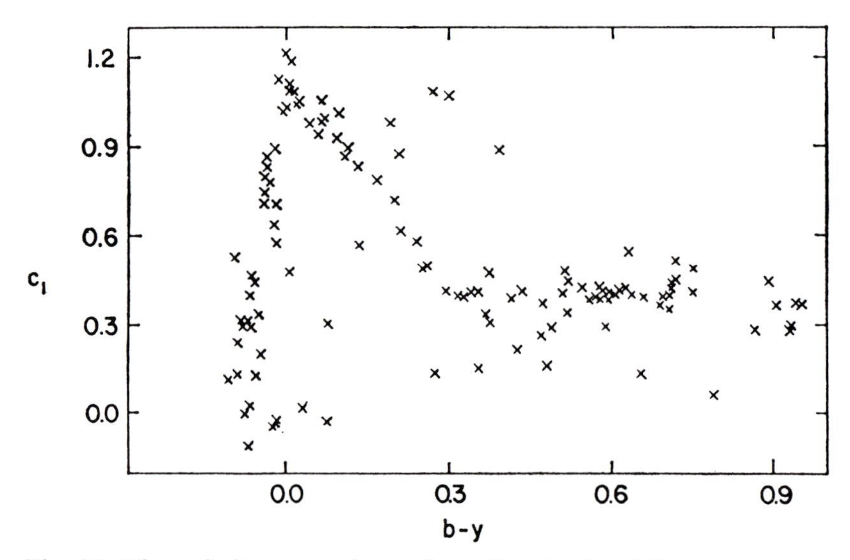

Fig. 4.7. The c_1 index versus $b - y$, from Crawford and Barnes.

luminosity and metallicity, although, in practice, the relationship of the variables to the sought quantities is not so direct, or even involves a fixed procedure over the whole range of stars. Temperature determination, for instance, can be made from several measured quantities; which is most suitable, in a given case, is decided on the basis of relative sensitivity of the available parameters and accuracy of the data.

Evolutionary status can also be derived from variations in luminosity and metallicity at a given spectral type, or $b-y$ value. The distribution of points representing a stellar sample in the planes of c_1, or m_1, against $b - y$ has been therefore studied. Generally one finds a clustering along particular loci in these diagrams, such as would be associated with the Main Sequence. A diffusion away from such a clustering can then be a measure of age (Figures 4.7 and 4.8).

The idea of correcting the measurements for the effects of interstellar reddening was embedded in the four-colour system from the outset. Along the lines of the preceding discussion of the Q-parameter, therefore, reddening-free parameters have been introduced, as follows:

$$\left.\begin{array}{rcl} [c_1] & = & c_1 - 0.20(b - y), \\ [m_1] & = & m_1 + 0.18(b - y), \end{array}\right\} \qquad (4.20)$$

which are verified on the basis of known colour excess forms, i.e. that $E(c_1) = 0.20E(b - y)$, $E(m_1) = -0.18E(b - y)$.

Another modified parameter of the system $[u - b]$ has also been in-

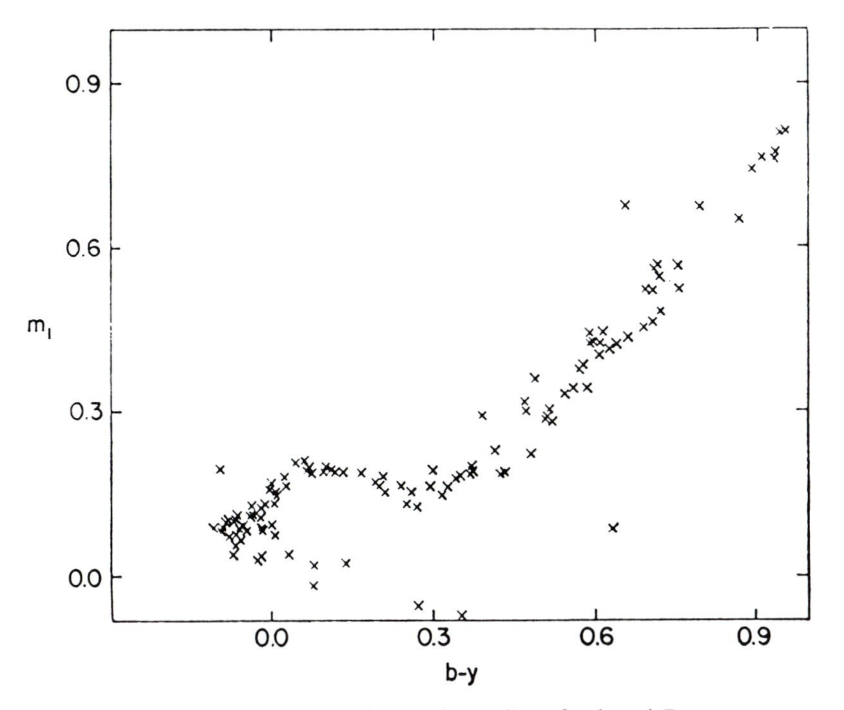

Fig. 4.8. The m_1 index versus $b - y$, from Crawford and Barnes.

vestigated. It can easily be found, since $[u - b]$ is defined to equal $(u - b) - 1.84(b - y)$, that

$$[u - b] = [c_1] + 2[m_1]. \tag{4.21}$$

The $[u-b]$ parameter has been found to be a useful measurer of temperature for very hot stars ($T > 11\,000$ K).

The *uvby* system has been adapted for more specialized studies of astrophysical effects, where detailed attention is paid to relative trends within restricted subgroups of stellar types. In addition, the photometry is often combined with narrowband photometry (next section) — usually of the H_β line. The investigations often seek enhanced sensitivity to a particular process. For example, the ages of certain groups of A type stars have been studied by means of an '*a*-parameter', which derives from a linear combination of $u - b$ and $b - y$, which, within a fairly restricted spectral range, becomes particularly sensitive to effective temperature.

uvby with β photometry applies its various parameter combinations to a number of taxonomic purposes. These may concern the showing up of composition peculiarities of certain stars, deciding whether stars relate to certain groups or clusters, or the comparison of metal abundances between clusters. Aside from composition, another factor which characterizes a star, even at zero age, is its state of axial rotation. Differing orientations of the rotation axis with respect to the line of sight, for otherwise similar stars, can produce measurable effects. Dispersions in the normal relationship between, say, c_1 and $b-y$ due to rotation induced effects have been calculated and the results compared with observations. The incidence of unresolved binarity is another factor which introduces dispersions from a single star trend in a two-parameter diagram. Yet another factor is the degree of 'microturbulence' in the atmosphere of the source object. This can influence line formation processes, and so change blocking levels, or m_1 values. Relative metal abundances derived from Δm_1 values have to take this into account.

4.4 Narrowband photometry

H_β, or simply β photometry as it has become known, centred on the second of the hydrogen Balmer lines, is a particularly prominent example of the somewhat different approach of narrowband work. The popularity of this particular line has probably resulted from its large absorption strength over a fair-sized range of stellar types, as well as its convenient positioning in relation to the sensitivity function of available photoemissive surfaces. For stars later than F type along the Main Sequence β photometry becomes relatively ineffective, however, due to the decline of the Balmer lines and the increasing and confusing influence of other lines.

Essentially, narrowband procedures involve comparison between flux measured in a narrow region centred on the selected spectral line, and the surrounding continuum; so that at least two optical filters are required. Ideally, the filters would have rectangular transmission profiles, such as produced by a monochromator.† In practice, interference filters are more convenient to use, and can be manufactured with the required transmission characteristics. Both filters are centred on the line of interest: the narrow filter would have a half-width $\Delta\lambda_n^{(1/2)}$ typically of order 20–30Å, and about one order of magnitude less than the corresponding half-width of the wide filter $\Delta\lambda_w^{(1/2)}$. The peak transmissions of the filters can be generally made to be $\sim 50\%$ or greater (Figure 4.9).

There is no loss of generality in using the notation of the H_β line, when we write for the equivalent width w_β

† A low-dispersion spectrometer with a controllable exit slit.

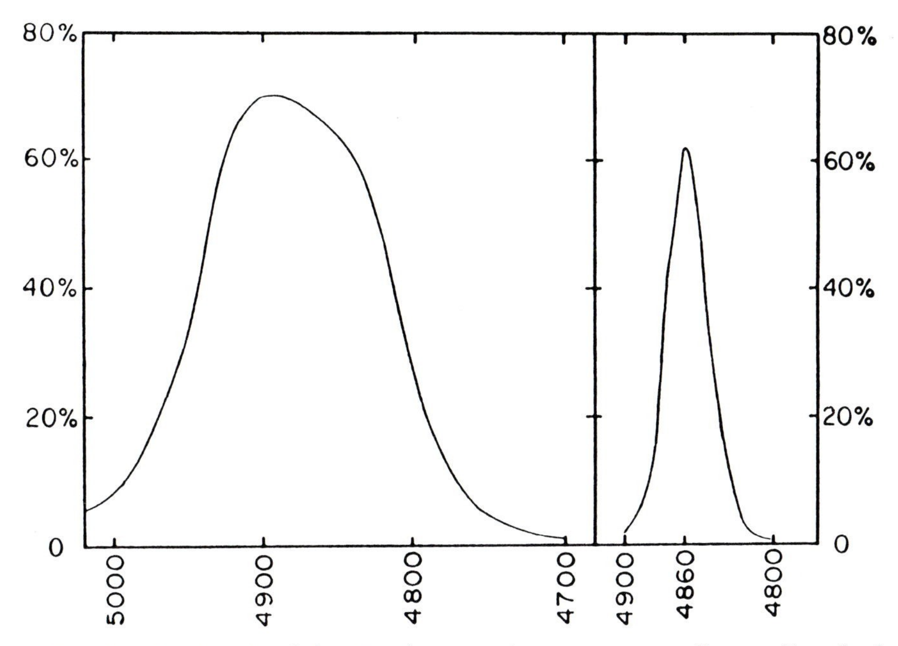

Fig. 4.9. Passbands of the H_β photometric system, according to Crawford and Mander.

$$w_\beta = \int_{\Delta\beta} \frac{l_c - l(\lambda)}{l_c} \, d\lambda, \tag{4.22}$$

where l_c is the mean continuum intensity, assumed independent of λ over the range $\Delta\beta$ of the line profile $l(\lambda)$. Typical values of w_β are of the order of a few angstroms for early type stars. The Balmer series lines are, of course, actually much broader than this, with wings which, for some stars, extend out to distances comparable to the edges of (though generally less than) the narrowband transmission window.

The flux received through the narrow filter is

$$f_n = k \int_{\Delta\lambda_n} l(\lambda) T(\lambda) \, d\lambda, \tag{4.23}$$

where multiplying the irradiance $l(\lambda)$ and transmission function $T(\lambda)$ is a scaling constant k. The range of integration corresponds to the region $\Delta\lambda_n$ for which T is non-zero. A similar expression holds for the flux through the wide filter f_w.

Assuming that in the region $\Delta\lambda_n$, $l(\lambda)$ is accounted for by the single line l_β, outside of which it reverts to the continuum value l_c, we can use the integral mean value theorem to show that the line index β is

$$\beta = -2.5\log_{10}\frac{f_n}{f_w} = -1.086\ln\frac{f_n}{f_w} =$$
$$= 1.086 w_\beta(\frac{1}{\Delta\lambda'_n} - \frac{1}{\Delta\lambda'_w}) + O(\frac{w_\beta}{\Delta\lambda'_n})^2 + \zeta_\beta, \quad (4.24)$$

where ζ_β is a zero constant. The quantities $\Delta\lambda'$ are given by $(\overline{T}_{\Delta\lambda}/\overline{T}_{\Delta\beta})\Delta\lambda$, and $\overline{T}_{\Delta\lambda}/\overline{T}_{\Delta\beta}$ is of order unity for normal transmission profiles. The overbar denotes an average value of the variable over the range indicated by the suffix.

Equation (4.24) establishes an approximate linearity between the measured line index and the equivalent width. Such a relationship is observable empirically, and Crawford published the relation $w_\beta = -19.1 + 38(\beta - 2.000)$ on the basis of H_β line profiles measured previously by Williams.

If more than one distinct feature was involved over the range of integration in (4.23), the linear nature of the integration operation would imply extra w terms, in addition to the β-suffixed ones in (4.24), appearing on the right hand side. The measured line index should take account of such separate contributions, in proportion to their equivalent widths. If the main feature is blended with some contaminating effect the derivation of (4.24) is not affected, since nothing was assumed about the form of $l(\lambda)$; however, differences in the behaviour of β for such cases would show up against comparison standards. This, indeed, provides a motivation for monitoring, particularly those stars showing emission effects in the Balmer lines.

The Balmer lines in stellar spectra are sensitive to 'pressure broadening' — the density-dependent widening out of lines in response to the perturbing effects of interparticle encounters on atomic energy levels. This effect provides a means of determination of stellar surface gravity (and thence size) if the variation of equivalent width for a line such as H_β can be suitably calibrated. In fact, the β index turns out to have a reasonably tight correlation with absolute magnitude, at least for early type stars — the form of variation of the β index with absolute magnitude is shown in Figure 4.10 for a range of such stars (B0–A1), where the effects of temperature, rotation and composition variation, at a given absolute magnitude, are seen not to introduce too strong a dispersion. Absolute magnitude can be confidently estimated to within 0.5 magnitude for a given β value on this range.

Actually, the higher quantum number n lines of the Balmer series are relatively more sensitive to pressure broadening than the lower ones, and indices γ and δ, corresponding to β in the foregoing, but for the H_γ and H_δ lines, have been efficiently used as luminosity measurers for early type

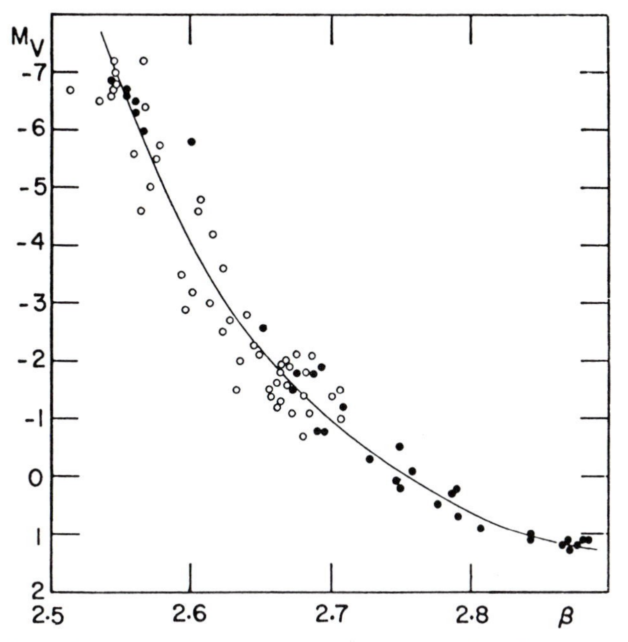

Fig. 4.10. Absolute magnitude versus the β index for early type stars, according to J. D. Fernie. Filled circles represent stars for which $B - V > -0.15$, and open circles have $B - V < -0.15$.

stars. Beyond H_δ isolation of a single strong line becomes more troublesome. On the other hand, the lower lines of the Balmer series are relatively more responsive to traces of emission contributions associated with circumstellar material. Plots of α against γ indices, for example, can therefore be useful in revealing stars which may have low levels of emission.

There are some very significant consequences of the *differential* basis of narrowband photometry: firstly, the line index is independent of any intervening broad extinction effect (interstellar or atmospheric), which affects both the wide and narrow measurements to the same extent. It is also possible to use an achromatic beamsplitter to enable simultaneous monitoring of both wavelength intervals. The time-dependent component of signal variation due to extraneous causes is thus minimized, and in this way a relatively high accuracy — of order 0.001 mag — is not an uncommon achievement.

For early type stars, β, measuring absolute magnitude, can be plotted against $[u - b]$, here measuring temperature. The resulting distribution of points maps part of the Hertzsprung–Russell diagram. Effects of evolution separate out as rightward trends in the appropriate region of the diagram. Such plottings have been placed over theoretical lines of constant age (isochrones), thereby determining stellar ages, or otherwise testing evolutionary models.

For stars later than about A0 type the β parameter is no longer so closely correlated with absolute magnitude, but instead shows much more of a single-valued dependence on temperature. Since it is inherently a reddening-free parameter, it then provides a means of estimating colour excesses[†] and therefore distances. In this dual role as luminosity tracer for early, and temperature tracer for late type stars, the β index has a similar, but opposite, behaviour to the Balmer discontinuity measurer c_1.

Lines, or narrow spectral regions, other than H_β have received attention in a variety of other filter-based photometric studies, where many of the preceding principles apply. For weaker lines, for example, where there is contamination of a chief line of interest with some others, the various relative contributions need be accounted for. The notion of a differential spectral measure is usually retained, either in the way described for H_β, or perhaps with a slightly different technique, as exemplified by 'Lindblad criteria' type parameters. Here there are three closely spaced and rather narrow monitorings; A, B and C, say, where B transmits maximally in the region of special interest. The mean of A and C, is compared with B, which is usually of somewhat narrower bandpass than the two outliers. The logarithm of the ratio, multiplied by –2.5, gives the sought index.

This approach has been followed in studies of photometric effects around the 'G band' — that broad mixture of line blends and molecular absorptions which becomes prominent for stars of type G and cooler. Attention has been directed to the CH band at around 4300 Å, the CaI feature at 4227 Å and the CN band at about 4200 Å. The magnesium triplet at 5170–5180 Å, as well as the sodium D doublet at 5900 Å, have also been studied in a similar way. For very cool stars, TiO bands offer comparable opportunities. Whether using separate wider filters to surround the spectral feature of chief interest, or the previously discussed narrow and wide concentred comparison, resulting indices are usually applied in pairs to provide a mapping of an H–R type diagram. This will be valid for some domain of (cool) stars, and normally permit the separating out of particular groups or evolutionary effects.

4.5 Photometry of extended objects

It has been mentioned that astronomical photometry often deals with the luminous flux from a celestial source overall, i.e. we express a certain mag-

[†] One would subtract an intrinsic, temperature-dependent colour (e.g. $(b-y)_0$), as calculated from the β value, from the corresponding observed colour.

nitude (e.g. $m_v = 0.0$) as equivalent to a certain number of lux (2.49×10^{-6}). Certainly for general stellar photometry, we are not usually directly concerned with the local surface brightness variation, or illumination geometry, though information on these may come from specific kinds of observation of variable stars. Stellar photometry, notwithstanding its historic role in the development of the subject, thus provides a special subset of conditions for analysis. The flux measured from the point-like source is essentially independent of the photometer's aperture, the main purpose of which is then to cut out extraneous input to the system from a background of no direct interest.

In the case of an extended object, though, such matters characterize the study from the outset, thereby introducing extra dimensions of potential complexity. From the observing point of view, greater stringencies in instrument positioning as well as aperture control are required. We continue our broad review of underlying themes by considering a few topics within extended object photometry: at first by some basic visual examples, followed by cases drawn from solar system, nebular and galactic contexts.

4.5.1 Visual photometry

A flux of 1.25×10^5 lx from the Sun was derived in Section 3.3. Dividing this by the 6.8×10^{-5} sr that the Sun subtends at the Earth, we obtain the intensity which, from (3.2), characterizes the radiation field in visual wavelengths, whether above the Earth's atmosphere or at the Sun's photosphere (averaged over the visible hemisphere and assuming the intervening space to be absorption free). The value is 1.84×10^9 nt. On a bright clear day with the Sun at the zenith the brightness would be reduced, by atmospheric extinction, to about 10^9 nt as an average over the entire disk, reaching a value of about 1.8×10^9 nt at its centre.

A similar brightness would characterize α Cen if the eye could resolve its disk. When the size of an object sinks below the eye's disk of resolution, though, the sensation of brightness no longer corresponds to intensity in the sense of (3.2), but rather with the net flux. When more than just one or two 'pixels' (picture cells) of the eye's retina become illuminated, the flux per pixel essentially corresponds to the surface brightness of the preceding paragraph. In this context, we meet Olbers' famous *gedanken* experiment, that if, in any direction, the line of sight sooner or later encounters a sun-like star, then the entire sky should be as bright as the surface of the Sun! To give full consideration to the solutions which have been offered for this paradox would take us quite beyond our present course, but the photometric premises on which it is based are straightforward enough.

With Rougier's value for the difference (14.29) in the magnitude of Sun and full Moon we derive the full Moon magnitude –12.46, which corresponds to 0.24 lx on the scale just cited. The Moon's apparent semidiameter varies, depending on its orbital position and where the terrestrial observer is located, but on average it is slightly less than that of the Sun at $\sim 940''$, or 6.59×10^{-5} sr. This implies a mean surface brightness of about 3000 nt for the terrestrial observer of the full Moon on a clear night. This is comparable to that of the mean daytime sky, which subtends a solid angle almost 10^5 bigger than the solar disk itself. With some inferences about the comparable fractions of sunlight scattered by the Moon's surface and atmospheric molecules, this comparability of brightness can be understood in a general way, but a more detailed exposition would be required to show why it is that the Moon can sometimes be bright enough to stand out against the daytime sky, and at other times not. The problem of first visibility of the lunar crescent, for instance, of ceremonial significance to Muslims, is potentially one of great complexity to treat in full detail.

The light from a $m_V = 0$ star spread over a patch of sky of size, say 1 'square degree' $= 3.04617... \times 10^{-4}$ sr, represents a number (8.17×10^{-2}) of lux per steradian, and is thus a unit of brightness. This unit is relevant for an object like the Milky Way, whose apparent brightness derives from the net effect of a large number of unresolved faint stars. A more generally applicable unit for faint sources, however, could be twentieth magnitude stars per square arcminute (mag 20 stars arcmin^{-2} $= 2.94 \times 10^{-7}$ nt).

The brightness of the daytime sky falls, on a clear moonless night in an isolated area, by a factor of several millions to, typically, $\sim 5 \times 10^{-4}$ nt, the residual light being due to the net effect of airglow, zodaical light, auroral and faint unresolved, or scattered, starlight contributions. Light 'pollution' in cities can enhance this by at least an order of magnitude, depending on the arrangement of local lighting. The faint light of the Milky Way has to be at least comparable to 10^{-4} nt in order to be distinguished, and typically turns out to be ~ 750 mag 20 stars arcmin^{-2}. If the Milky Way covers around $\sim 10^7$ arcmin^2 of sky, we quickly form some feeling for what numbers of sun-like stars are required to account for its light in this simplistic way, keeping in mind that $m_V = 20$ compares with what the Sun would be at a distance of the centre of the Galaxy, and neglecting the effects of interstellar extinction.

Standing out against the background of the Milky Way are a few clusters, for example h and χ Persei, both of area around 100π arcmin^2. Their total illuminances are comparable to that from a fourth magnitude star, which implies a brightness of about 1 mag 10 star arcmin^{-2}. The clusters are known to contain a few hundred bright stars, but they are located at distances of somewhat over 2000 pc. We deduce directly, therefore, that these stars must be of essentially greater visual luminosity than the Sun, by around 6 mag, on average.

An alternative to the number of stars of given magnitude per square angular unit is to put one star of equivalent light output. Thus 750 mag 20 stars arcmin^{-2} = 12.81 mag arcmin^{-2}, or 21.70 mag arcsec^{-2}. This latter representation is frequently used to mark contour levels in mapping the surface brightness distribution of galaxies, i.e. a given locus corresponds to μ mag arcsec^{-2}.

Galaxies external to the Milky Way, like the Magellanic Clouds, or the Andromeda Galaxy, which are visible as extended objects, still appear to have a comparable brightness to the Milky Way, despite their much greater distances. This can be easily understood if the galaxies are composed of comparable column densities of stars in a similar inherent luminosity distribution. The decline in received flux from each individual star type with increasing distance squared is then counterbalanced by a proportional increase in numbers per unit area of sky. Regions of relatively greater stellar (column) density, like the bar region of the Large Magellanic Cloud, appear correspondingly brighter.

4.5.2 Solar system photometry

The solar system also presents a particular subset of conditions for photometry. Thus, there is only one predominating source of illumination, i.e. the Sun, and distances and angles involved in the configuration can be determined with precision. The illuminating pencil may be regarded, with sufficient accuracy, as a parallel beam. Attention is then directed to the nature of the reflecting surface: its 'albedo', or the laws characterizing the reflected light in dependence on either wavelength or geometry (phase).

Consider, for example, the illumination of a resolvable spherical planetary body as represented in Figure 4.11. x, y, z ($\equiv$ **x**) represents a coordinate scheme, conveniently oriented for the terrestrial observer. The x, y plane is perpendicular to the line of sight z. The unit of distance in **x** is taken, for convenience, to be the radius of the planet. The coordinate scheme **x** can be related to another rectangular system ξ, more appropriate for the illumination geometry. The relationship between **x** and ξ is established by a sequence of rotations **R**, involving known angles, in the manner of Section

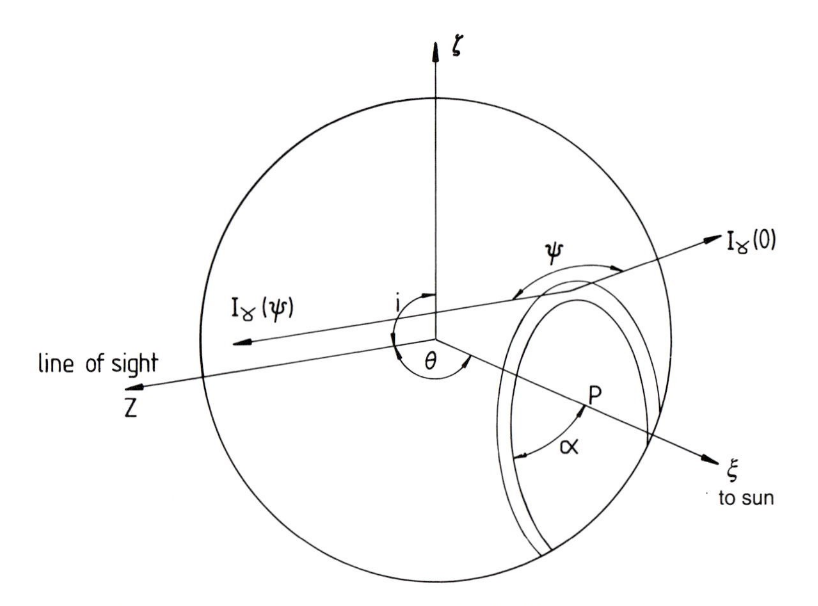

Fig. 4.11. Scattering from a spherical body illuminated by a remote source.

4.1.2. We thus write,

$$\xi = \mathbf{R}\cdot\mathbf{x}\,. \tag{4.25}$$

The total amount of reflected light then becomes

$$\mathscr{L}_r = k \iint_A I_\gamma(\psi)\hat{\mathbf{z}}\, d\mathbf{s}, \tag{4.26}$$

where $I_\gamma(\psi)$ is the intensity of reflected light at angular distance γ from the subsolar point, and in the direction ψ to the surface normal. The integration is carried out over the illuminated area A. We can put

$$\hat{\mathbf{z}}.d\mathbf{s} = ds \cos\theta \ (\equiv dxdy), \tag{4.27}$$

where $\hat{\mathbf{z}}$ is the unit vector in the direction parallel to the line of sight, and ds is an increment of area on the surface of the planet. The constant k allows the units of $\mathscr{L}_r$ to be appropriately expressed in (4.26). The simplest reradiation formula is that of completely uniform diffusion of the reradiated (scattered) flux ('Lambert's law'). In this case $I_\gamma(\psi) = I_\gamma = \text{const.}$

More generally we can use a limb-darkened (cosine law) form, i.e.

$$I_\gamma(\psi) = I_\gamma(0)(1 - u + u\cos\psi), \tag{4.28}$$

where u is the limb-darkening coefficient. $I_\gamma(0)$ can be written in the form

$$I_\gamma(0) = I_0 f(\xi). \tag{4.29}$$

Spelling out the first row of the direction cosine matrix **R**, we have

$$l_1 x + l_2 y + l_3 z = \xi, \tag{4.30}$$

where we set ξ to lie in the direction of the centre of the Sun. We choose the x, y orientation such that $l_2 = 0$, and then l_1 and l_3 are $\sin\theta$ and $\cos\theta$, respectively, where θ is the angle indicated in Figure 4.11. Alternatively, we could write $l_3 = \cos\phi \sin i$, where ϕ measures the angle between Earth and Sun in the plane perpendicular to the axis ζ (phase), which is inclined at angle i to the line of sight.

We then combine (4.30) and (4.29), and insert into (4.26), using also (4.27) and (4.28), and we find,

$$\mathscr{L}_r = kI_0 \iint_A \{C_1 f(l_1 x + l_3 z) + C_2 f(l_1 xz + l_3 z^2)\}\, dx\, dy, \tag{4.31}$$

where C_1 and C_2 are constants, dependent on u, which normalize the integration, so that the net returned light remains constant with different assigned values of u. The suffix A denotes the projection of the illuminated area in the line of sight.

Writing $I_0 = L_\odot/\pi$ for the intensity of the illuminating beam, the appropriate value for k becomes $r_p^2 R_{se}^2/(R_{sp}^2 R_{ep}^2)$, where r_p is the planetary body's radius, R_{se} the distance between Sun and Earth, R_{sp} the solar–planet distance, and R_{ep} that between the planet and the Earth. For convenience, we make the Sun–Earth separation to be the unit of distance (i.e. 1 a.u.), so that R_{se} drops out of the formula for k.

At least for a conservative bolometric radiation field, we now have suitably arranged formulae to scale the brightness seen by an observer at the Earth to the solar luminosity. For, if there is no build up or loss of radiative energy in the considered process, all incoming radiation will be reradiated by the illuminated surface (though usually with some redistribution in wavelength) and the bolometric albedo a_{bol} is unity. Over some particular wavelength range, however, this does not normally hold, and k should include a non-unity albedo a_λ. This can be regarded as an empirical scaling factor multiplying the returned sunlight, in the given wavelength range, which makes the calculated reflected light agree with the observed level, after appropriate allowance for directional effects in the scattering.

This observed level is studied in its dependence on the orbital angular argument — the 'phase law'. Since the integral in (4.31) takes different values at a given phase in dependence on the form of $f(\xi)$, however, the form of the phase law will be similarly dependent. Hence, we must perform the integration (4.31) in order to study such phase laws for different $f(\xi)$. This integration introduces a class of integral of the following type:

$$\pi \imath_n^m = \iint_{A_c} x^m z^n \, dx dy, \tag{4.32}$$

where A_c represents the area A enclosed by some perimeter c on a sphere. The integrals fall into three classes, depending on whether the perimeter curve lies entirely within ('annular') the circular boundary of the sphere, as seen from a great distance along the z-axis; whether it intersects with this outer circular boundary ('partial'); or whether this outer boundary itself delimits the area over which the double integration is carried out ('total').

In this book we meet two types of integral of this kind, whose difference lies in the different perimeter curves c. The presently arising kind we identify as σ_n^m. In this case, the curve c is a circle formed by the intersection of a plane, at an arbitrary orientation, with the sphere. Seen from a distance along the z-axis this circle appears as an elliptical curve, unless it crosses over the outer boundary (partial case). The other kind of integral we denote α_n^m. In that case, the perimeter is formed by the sphere's intersection with a cylinder whose axis is parallel to the z-axis. Its outline, in the line of sight, then appears circular.

In the case of a near parallel illuminating beam an appropriate form for $f(\xi)$ is $f(\xi) = \xi$ $(\xi \geq 0)$, and $f(\xi) = 0$ $(\xi < 0)$. We can then find $C_1 = 3(1-u)/(3-u)$ and $C_2 = 3u/(3-u)$. Using the σ-integral notation, we rewrite (4.31) as

$$\mathscr{L}_r(\phi) = \frac{a_\lambda r_p^2}{(R_{sp}^2 R_{ep}^2)} L_\odot \{C_1(l_1\sigma_0^1 + l_3\sigma_1^0) + C_2(l_1\sigma_1^1 + l_3\sigma_2^0)\}. \tag{4.33}$$

The σ-integrals in this expression will be defined in terms of two parameters: the angular extent γ_m of the illuminated region, and the distance z_0 of the subsolar point above the plane $z = 0$. In the geometrical arrangement under consideration an appropriate value for γ_m is $\pi/2$. The value of $z_0(= l_3)$ is given by

$$z_0 = \cos\phi \sin i, \tag{4.34}$$

while $l_1 = \sqrt{1 - z_0^2}$. Once values of γ_m and z_0 are provided, the whole range of σ_n^m for integral values for $m \geq 0$, $n \geq -1$ forms a determinable set, involving functions no more complicated than inverse trigonometric

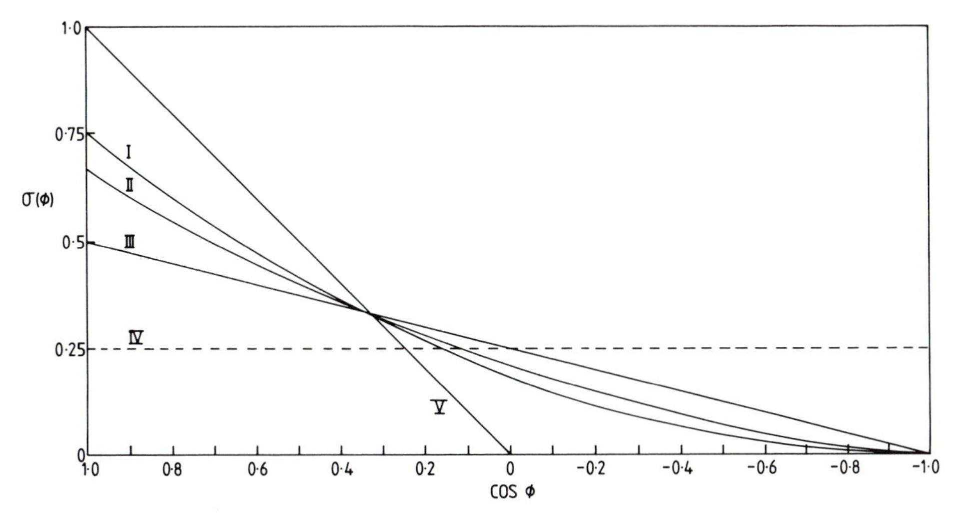

Fig. 4.12. Phase variation of alternative reflection laws for parallel beam illumination. The forms correspond to: a linear limb-darkened scattering with coefficient $u = 1$ (I); $u = 0$ (Lambert's law) (II); the 'area' law (III); the 'specular' law(IV); and the uniform flat disk (V). For simplicity we have set $i = \pi/2$. Observe that $2\int_{-1}^{1} \sigma(\phi)\, d\cos\phi = 1$ in each case.

functions and algebraic expressions. For $\gamma_m = \pi/2$ it can be shown that (Chapter 9)

$$\sigma_0^0 = \frac{1 + z_0}{2}; \quad \sigma_1^0 = \frac{2(\arccos(-z_0) + z_0\sqrt{1 - z_0^2})}{3\pi}.$$

In general, we can write $\sigma(u, \phi)$ for the phase law, corresponding to the expression in curled parentheses in (4.33). The phase laws with $u = 0$ (II) and 1 (I) are shown in Figure 4.12. Also shown for comparison are some other simple reflection laws: the area law (III), for which $f(\xi) = \frac{1}{2}$ in (4.29) and $u = 0$, so that $\sigma(\phi)$ reduces to,

$$\sigma(\phi) = \frac{1}{2}\sigma_0^0, \tag{4.35}$$

the specular reflection law (IV), for which

$$\sigma(\phi) = \frac{1}{4}, \tag{4.36}$$

and the uniformly diffusing flat disk (V),

$$\sigma(\phi) = z_0. \tag{4.37}$$

The phase laws shown in Figure 4.12 satisfy $2\int_{-1}^{1} \sigma(u, \phi)\, d\cos\phi = 1$ due to the normalizing action of the coefficients C_i in (4.31). In some treatments, however, a phase law $q'(\phi)$, proportional to the foregoing, is arranged with a starting value $q'(0) = 1$, and the integral $q = 2\int_0^\pi q'(\phi)\sin\phi\, d\phi$ is scaled accordingly. This then makes for a second coefficient p, where pq is set equal to the albedo a_λ.

We can rewrite (4.33) in magnitudes as

$$m_p = m_\odot - 5\log[r_p/(R_{sp}R_{ep})] - 2.5\log[a_\lambda \sigma(u, \phi)] \qquad (4.38)$$

to express the overall apparent magnitude variation of the planetary body.

The phase laws observed for real cases are generally more peaked towards low phase angle than even the $u = 1$ form shown in Figure 4.12. This is particularly true for those planetoids with minimal atmospheres and rocky surfaces. Surface scattering properties can be more fully investigated, however, by also taking into account the polarization of the returned radiation (next section).

While the foregoing procedures allow first order models of scattering to be tested, thereby checking general properties, currently available practical techniques are able to reach well beyond this, in particular with high resolution areal imaging. Departures from the smooth forms of the functions we have considered, i.e. as $I_\gamma(0)$ (4.29), are thus thrown into sharper relief. More intensive studies of the nearer planets, which can be expected in an era when actual samples are recovered, imply that Earth-based photometry plays increasingly a supportive role to more direct work.

4.5.3 Photometry of galactic nebulae

The word nebula (Lat. *mist*) is used to represent a wide range of extended celestial sources, involving a variety of different physical situations. The bright, glowing nebulae range from quite symmetrical, quasi-spherical 'planetary' sources, to irregular, clumpy or fragmented clouds which may spread, in extreme cases, to over a hundred parsecs or more in linear extent. There are also dark nebulae, conspicuous by the special extinction they produce in the light from stars behind them. The topic is thus big and complex, and data are often imprecise, or, at best, lacking a fully comprehensive quality (of necessity). Nevertheless, the astrophysical problems posed by these objects are deep and exciting, and continue to raise new challenges.

Within the present broad review we can raise at least some of the more obvious questions, such as where does the light come from — or how is a bright nebula's energy budget balanced, to put it in a more structured way. Among the primary photometric tasks is then to determine the net radiative energy losses. The well-known Crab nebula (Messier 1), in the constellation of Taurus, provides an instructive general example.

The integrated magnitude of the ~5 arcmin diameter (continuum) source is about $m_V = 8.6$. The distance of the source has been determined to be about 2200 pc, implying a linear size of ~3 pc. The large distance also results in a significant interstellar extinction (Section 4.1.3), so that the intrinsic apparent magnitude of the source would be around $m_V = 7.1$. Using the calibration of Section 3.6, this yields some 2.4×10^{-6} W m^{-2} at the source, with the foregoing size and distance, or $\sim 7 \times 10^{28}$ W altogether in the visual region. This is already more than 100 times greater than the Sun's bolometric output, but it is not all. The non-thermal continuum power of the nebula declines with a low negative power of the frequency, extending from the radio up to the X-ray region, where individual photon energies are two (or more) orders of magnitude greater than in the optical. An appropriate integration will show $\sim 10^{31}$ W to be a fair estimate for the entire rate of radiation loss of the Crab in its predominating synchrotron continuum emission.

Many bright nebulae shine mainly in discrete emission lines rather than a continuum. Two-dimensional photometry, particularly with high resolution scanners or array type detectors, are increasingly revealing more complexity of structural detail in such sources, requiring more fine tuning of the many-parameter models which address themselves to the data. As a broad illustration of how photometry can approach this process, we consider a procedure for evaluation of the effective temperature of the central star of a planetary nebula, along the lines proposed originally by H. Zanstra in the 1920s.

We suppose that the radiative emission from the hydrogen atoms, of which the nebula is largely composed, arises from a cascade recombination process, giving rise, in the emergent radiation field, to the Balmer lines, together with Lyman α — the first level to ground transition of atomic hydrogen associated with photons of wavelength 1215 Å. The Ly$_\alpha$ transition's upper state is supplied from Balmer line recombinations through the nebula. Attention then fixes on these Balmer lines, which derive, in this picture, from an input photon stream coming originally from beyond the Lyman limit of the central star's radiation field.

A quantity which can be observationally determined is νA_ν, the ratio of the net illuminance from the nebula L_ν in a Balmer line to the spectral irradiance from the central star H_s, at frequency ν_1, say, in a given line-free region of its optical spectrum. H_s can be written as a fraction of the entire illuminance beyond the Lyman limit (912 Å), to an extent which depends on the temperature of the photosphere — a simple Planckian form was assumed in the original presentations. It is more relevant, however, to relate H_s to the integral of the distribution of *numbers* of photons per second with wavelengths beyond 912 Å, using $E_\nu = h\nu$ (Chapter 2). This is because the main underlying premise is that the total rate of photon production in all the Balmer lines of the nebula is equal to this integral. The sum S_A of the values of A_ν for all the Balmer line images of the nebula then results in an implicit equation for the temperature. This is of the form

$$S_A = \int_{x_0}^{\infty} \frac{x^2\,dx}{e^x - 1} \Big/ \frac{x_1^3}{e^{x_1} - 1},$$

where $x = h\nu/kT_\star$, x_0 corresponds to the Lyman limit, and x_1 to ν_1.

The A_ν values are probably best determined using high-dispersion spectrophotometric techniques and areal detectors. Single-beam photometry could be considered, however, by means of a photometer equipped with a controllable 'iris' type entrance aperture. The stellar continuum flux is measured with the iris stopped right down around the central star, using a filter which avoids nebular lines as much as possible, but is of sufficient width to allow a good continuum signal. The total light from the nebula through each of a set of narrowband filters centred on the principal Balmer lines is then measured. The initially small aperture is progressively widened until the recorded level reaches a constant limit taken to be that of the nebula as a whole. The observed photon incidence rates in the various lines are converted into above-the-atmosphere values, by taking account of the net transmission and detector response functions, i.e. the relevant η and k-loss terms referred to in Section 3.6. A similar procedure applies also to the continuum measurement, though this has to be divided by the frequency bandwidth of the filter to obtain the spectral irradiance at ν_1.

It seems likely that only a few Balmer line narrowband filters would be available for a convenient and practical arrangement — and, fortunately, this usually suffices, since the relatively rapid decline of intensity of the Balmer lines with n for many planetary nebulae permits an acceptable extrapolation for the higher lines of the series (which can be theoretically checked), while the Balmer continuum is usually neglected in the method. The corrected S_A value then yields the photospheric 'Zanstra temperature' $T_\star$ of the central star.

This is a very simple model with very limiting assumptions built in, but it illustrates the procedure of calibrating 'surface photometry', in emission line radiation, by simultaneous observation of an adopted reference source. A more complete theory would describe how a nebula cools to give rise

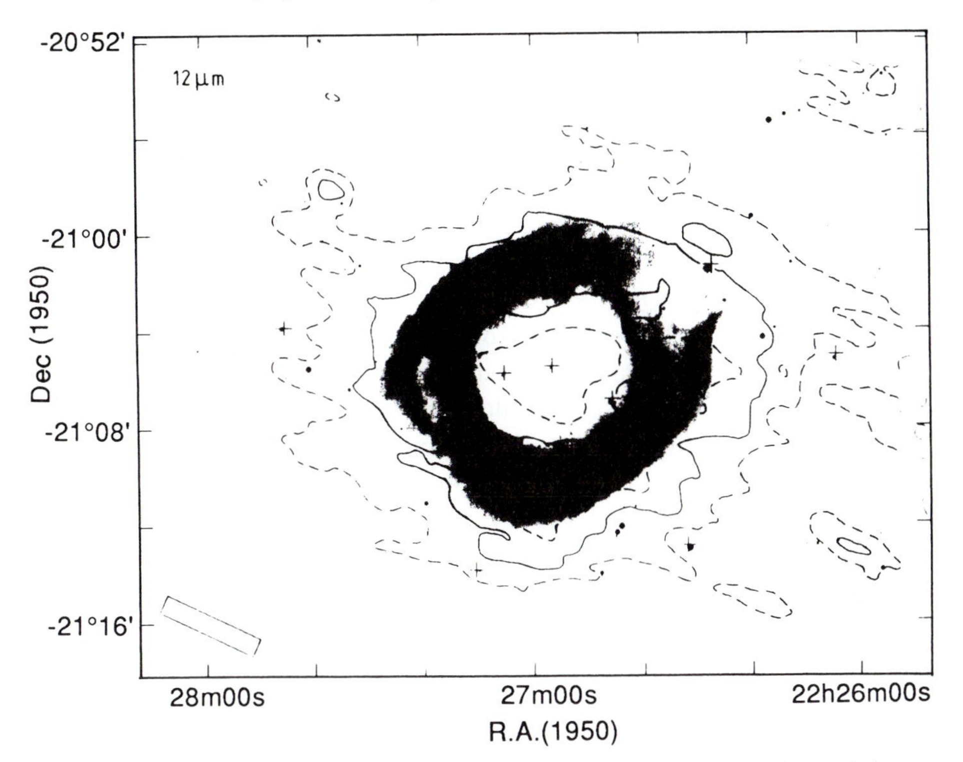

Fig. 4.13. Surface photometry of NGC 7293 (the Helix Nebula). Brightness contours in the infra-red are superposed on an optical image. Stars are indicated as crosses. The outer and inner (dashed) contours about the nebula correspond to irradiance levels of 3.2×10^{-16} W m^{-2} μm^{-1} arcmin^{-2}. The continous curves are at 6.4×10^{-16} W m^{-2} μm^{-1} arcmin^{-2}, within which some brighter spots (dashed) 9.6×10^{-16} W m^{-2} μm^{-1} arcmin^{-2} are found.

to a given flux distribution depending on densities, temperatures, relative abundances, velocity fields, and perhaps other agencies besides. The more detailed information that is brought together, such as by surface photometric mapping in selected wavelengths of emissions (Figure 4.13), the better the chances of clarifying and refining the theoretical model.

4.5.4 Photometry of galaxies

Specification of the brightness contours of galaxies in units of mag arcsec^{-2} gives rise to a numerical way of representing galaxies. It enables a quantitative approach to their morphological classification. Hubble, early on in the development of this subject, for instance, wrote that elliptical galaxies tend to show a 'light curve' along their principal axes of the form $I = I_0/(r+a)^2$, where I measures surface brightness on a linear scale, r radial distance along an axis, and a is a constant.

Later, however, it was realized that the apparent central flattening of brightness which such a function indicates is critically related to the resolution of the imaging facility — photographic plates in the time of Hubble, and for a long time after. The foregoing brightness unit thus suggests a potential resolution of surface features which would have been impractical on all but the best nights and with the largest telescopes. But as more detailed information became available it was realized that the light curves of elliptical galaxies are more peaked towards the centre than the $(r + a)^{-2}$ law indicates, and the subsequent decline with distance r tends to be more of a power of magnitude than linear intensity. The central peaking of the luminosity function, coupled with resolution limitations, means that referring the light distribution to the central intensity is not the most stable or satisfactory method. Instead an isophotal radius, enclosing half the light of the galaxy, proves to be a useful reference contour, with which other light levels at different distances may be compared, and provides a suitable unit with which to scale the radial distance r.

From such a basis de Vaucouleurs empirically derived what has become established as the '$r^{\frac{1}{4}}$ law' for elliptical galaxies. This can be presented as

$$\mu = a + br^{\frac{1}{4}}, \tag{4.39}$$

where a and b are constants and μ was introduced in Section 4.5.1. The formula has been found to be applicable to a wide range of normal ellipticals, out to a relatively large radial extension, as well as the spheroidal bulge component of galaxies with disks. The disk component in spiral galaxies satisfies the more steady 'exponential' decline, i.e.

$$\mu = c + dr. \tag{4.40}$$

More detailed study has shown that disks can be further decomposed into young (arm) and old (bulge) components, and indeed, further photometric morphological subclassification may refer to features such as bars, lenses, inner and outer rings, envelopes and asymmetries. A systematic, non-subjective way to characterize such morphological variation is provided by taking moments of the two-dimensional intensity distribution. In principle, this is like summarizing the salient transmission properties of photometric filters, in terms of the zeroth (extent), first (centroid locator) and second (shape or concentration descriptor) moments (Section 3.6).

The magnitude surface density of a galaxy, μ, is related to the derivative of the 'magnitude–aperture' curve, in that the latter represents a mean value

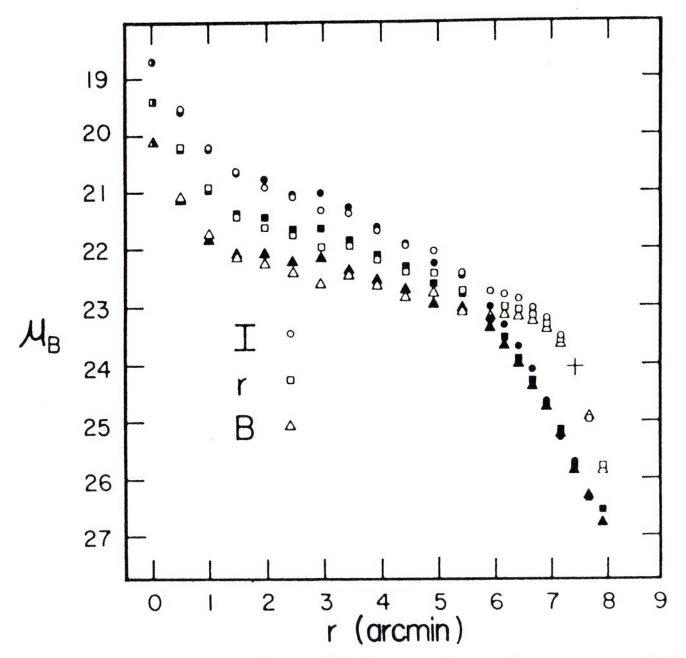

Fig. 4.14. Surface (major-axis) three-colour photometry of the edge-on spiral galaxy NGC 4565, after Jensen and Thuan (1981). Full symbols are for the southeast side, open for the northwest. The *I* and *r* curves are normalized to *B* at the cross. The 'exponential' decline can be seen along most of the disk, with a noticeable blueing of the galaxy before its sharp cut-off at the end of the profile.

of the former averaged over a circular ring concentric with the aperture centre. This magnitude–aperture curve provides an alternative approach to the photometry of extended objects, as already encountered with the iris photometry of planetary nebulae. The total or integrated magnitude of a galaxy can be approached directly in this way. Such data also allow some methods of classification. The galaxy is measured through increasing apertures until it reaches a constant total value. This measurement with broadband filters implies a continuous process of sky subtraction, which could be carried out using data from a neighbouring patch of sky, as with conventional stellar photometry. The information retrieved from such an operation is not just the total magnitude, and colours, of the galaxy, but a plot of the magnitude and colour aperture curves.

The colour of most galaxies becomes more blue as the aperture is increased. The effect is more marked for normal spirals than ellipticals, and is probably related to young star formation in spiral arms, but would also be influenced by the relative amounts of dust in different regions (Figure 4.14). There are, however, a few peculiar galaxies which show a reversal of this, i.e. blueness increasing towards the nucleus, and these include galaxies with strong radio emission. Colour variation with aperture, in general, is more marked for spirals than ellipticals.

The recent growth of high-sensitivity areal detectors allows investigators to build on the systematics of previous techniques, but with the advantages of linear detection, digital recording and computer-based image reduction facilities more detailed features of galactic morphology are opened up.

4.6 Photopolarimetry

Conventional photometry, concerning itself with the intensity I of an incident light beam, addresses only part of the information that the light contains. The wave properties of light can be summarized by referring to four 'Stokes parameters' I, Q, U and V. These contain information not only about the intensity I, which is closely related to the subject matter hitherto, but also about the two perpendicular components into which the electric vector of an electromagnetic vibration is resolvable: their relative magnitudes and orientation with respect to a frame of reference specified by the observer, and their separations along the axis of propagation (phase difference). An input stream of wave packets from a source normally shows some prevailing tendencies with these properties, though we also find a good deal of randomization when the wave packets have passed through turbulent inhomogenous media and are collected over a wide range of frequencies.

The omission of data on three of the four basic parameters characterizing the input could be seen as rather a negligence, were it not for the usually much greater information yield, for a given amount of observational effort, from I than the other three parameters. In order to obtain reliable data on representative values for Q, U and V, along with I, particularly in stellar optical photometry, very significant improvements in equipment and operation are necessary. Nevertheless, it is appropriate to include a brief discussion of this subject, which has developed as an offshoot from conventional photometry.

We approach the Stokes parameters via the solutions of Maxwell's equations for a single electromagnetic wave propagation involving components E_x and E_y of the electric vector, i.e.,

$$\left.\begin{aligned} E_x &= E_1 \cos(\omega t - 2\pi z/\lambda + \delta_x), \\ E_y &= E_2 \cos(\omega t - 2\pi z/\lambda + \delta_y), \end{aligned}\right\} \qquad (4.41)$$

where we have put the vector components in the x and y directions at time t, for angular frequency ω, and wavelength λ, at position z along the axis of propagation (z-axis), where there are phase advancements δ_x and δ_y at $z = 0$.

We now introduce the Stokes parameters I, Q, U, V, as

$$\left.\begin{array}{rcl} I & = & E_1^2 + E_2^2, \\ Q & = & E_1^2 - E_2^2, \\ U & = & 2E_1E_2\cos\delta, \\ V & = & 2E_1E_2\sin\delta, \end{array}\right\} \quad (4.42)$$

where δ expresses the phase difference $\delta_y - \delta_x$, and E_1 and E_2 are the amplitudes of the wave motions in the x and y directions.

The foregoing expressions (4.41) for E_x and E_y satisfy

$$\left(\frac{E_x}{E_1}\right)^2 + \left(\frac{E_y}{E_1}\right)^2 - \frac{2E_xE_y}{E_1E_2}\cos\delta = \sin^2\delta, \quad (4.43)$$

which is the equation of an ellipse in the E_x, E_y plane. The sum of squares of the principal axes of this ellipse is seen to equal the sum of the amplitudes $E_1^2 + E_2^2$, which is invariant against a rotational transformation of the axes x, y; and is thus I by the preceding definition. The (signed) product of these principal axes, $E_2E_1\sin\delta$, is a similar invariant, proportional to the area of the ellipse, and turns out to be $\frac{1}{2}V$ in the foregoing notation. The ratio V/I is thus a measure of the nearness of the ellipse to a circle — the 'circular polarization' of the wave, sometimes denoted by q, which varies on the range $-1 \leq q \leq 1$.

The remaining two quantities relate to the difference of the squares of the principal axes. Averaged over a suitable time interval, to allow representative statistics for a large number of received wave packets, they indicate any tendency towards a linear polarization prevalence in the radiation — the two quantities fix not only the 'degree of linear polarization' p as

$$p = \frac{\sqrt{Q^2 + U^2}}{I}, \quad (4.44)$$

but also the orientation ψ of the major axis with respect to the x-axis as

$$\psi = \frac{1}{2}\arctan(U/Q). \quad (4.45)$$

A specification of any systematic polarization in the illuminance in terms of the four Stokes parameters calls for the insertion of an 'analyser' into the optical train, which monitors the linear polarization in at least two orientations, as well as a 'depolarizer', to account for any net instrumental effects having nothing to do with the source under study. A 'quarter-wave plate' retarder placed at different orientations in the wave train introduces an effect in the photometric output whose coefficient is V. It is thus possible to have a suitable arrangement of rotatable optical components — a 'Stokes-meter' — where all four parameters are monitored together. The relative accuracy of determination is, however, then spread more thinly, and in practice many observers, particularly in the stellar context, have settled for concentrating on the linear polarization in their data.

The effect of any particular piece of polarizing, or optically 'active', material on a given beam, specified by a vector $\mathbf{S}$ with four components corresponding to the four Stokes parameters, can be studied by means of a 'Mueller matrix' $\mathbf{M}$, whose form specifies the action of the optical agent. The output beam $\mathbf{S}'$ satisfies

$$\mathbf{S}' = \mathbf{M}\cdot\mathbf{S}\,. \tag{4.46}$$

The net effect of a combination of n components: analysers, retarders or whatever; is examined as the result of the sequence of operations $\mathbf{S}_{\text{out}} = \mathbf{M}_n\cdot\mathbf{M}_{n-1}...\mathbf{M}_1\cdot\mathbf{S}_{\text{in}}$. Operating with such a Mueller calculus one gains a fuller understanding of the design and action of a photopolarimeter. Two basic forms of such matrices can be set out. For a perfect analyser $\mathbf{M}_a$, we have,

$$\mathbf{M}_a = \begin{vmatrix} 1 & \cos 2\theta & \sin 2\theta & 0 \\ \cos 2\theta & \cos^2 2\theta & \frac{1}{2}\sin 4\theta & 0 \\ \sin 2\theta & \frac{1}{2}\sin 4\theta & \sin^2 2\theta & 0 \\ 0 & 0 & 0 & 0 \end{vmatrix},$$

where θ measures the orientation of the principal plane of the analyser with respect to the initial axis. For a perfect retarder of retardance τ with the axis of optical activity at position angle ϕ, we write $\mathbf{M}_r$ as

$$\mathbf{M}_r = \begin{vmatrix} 1 & 0 & 0 & 0 \\ 0 & G + H\cos 4\phi & H\sin 4\phi & -\sin\tau\sin 2\phi \\ 0 & H\sin 4\phi & G - H\cos 4\phi & \sin\tau\cos 2\phi \\ 0 & \sin\tau\sin 2\phi & -\sin\tau\cos 2\phi & \cos\tau \end{vmatrix},$$

where $G = \frac{1}{2}(1 + \cos\tau)$ and $H = \frac{1}{2}(1 - \cos\tau)$.†

We consider, as an application of photopolarimetry to observational astrophysics, the case of the 'Q, U curve' of Algol binaries.

Let us first note the simple Mueller matrix $\mathbf{M}_t$ for a rotational transformation through an angle φ of the axes of reference, which follows from the

† For more details on this, the reader is referred to Shurcliff and Ballard's *Polarized Light*, van Nostrand, Princeton, N.J., 1964.

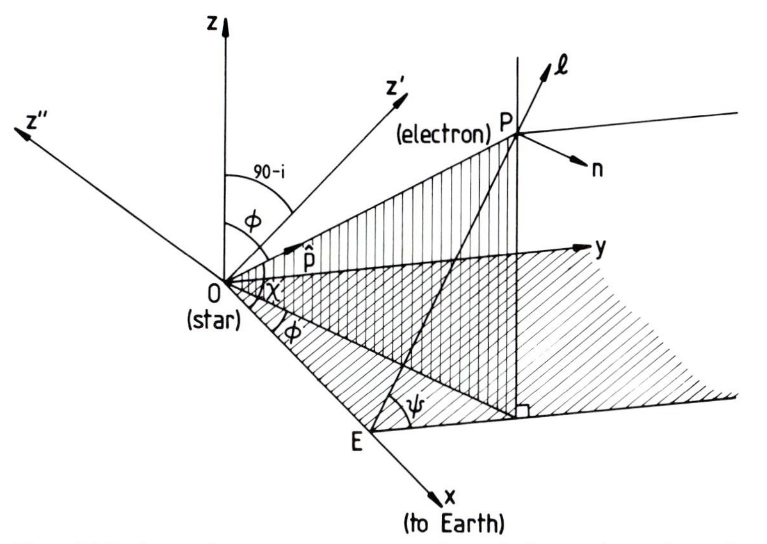

Fig. 4.15. Scattering geometry — adapted from that given by Brown, McLean and Emslie (1978).

discussion following (4.43) as

$$\mathbf{M}_t = \begin{vmatrix} 1 & 0 & 0 & 0 \\ 0 & \cos 2\varphi & \sin 2\varphi & 0 \\ 0 & -\sin 2\varphi & \cos 2\varphi & 0 \\ 0 & 0 & 0 & 1 \end{vmatrix}$$

With the aid of $\mathbf{M}_t$, we can transfer from any set of axes convenient for the instrument, to an astronomically useful orientation in the plane of the sky.

Definite linear polarization effects for Algol binaries have been known about at least since the work of Shakhovskoi and Shulov in the early sixties. The effects are slight — in general less than $\sim 1\%$ — and the preferred mechanisms for explanation generally refer to singly scattered photons from electrons in optically thin material in the vicinity of the binary. The geometry of the scattering is shown in Figure 4.15. We presume that light from the early type primary (typical of Algols) is the primary source of originally unpolarized, scattered light, and this is centred at O in the diagram. We neglect the finite size of the source in this simplified treatment. The plane of scattering is then OPE, and, with standard electron scattering, light scattered in the direction OE, through the scattering angle EOP (χ), will be linearly polarized in the plane perpendicular to OPE, in such a way that:

$$\left.\begin{aligned} I &= \mathscr{K}(1+\cos^2\chi), \\ Q &= \mathscr{K}(\sin^2\chi\cos 2\psi), \\ U &= \mathscr{K}(\sin^2\chi\sin 2\psi), \end{aligned}\right\} \tag{4.47}$$

where $\mathscr{K} = I_0\sigma_0 \int_V (n/r^2)\, dV$ is regarded as an integrating operator containing an effectively constant component (optically thin condition) of the scattered illuminance, deriving from a source of intensity I_0. σ_0 represents the coefficient of Thomson scattering $\frac{1}{2}(e^2/m_e c)^2$ from electrons whose local number density is n, and the volume integration extends over the whole illuminated volume V scattering light towards the Earth.

Now the coordinate system indicated in Figure 4.15 has its y– and z-axes located in the plane of the sky, but it will be advantageously oriented when the x, z plane contains the axis parallel to that of orbital revolution of the binary passing through the source at O. The angle between this axis Oz' and the line of sight Ox is the inclination i.

A unit vector $\hat{\mathbf{p}}$ in the direction OP has coordinates λ, μ, ν when referred to the x-, y-, z-axes. We may rotate Ox into alignment with OP by the following sequence of operations, (cf. Section 4.1.2)

$$\hat{\mathbf{p}} = \mathbf{R}_y(\theta - 90)\cdot\mathbf{R}_z(\phi)\cdot\boldsymbol{\lambda}. \tag{4.48}$$

Alternatively, we can write,

$$\hat{\mathbf{p}} = \mathbf{R}_z(\chi)\cdot\mathbf{R}_x(\psi)\cdot\boldsymbol{\lambda}, \tag{4.49}$$

and hence see the equivalence of the two separate matrix products. Reverse the sequence of rotations, multiply out, and we find

$$\left|\begin{array}{c} \lambda \\ \mu \\ \nu \end{array}\right| = \left|\begin{array}{c} \cos\chi \\ \sin\chi\cos\psi \\ \sin\chi\sin\psi \end{array}\right| = \left|\begin{array}{c} \sin\theta\cos\phi \\ \sin\theta\sin\phi \\ \cos\theta \end{array}\right|, \tag{4.50}$$

which allows us to rewrite the integrations (4.47) in terms of the angles θ, ϕ, thus:

$$\left.\begin{aligned} I &= \mathscr{K}(1+\sin^2\theta\cos^2\phi), \\ Q &= \mathscr{K}(\sin^2\theta\sin^2\phi - \cos^2\theta), \\ U &= \mathscr{K}(\sin 2\theta\sin\phi). \end{aligned}\right\} \tag{4.51}$$

Matters are considerably simplified by assuming that the scattering material is symmetrically distributed about the orbital plane and is also axisymmetric about a radius vector emanating from the primary source at O, such as the line joining the centres of the two binaries. This would certainly hold for the illuminating pencil giving rise to the well-known 'reflection effect', which is particularly noticeable in Algol binaries. Although reradiation from the secondary photosphere will produce some polarization, multiple scattering and absorption processes are also involved, and it is generally considered that a more effective source for the observed linear polarization is optically thin, highly ionized plasma distributed over a region of space appreciably sized in comparison to the interbinary distance scale. A 'gaseous stream', as evidenced in a good number of Algol systems, appears a good candidate, and although this is displaced somewhat from the line of centres, axisymmetry about a radius line provides a tolerable first approximation. The material can also be safely assumed to lie close to the orbital plane.

Under these circumstances, the angular integrands involved in (4.47) can be replaced by δ-functions $\delta(\theta_r)$, $\delta(\phi_r)/\sin\theta$; whose arguments specify the direction of the radius of symmetry, with the density function reducing to a one-dimensional variable expressing the appropriate value for each areal section of the scattering volume as one moves along the radius $\theta = \theta_r$, $\phi = \phi_r$. The $\mathscr{K}$ operation then only affects this density function, and becomes effectively a constant, separated from the angular coordinates. The form of the density function will undoubtedly affect the scale of the observed polarization, but the shape of the Q, U curve now depends only on the changing aspect of radius of symmetry as seen from the Earth. For example, if the orbital plane happens to coincide with the x, y plane in Figure 4.15, the effective value for θ in (4.48) is $\pi/2$, and so the U parameter would remain zero. The polarization would vary with orbital phase, but would always be oriented perpendicular to the orbital plane, which projects as a very short straight line on the sky.

Now θ and ϕ, for the radius of symmetry, satisfy

$$\left.\begin{aligned} \cos\theta &= -\cos\alpha\cos i, \\ \tan\phi &= \tan\alpha\csc i, \end{aligned}\right\} \tag{4.52}$$

where α expresses the phase, and linearly increases with time for Algol binaries.

The relations (4.52) can be used to rewrite (4.51). We find, after a little manipulation, that

$$\left.\begin{aligned} I &= \mathscr{K}(1 + \sin^2 i \cos^2\alpha), \\ Q &= \mathscr{K}(\sin^2\alpha - \cos^2\alpha\cos^2 i), \\ U &= \mathscr{K}(-\sin 2\alpha\cos i). \end{aligned}\right\} \tag{4.53}$$

These Stokes parameters apply only to the small proportion of singly scattered light. They are normally scaled by the predominating, direct unpolarized intensity of the source. The mean value of U over the orbital cycle is zero, in any case, but Q has a non-zero average, which depends on the orbital inclination. This was seen, originally, as giving a way to find inclination values, though this method is critically dependent on the model and its scaling factors.

If we rewrite Q as $\frac{1}{2}(1 - \cos 2\alpha)(1 + \cos^2 i)$, we see that U, and Q with the removal of a constant, are terms of different amplitude in $\sin 2\alpha$ and $\cos 2\alpha$, respectively. These variables therefore trace out an ellipse, in the Q, U plane, twice, as the radius of symmetry sweeps once around the orbit. The eccentricity of this ellipse is $\sin^2 i/(2 - \sin^2 i)$, which thus furnishes a way of finding the inclination independently of the details of the operation $\mathscr{K}$. This is a result of potentially far-reaching significance. Previous to its discovery, the only way to determine the orbital inclinations of close binary systems was from the analysis of the light curves of those systems which happened to eclipse.

The foregoing analysis does not depend on the circumstances of eclipses; in fact, these rather complicate the issue — but as the eclipse effect, which for Algols is usually marked for only the primary star, takes place over only a limited range of phases, satisfactory coverage for the eclipse-free form of the Q, U curve can be derived from observations distributed throughout the rest of the light cycle.

In Figure 4.16 data points taken from Rudy and Kemp's observations of the polarized light from Algol itself have been plotted in the Q, U plane. The generally double elliptical form of the polarization curve is immediately discernible (with some additional complications). The Q, U data plotted are in the instrumental system, and so require a further rotation $\mathbf{M}_t$ to bring the Q-axis into alignment with the long axis of the average ellipse. This, in principle, then allows a determination of the orientation of the plane of the eclipsing pair in the sky. Such a rotation preserves the eccentricity of the mean ellipse and thus the derived inclination. The dashed ellipse, shown for comparison, corresponds to the generally accepted inclination value of 82 deg, determined from photometric analysis of the eclipses. It is evident that the double quasi-elliptical track of the Q, U values with phase approximately confirms this inclination. The scale of the locus is somewhat greater in the second half of the orbital cycle, which may relate to the inherent asymmetry of the stream, as expected for Algols, and the consequent difference in the relative visibility of scattered light from near the two stars. In other words, the effect is consistent with the expected greater visible aspect of the stream in the second half of the orbital cycle.

The sense in which the ellipse is traced out also is given in Figure 4.16. This is anticlockwise — a positive sense, by convention. In the foregoing analysis, both Q and U, as terms in $\cos 2\alpha$ and $\sin 2\alpha$, had the same sign,

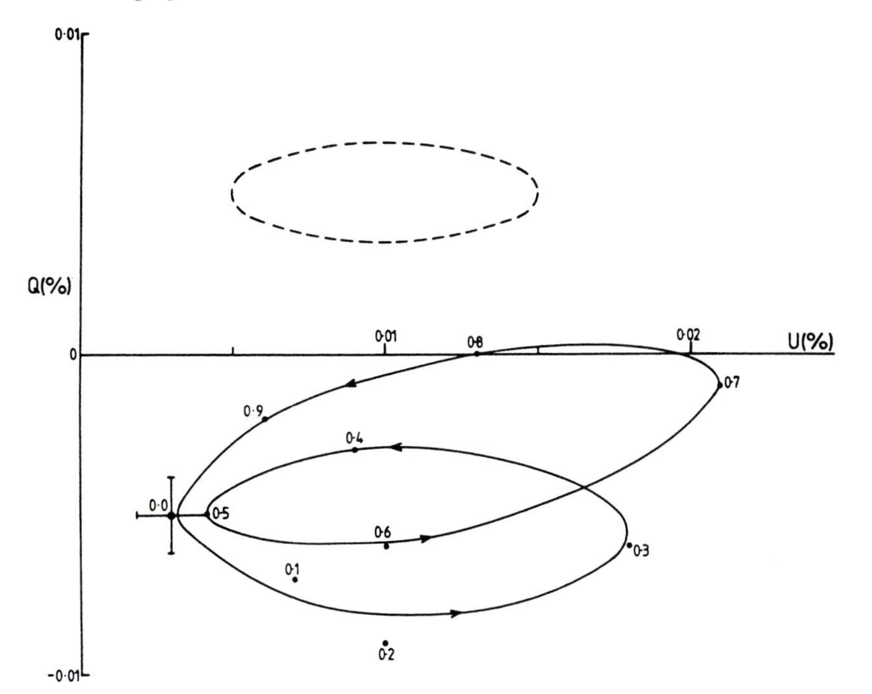

Fig. 4.16. The polarization curve of Algol, according to Rudy and Kemp (1980).

corresponding to an inclination as shown in Figure 4.15, i.e. north pole visible at the Earth. Thus polarization data have the potentiality to reveal full details on the geometry of the orbit, free from the sign ambiguities which usually beset the task of the astronomical analyst.

4.7 Bibliographical notes

The treatment of atmospheric extinction is standard, and a summary is given in M. Golay's *Introduction to Astronomical Photometry*, Reidel, 1974, Dordrecht, Chapter 2. Formula (4.7) is attributed to Bemporad (cf. E. Schoenberg, *Handbuch der Astrophysik*, J. Springer, Berlin, **2**, 268, 1929).

Much more background on orthogonal matrix transformations and their application to the rotational transformation of coordinate axes can be found in a standard reference such as *Elementary Matrices*, R. A. Frazer, W. J. Duncan and A. R. Collar, Cambridge University Press, Cambridge, 1946, Chapter 8.

A summary on interstellar extinction is provided in Chapter 4 of C. and M.

Jaschek's *Classification of the Stars*, Cambridge University Press, Cambridge, 1987. The mean formula

$$A(r, \lambda) = A_0 \sec b \left[1 - \exp\left(\frac{-r \sin|b|}{\beta}\right)\right],$$

where A_0 is a constant, b is the galactic latitude and β the thickness of the extinction providing layer, is provided there, though clear evidence is also given to show the marked local variations which occur in particular directions. The of form of the interstellar extinction law (Fig. 4.2) is also discussed in various places in the Symposium to honour K. Nandy, *Q. J. R. Astr. Soc.*, 1987, **28**, 207 (*et seq*).

The purposes behind the choice of optical filters have stimulated many papers and discussions. Significant sections of *IAU Symp. 24* (ed. K. Lodén, L. O. Lodén and U. Sinnerstad) Academic Press, London, 1966, and *50* (ed. C. Fehrenbach and B. Westerlund) Reidel, Dordrecht, 1973, were dedicated largely to this subject. The general forms shown in Figure 4.3 appear in many places: they accord with the typical numerical values tabulated in C. W. Allen's *Astrophysical Quantities*, Athlone Press, London, 1974. The Q parameter of equation (4.11) appeared in Johnson and Morgan's original paper *Astrophys. J.*, **117**, 313, 1953.

The respective merits of *UBV* and *RGU* (W. Becker, *Veröff. Univ. Sternw. Göttingen*, No. 80, 1946), as well as various other, filter systems were thoroughly reviewed in Golay's *Introduction to Astronomical Photometry*. Figure 4.4 comes from J. Stebbins and A. E. Whitford *Astrophys. J.*, **98**, 20, 1943. H. L. Johnson introduced the *RJKLMN* extension to the *UBV* system in *Astrophys. J.*, **141**, 923.

The original papers on the intermediate-band *uvby* set were B. Strömgren's in the *Annu. Rev. Astron. Astrophys.*, **4**, 443, 1966, and D. L. Crawford's article in *IAU Symp. 24*, p170. Some of the basic diagrams in this photometric system have been shown in the text. Thus, Figures 4.7 and 4.8 come from D. L. Crawford and J. V. Barnes' article in *Astron. J.*, **75**, 978, 1970, which also discusses the reduction of observations in this system, as well as presenting lists of standard stars. Figure 4.6 was shown in Crawford's pedagogic summary in *Stellar Astronomy* (ed. H.-Y. Chiu, R. L. Warasila and J. L. Remo) Gordon and Breach, New York, 1969. Useful programs for extracting stellar parameters from Strömgren photometry were published by T. T. Moon, Commun. Univ. London Obs., No 78, 1985.

The H_β transmission diagram is from D. L. Crawford and J. Mander in *Astron. J.*, **71**, 114, 1966. The analysis which is followed through in equations (4.22-24) was given by E. Budding and N. Marngus, *Astrophys. Space Sci.*,

67,477, 1980. J. D. Fernie's paper in *Astron. J.*, **70**, 575 1965, is referred to in Figure 4.10, while a $[u - b], \beta$ plot can be found in B. Strömgren's cited review of 1966.

Work in extended object photometry is diffused throughout a wide literature, of which Section 4.5 can give only the briefest sketch. The well-known paradox of H. W. M. Olbers first appeared in the *Berliner Astron. Jahrb. Jahr 1826*, and general discussions of it abound. The problem of assessing the various contributions to the background light of the night sky is convoluted with many issues. A short critical review was given by A. S. Sharov and N. A. Lipaeva in *Sov. Astron.*, **17**, 69, 1973. A more recent paper discussing newer satellite-based photometric data, but referring to previous sources, is that of G. Toller, H. Tanabe and J. L. Weinberg in *Astron. Astrophys.*, **188**, 24, 1987.

The quoted value of the magnitude of the full Moon comes from G. Rougier, *Ann. Obs. Strasbourg,* **3**, 257, 1937. More recent studies of the photometric and polarimetric properties of the Moon, and other members of the solar system, are strongly influenced by the results of space missions, as mentioned in the text. Specialist workshops and colloquia are convened to review advances on individual planets. Some source material can be found from the bibliographies of journals such as *Earth, Moon, Planets* (ed. Z. Kopal, M. Moutsoulas and F. B. Waranius) **44**, 47, 1989. One paper dealing with problems of modern lunar photometry is that of L. K. Akimov in *Kinematics Phys. Celest. Bodies* (original verion **4**, 10, 1988). A good overview of the theory of planetary photometry was given by T. P. Lester, M. L. McCall and J. B. Tatum in *J. R. Astron. Soc. Can.*, **73**, 233, 1979. A recent article discussing similar problems was that of W. J. Wild in *Publ. Astron. Soc. Pac.*, **101**, 844, 1989. A more general article addressing the issue of planetary photometry was by S. P. Worden in *I.A.P.P.P. Commun.*, **9**, 120, 1983.

A classic background on Zanstra temperature determination for the central stars of planetary nebulae was given by Lawrence Aller in his *Gaseous Nebulae*, Wiley, New York, 1956. For a recent review of techniques see e.g. J. A. Kaler's paper in *IAU Symp. 131: Planetary Nebulae* (ed. S. Torres-Peimbert) Kluwer, Boston, 1989, p229. Optical morphologies were reviewed by N. K. Reay in *IAU Symp. 103: Planetary Nebulae* (ed. D. R. Flower) Reidel, Dordrecht, 1983, p31. Figure 4.14 comes from A. Leene and S. R. Pottasch's study of IRAS data in *Astron. Astrophys.*, **173**, 145, 1987. Practicalities of absolute photometry of H II region type nebulae were summarized by J. Caplan and L. Deharveng in *The Messenger*, **32**, 3, 1983.

Problems of the photometry of galaxies were the subject of the workshop on *Photometry, Kinematics and Dynamics of Galaxies* (ed. D. S. Evans) University of Texas at Austin, 1981, where formulae like (4.39) and (4.40) are discussed at some length. This book also contains the paper of E. B. Jensen and T. X. Thuan (p113), which is referred to in Figure 4.15. An example of a detailed study of the photometric properties of an individual galaxy is that of M. F. Duval and E. Athanassoula in *Astron. Astrophys.*, **121**, 297, 1983. The subject was recently reviewed by J. Kormandy and S. Djorgovski in *Annu. Rev. Astron. Astrophys.*, **27**, 235, 1989.

Basic definitions of the Stokes parameters can be found in K. R. Lang's *Astrophysical Formulae*, Springer Verlag, Berlin, 1980, p11, while the Mueller calculus is well written up in the cited text of W. A. Shurcliff and S. A. Ballard *Polarized Light*, van Nostrand, Princeton, N.J., 1964. As also indicated in the caption of Figure 4.15, J. C. Brown, I. S. Mclean and A. G. Emslie's paper in *Astron. Astrophys.*, **68**, 415, 1978, provides a much fuller version of the material discussed through equations (4.47)–(4.53), though the present text's derivation of the inclination formula is original.

5
Practicalities

5.1 Overview of basic instrumentation

What is an astronomical photometer? In answering this natural introductory question we refer to Figure 5.1, where the basic optical photometric system in its observatory setting is schematized.

At the heart of the system is the detector. There has been a continued trend towards a more reliably linear and efficient photoelectric responder

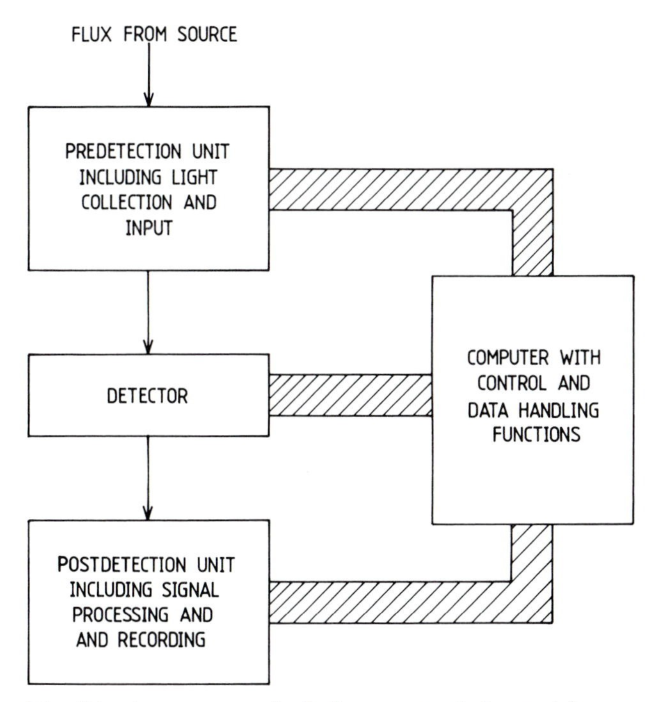

Fig. 5.1. An astronomical photometer (schematic).

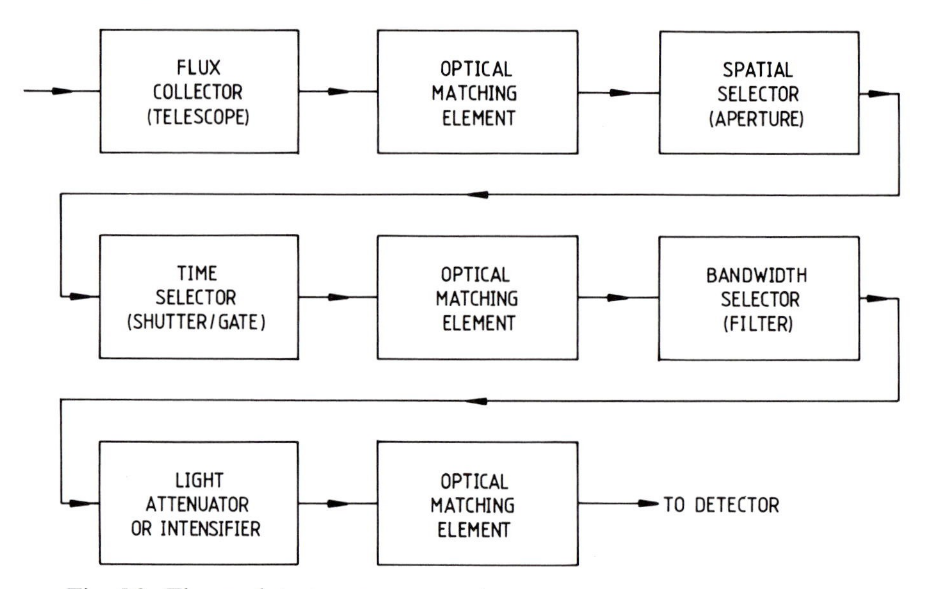

Fig. 5.2. The predetector arrangement.

here. Photocathodes in photomultiplier tubes have played an important part in this. The signal which such a responder emits varies in direct proportion to the flux of received light on a specially arranged target area. The photocathode tube, as a single-channel, spatially non-resolving detector, represents the simplest type of detector in common use, and forms a natural basis for discussion, especially where we deal with the relative brightnesses of stars.

Initially, light has to be gathered and optically processed before passing to the detector. This is indicated in the expanded version of the predetector unit shown in Figure 5.2. With reference to the flux collector — reflectors are usually preferable, since a wider range of received wavelengths (particularly ultra-violet), are thereby available. Since high spatial resolution is not essential for general stellar photometry, the optical figure of the flux collector need not be at so high a standard as for a telescope intended for accurate positional work, though there is always some potential information advantage at higher resolution.

There are more basic considerations about what kind of photometric system is to be implemented, relating to where it is situated and the wavelength range of the receivable flux. Thus, ultra-violet reception is better at higher altitudes, infra-red observations are usually more successfully made from arid locations, and so on. The emphasis in what follows, though, will be on instrumentation for an already collected flux. When originally describing the requirements for definitive *UBV* photometry, Johnson made specific recommendations about telescope type and location.

Not all the components indicated in Figure 5.2 are necessarily employed in a real photometer, nor need they be in the order shown. Those after the flux collector are usually contained together in a single assembly, containing also the detector, and perhaps initial parts of the postdetection electronics. This is what would normally be called 'the photometer', or 'photometer head'. It is often separated into optical, detector and preamplifier boxes. The remainder of this section concentrates on the optical compartment. A practical arrangement, which bolts on to a telescope's backplate, is shown in Figure 5.3.

The optics might start with a component to narrow down the input pencil. To allow ultra-violet access, lenses, or other windows, are usually made from a transmissive glass such as crown borosilicate or quartz. A longer focal ratio allows more convenient spacing in the optical box, so practical photometers are commonly arranged for the Cassegrain focus. The aim is to select an object of interest, and direct its light to the input aperture, free from the confusing effects of other sources, or the background sky, whose light contribution rises in inverse proportion to the focal ratio.

In a conventional single-channel photometer, the aperture takes the form of a round 'pin-hole', though more specialized designs are sometimes advantageous. Several apertures of varying diameters, usually in the range 10–60 arcsec, are commonly set on a sliding bar or wheel. During observations, one of these would be centred on the image at the principal focus. The choice depends on factors such as atmospheric 'seeing' conditions, background sky brightness, optical quality of the flux collector, accuracy of its tracking, or other special circumstances of the observation. For normal photometry of not-too-faint stars 20–30 arcsec is a typical diameter.

The single-channel input may be subsequently analysed by more than one detector in a 'multiple-beam' arrangement (Figure 5.4). This is distinguished from 'multi-channel' photometry, where two or more apertures are used. With a single flux collector more sophistication in the fore-optics or image directing mechanics is then required. Subsequent to the apertures the various channels perform as separate units — though designs also exist where such channels are directed to one detective surface, which scans them alternately. Another form of controllable aperture is of the iris type. In stellar photometry it can be adjusted to suit varying seeing or background levels. It can also be applied to extended sources (Section 4.5).

Manual centring and guiding frequently involves two separate eyepieces; one in front of, and one behind the main aperture. It is desirable for such

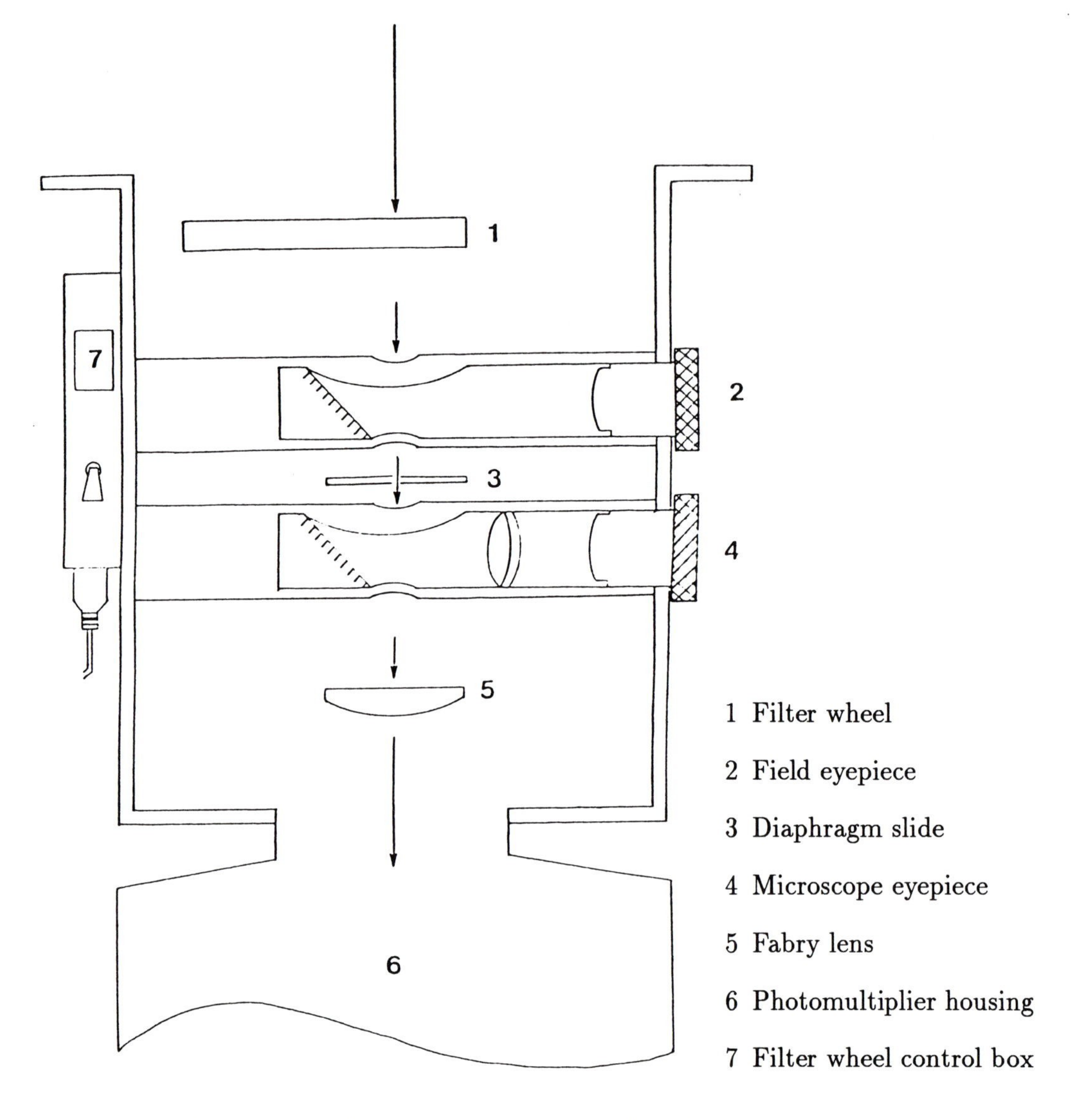

Fig. 5.3. Optical arrangement of a typical working photometer.

viewing arrangements to be simple, unobstructive, light-tight and robust. A suitably marked graticule superposed by the field eyepiece assists object location, particularly if 'offset' guiding is required (i.e. a brighter star at a known separation, distance and angle, is used to set and guide on the target). Various alternatives exist for viewing facilities, e.g. mirror mounted on the end of a retractable eyepiece (Figure 5.3), mirror with a central hole (Figure 5.4), or flip mirror (Figure 5.5).

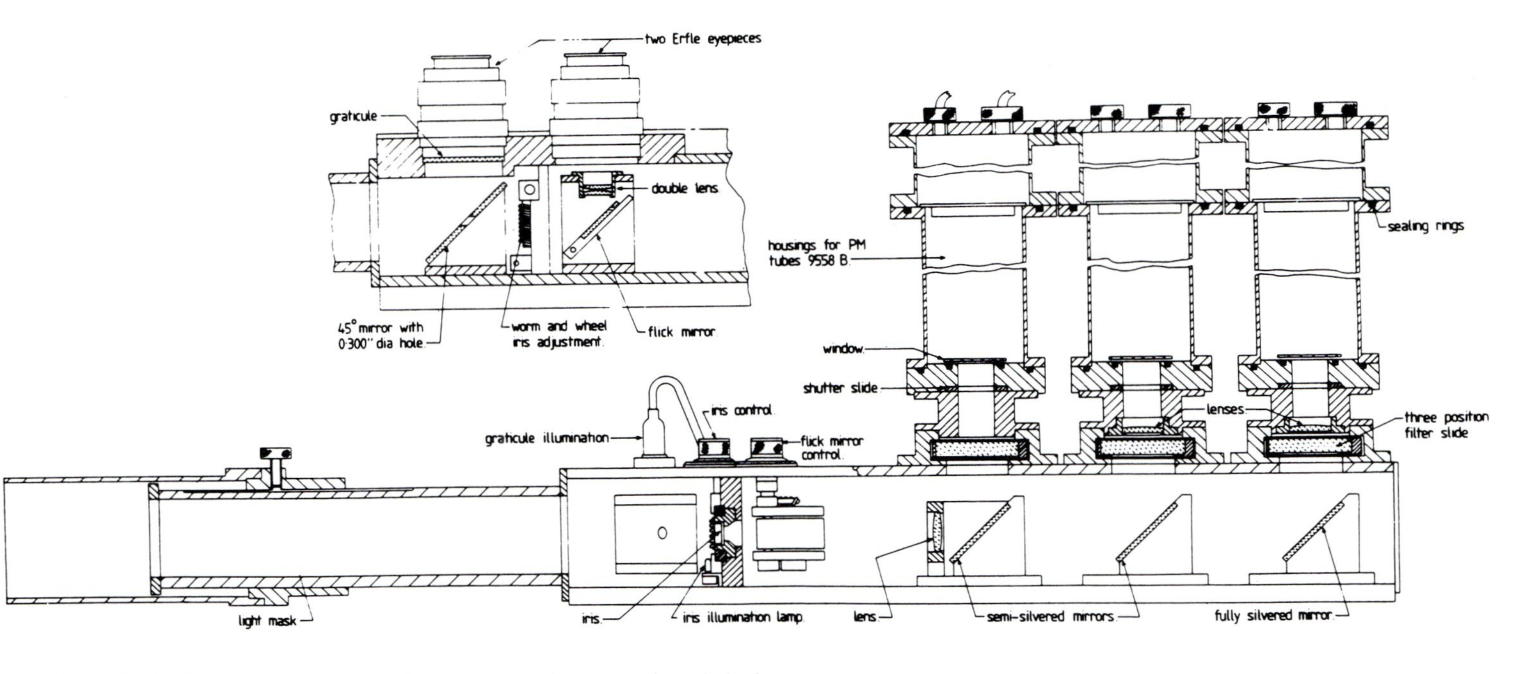

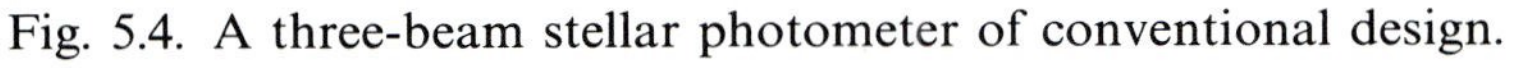
Fig. 5.4. A three-beam stellar photometer of conventional design.

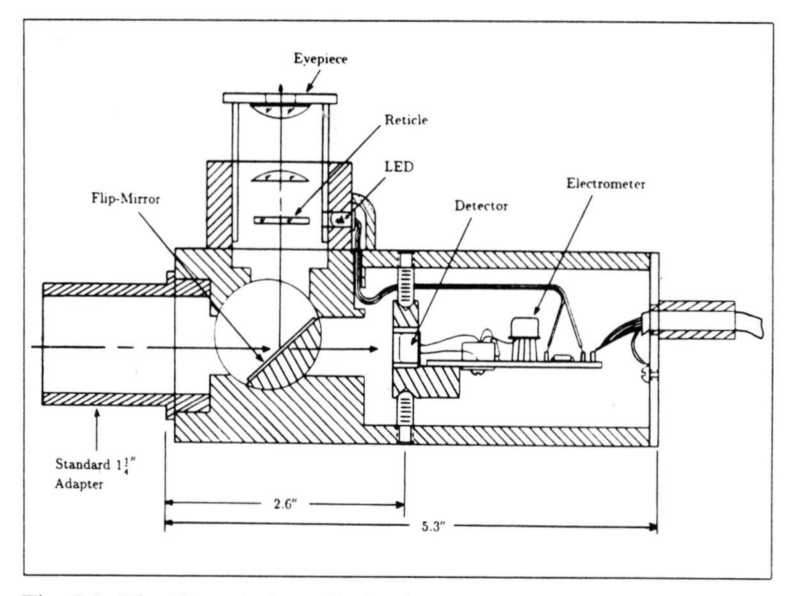

Fig. 5.5. The 'Optec' photodiode photometer.

With the advent of widely available, versatile control electronics, at relatively low costs, more varied sky masking and viewing arrangements are possible; television, for example, replacing the traditional eye to eyepiece. Modern areal detectors allow sky-subtraction and comparison of sources to be organized from an on-line displayed image, with great improvements in time resolution and general observing convenience, at the cost of some restrictions on choices of comparison stars, and vulnerability to technical mishap.

The exposure duration is usually controlled by an electronic gating system, certainly if the postdetector unit involves photon pulse counting ('PC' method). Since the underlying measurement is one of power, a division of the photon count by the precisely measured counting time interval is implied. On the other hand, the measurement itself may be a rate (or current – 'DC' method), in which case exposure timing is less stringently defined, and limited by how the steady, representative rate of photon events is determined. If the interval is too short the required averaging may be too strongly influenced by irrelevant noise contributions; if too long, variations of real interest in the source may be lost in the smoothing. The determination of an optimum gate becomes a critical matter in 'fast' photometry, where sources may be both rapidly varying and intrinsically faint. There are other tasks, however, e.g. (bright) standard star calibration, where the selection of a gate time interval becomes less important. The necessary time of day specification can be conveniently signalled by the exposure starter. A controlling computer would usually provide a stable and accurate reference clock, but a gate timer can be modified to yield elapsed time after a given initial signal, while a DC recorder can be similarly calibrated against an external clock.

Frequently adopted integration times for general work are of the order of several seconds, though in fast photometry the interval may drop to f_s^{-1}, where f_s is the rate of measurable photon events per second divided by a quantity related to the desired signal to noise ratio (i.e. $(S/N)^2$ for poissonian statistics). Values of f_s^{-1} as little as 10^{-3} seconds are found with bright stars and intermediate-sized telescopes. Of course, such a rate of data collection quickly puts large demands on storage, and would usually only occur where rapid events are being monitored (e.g. lunar occultations of stars).

Optical matching beyond the aperture means, for example, beam collimation, since narrowband filters are designed to work with a parallel beam. Beam-splitting can also be performed here. The time resolution of narrowband photometry is optimized by simultaneous monitoring in the different spectral regions, i.e. two-beam photometry. The same argument also applies to photopolarimetry, where simultaneous comparison of different polarization components of the light input is highly desirable, in view of the usually small scale of effects.

Recommendations on bandwidth selection, i.e. filter designs, are normally provided in the original papers of system initiators, though, of course, the net response functions are products of those of all agents involved in the light throughput, including the atmosphere, flux collector, and, particularly, the detector's own spectral efficiency curve. In the case of well-used sets, like the *UBV*, a great body of additional experience and alternative filter/detector combinations now exist. Recent developments in solid state detectors, for example, with their enhanced red sensitivity, have required extensive redesign of filters, notably for the *U*, where the well-known red leak introduces complications. Filters commonly come in sets of at least three units, and often the number is more like five or six. For ease of computer control and light-tight optics, an internally mounted wheel, equipped with a position-registering switch, or encoder, looks good. Six standard sized (25 mm) filters are easily accommodated on a wheel of order 10 cm in diameter.

Apart from the different types of filter already discussed a neutral density filter is sometimes employed, for example, to reduce the signal from bright standard stars to manageable proportions compared to a programme star, or to test linearity of the measurement system. Conversely, it is also possible to arrange for photoelectronic image intensification prior to the main measuring component for very faint objects. Care is required with either of these introductions, which may lead to loss of accuracy. Light diminution, though perhaps convenient in some operations, must inevitably reduce the signal, and though the gross photon input S/N ratio may still be acceptable, some component of potential interest may not be. Moreover, the signal reduction, or the form of its dependence on wavelength, may be imprecisely known. On the other hand, intensification, which essentially means reimaging and therefore implies areal detection, usually brings in some extra noise and perhaps loss of linearity, focusing, or other disadvantage. It would usually be implicit, however, that such disadvantage is offset by sky noise (or other) variation during the long integration period required for the measurement of unenhanced very faint sources.

The final optical matching element is the Fabry lens. Its purpose is to image the entrance pupil of the flux collector on the detector surface (Section 2.2). A fixed patch of illumination, is then maintained on this surface. A wandering image on the detector surface, even if containing a constant flux, would only produce a constant signal if the response function were perfectly constant over the surface. This is difficult to achieve for real detectors. In practice, of course, the presence of the atmosphere means that a star's light, though largely concentrated within the seeing disk of perhaps a few arcsec diameter, is actually scattered, in principle, over the whole sky. It is therefore desirable to centre the image in the aperture, whereupon the Fabry lens minimizes the combined effects of image 'dancing', or small drifts from other causes, and scattering. Rigidity of the optical system is also essential for accurate monitoring.

There are built-in dilemmas to general photometer design. On the one hand, one wants a light-tight, electronically noise-free box, which can stay, with its components, in rigid attachment to the telescope and be convenient for viewing and guiding. On the other hand, one may seek portability, and the ability to change plans or components rapidly in response to varying conditions, or target availabilities. In selecting choices of layout and component facilities it is difficult to satisfy aims somewhat at variance with each other, and some compromises almost certainly have to be tolerated.

5.2 Detectors

5.2.1 Detective processes

The underlying principle whereby the energy of an electromagnetic wave packet, or photon, is transferred to a registrable electron event can be intuitively appreciated fairly directly. The matter of just how some particular electron energization — an intrinsically very minute process — is actually registered, when we suppose the apparatus to contain myriads of electrons, each subject to the likelihood of energizations from other photons and by continuous quasi-random interactions with surrounding atomic particles, is by no means obvious.

The familiar photocathode responder depends on *photoemission* — a process associated with the action of light on a clean-surfaced, heavy or alkaline metal compound or alloy subject to a strong electric field gradient in a near vacuum container with a transparent entrance window. The cathode is usually physically attached to, or forms part of this window. A photon with energy greater than the photoemissive work function (Section 2.4) has a finite probability of releasing an electron from such a cathode, whereupon, under suitable amplification conditions, it can trigger enough subsequent energy release to enable registration. This process is not so straightforward, however. Electron emission is intrinsically a probabilistic behaviour at the quantum level, and, for example, the direction in which an electron energized by a particular photon will start to move cannot be predicted. This implies that only some proportion of primary photons, in the right energy range, result in registrable electron events. In fact, the process is relatively inefficient — at most, about a quarter of the incident photons trigger electron pulses in practical photomultiplier tubes, and usually the probability is much less than this over the range of accepted wavelengths.

Apart from photoelectric emission, matter can respond to irradiation by changes in the relative amount of charge displaced through it by the action of an electric field, in other words, a decline in its electrical resistance may occur. The effect is relatively marked in semiconductors, i.e. where there is an energy gap between the upper limit of the distribution of energy levels for unexcited material, the valence band, and the band of higher levels, forming the conduction band (Figure 5.6). This *photoconductive* effect was used in the astronomical bolometer of Pettit and Nicholson (Chapter 3). The energy gap is comparable to that of infra-red photons for semiconductor material such as silicon and germanium.

The rare population of conduction levels for non-metals means that electrons energized to such higher states can travel a further distance before a close encounter with some similarly energized particle, i.e. they have a longer mean free path than similarly energized electrons in metals. This is an advantage for practical photoemitting material, since it leads to a greater probability of surface escape by electrons, which have been excited at some depth by incident photons. Semiconductor material is thus suitable for the construction of practical photocathodes.

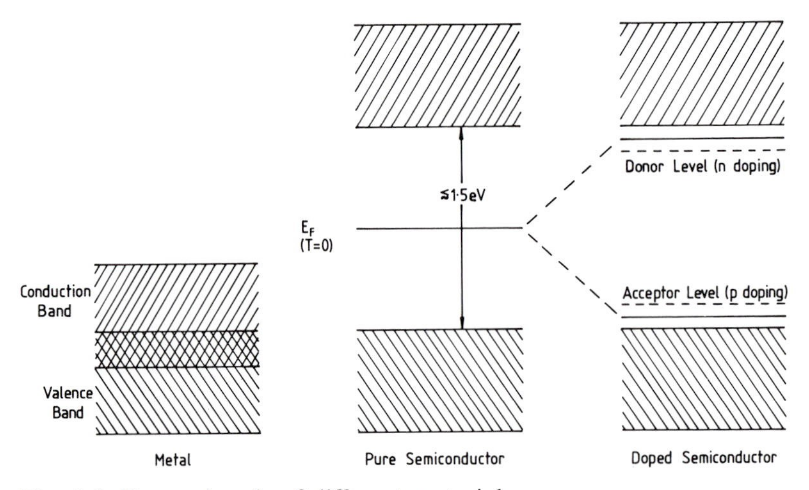

Fig. 5.6. Energy bands of different materials.

In order to escape from the surface of the cathode material electrons must be energized not just across the 'band gap' separating valence and conduction bands, but also overcome the 'electron affinity', deriving from the attraction of residual positive charge on the material from which the electron has escaped. The work function for photoemission is thus the sum of band gap and electron affinity potentials. This can be distinguished from a 'thermionic' work function, associated with the diffusion of electrons at levels above the zero temperature valence band limit from the surface in the presence of ambient thermal phonon energization. Such an emission of electrons can derive from lower photon energies (i.e. infra-red) than normal photoemission, but it requires a sufficient population of electrons at enhanced levels, i.e. it is a function of the ambient temperature.

The detailed nature and scale of the response of semiconductors to incident radiation is very strongly influenced by any proportion of impurities in the material. A very advanced technology now exists whereby semiconductors are 'doped' with selected impurities, which can raise the Fermi level E_F (where the probability of a state being occupied $= \frac{1}{2}$) towards the conduction band ('donor' impurity) or lower it towards the valence band ('acceptor' impurity). This is intimately related to current research and understanding of photodetective materials and methods, of a variety of types.

When two such differently doped semiconductors are placed into physical contact an electromotive force develops at the junction. The static level of this potential difference can vary in response to incident radiation (the *photovoltaic* effect). More commonly found practical light detectors, however, employ the conductive properties of such a semiconductor diode.

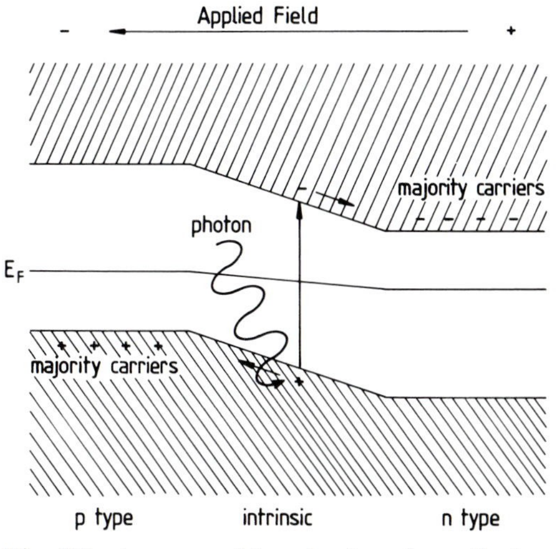

Fig. 5.7. A reverse biased p-i-n photodiode.

The normal mode of operation is with an external voltage applied in series with a load resistor across the diode. The sense of the voltage is to oppose free diffusion of electrons across the junction from donor to acceptor material — a trend which would naturally occur in the absence of this voltage, until some equilibrium distribution was arrived at. The excess of electrons on the acceptor side would then set up an intrinsic electric field in the same sense as that to be applied. This is called reverse biasing (Figure 5.7). There is no significant current in the absence of any illuminance.

When the diode is illuminated and electrons, in the vicinity of the junction, are energized into the conduction band, they can be pulled through to the positive side of the applied field, in effect forming part of a current whose value is very linearly related to the incident flux of photons with energies above that of the band gap. 'Holes' in the microscopic lattice arrangement of atoms in the solid play an equal and opposite role to that of electrons, i.e. an electron–hole pair is what is created by the energizing photon: the migration of electrons in response to the field is matched by a similar migration of holes in the opposite direction.

The inclusion of a layer of intrinsic or undoped semiconductor material between acceptor ('p type') and donor ('n type') regions results in a 'p-i-n' diode. This has an effectively extended energization cross-section over the simple junction, particularly at longer wavelengths, as well as a relatively faster response time. Such diodes, as robust, light-weight and inexpensive detectors, not requiring high voltages, have been usefully applied to astronomical photometry. However, though they give better access to the near infra-red, they tend to be relatively noisy as visible region detectors compared with photomultiplier tubes. This is essentially because of the inherent information advantage produced by the secondary emission cascade gain.

Practical photocathodes are often made of p type material, i.e. with the Fermi level closer to the valence band, which now has additional acceptor levels above it. The population of these acceptor levels can give rise to a noticeable red-leak in the cathode response function. This population is strongly temperature-dependent, as well as critically related to impurity concentration. This entails individual variation in red response from tube to tube depending on operating temperature — a point of significance to photoemission-based photometry at longer wavelengths, and its reduction procedure.

Electron energization in semiconductor material is also basic to areal detection, or imaging. This may take the form of a suitably contrived array of small photodetection sites formed on a demarcated and controllable substrate, as in a charge coupled device (CCD), or it may involve the rapid sequential scanning of the electric potential distribution on a photoresponsive target, as with a television camera. In either case one seeks a low-noise mapping of an areal flux distribution into a corresponding, linearly dependent distribution of electron concentrations, which can be faithfully read out. Although considerable complexity may then be introduced in dealing with the greatly increased amount of data, the underlying detection principles remain essentially similar to those of single-channel units.

5.2.2 Detector characteristics

The practicalities of astronomical (weak) signal detection raise the question of *detector efficiency*. This may be defined as,

$$Q = \langle \text{number of registered events/incident photon} \rangle \tag{5.1}$$

We commonly find a poissonian distribution of registrations, so that the noise in either input or output signals is proportional to the square root of the signal, which is thus also proportional to the signal to noise ratio (S/N). Another useful parameter of performance may be introduced, which we call the 'information transfer efficiency' E, given as

$$E = (S/N)^2_{out}/(S/N)^2_{in} \tag{5.2}$$

In an ideal situation, one would find $(S/N)_{out} = (QS/\sqrt{QS})_{in}$, and so $E = Q$ for the detective process. In practice, $E \leq Q$, because additional processes, irrelevant to detection, add into the output noise N_{out}, and correspondingly reduce $(S/N)_{out}$. A reduction of E from Q also results from non-uniformity in the weighting of registered photon events from a uniform input.

For convenient detection one seeks a linear responder. Although the photoelectric effect is at the root of the silver halide grain darkening which takes place in photographic emulsions, the variety of additional processes leading to the final state of a photographic plate or film tend to make the net registration very non-linear. In general, we characterize linearity as the slope of the curve of accumulated registered events versus the number of accumulated incident photon events at the same surface. Where there is a continuous, quasi-instantaneous registration of events (e.g. a photocathode) any particular registration will be independent of previous ones, and thus produce a linear detection. Where an accumulation of events can interact at a detection site prior to final registration (as with a photographic emulsion), the likelihood of non-linearities increases.

This raises the matter of 'event capacity' at a detection site, i.e. how many events can be usefully accumulated before, for example, saturation effects start to impede linearity? Also there is the 'dynamic range' — essentially the same as event capacity for a noiseless accumulative detector, but more generally interpreted as the difference between useful maximum and minimum event counts.

For areal detectors, such as CCDs, one generally seeks information on pixel size (which should be comparable to telescopic resolution in the focal plane), and the total number of pixels and their geometric organization by the controlling software. A distinction can be made made between pixel size, as determined by the software, and detection-site size, which is $\leq$ pixel size.

5.2.3 Photomultiplier tubes

The advance in design of photomultiplier tubes over pre-existent cell types essentially concerns the multiplier system of secondary emission electrodes, mounted in the same near-vacuum container as the cathode and anode. The secondary 'dynodes' are arranged in a series, each one at an electrostatic potential of around 100 V higher than its predecessor. The first dynode, at which special arrangements are sometimes included to maximize photometric efficiency, is at a similar potential above the cathode. It attracts electrons released from the cathode, but after impinging on the dynode surface with sufficient kinetic energy each such electron releases several more toward the following dynode. The process is repeated perhaps a dozen times along the dynode chain. In this way, large numbers of electrons are produced as pulses in response to photon incidence, and are thus significantly more

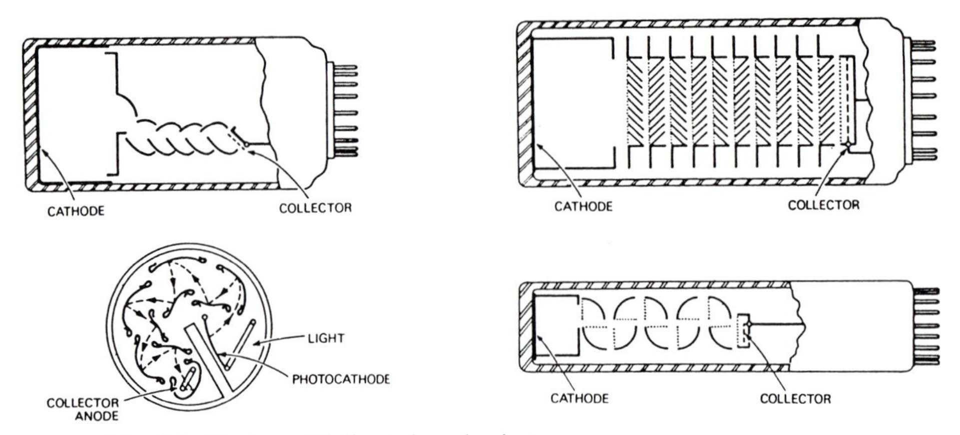

Fig. 5.8. Photomultiplier tubes: basic types.

measurable against the background of quasi-random thermal energizations in the measuring system. Some examples of different designs of photomultipliers are shown in Figure 5.8.

The voltages along the dynode chain are maintained by a 'divider', usually in the form of a series of high-quality resistors mounted across relevant pins of the photomultiplier base. This resistive chain, in parallel with the dynamic resistance of the tube itself, involves a compromise on the choice of component values. Too high resistances in the chain would mean that the current could start to become comparable to that of the tube at high photon incidence. Hence the dynode voltages would be reduced somewhat, leading to loss of linearity. Too low resistances, on the other hand, could start to introduce i^2R heating around the tube base, enhancing undesirable thermionic effects. The former consideration is more important in the DC method, where resistance values are typically 100 kΩ. For the PC method higher values of resistance can be tolerated, since this method is stable against variations in value of the output current. In that case megohm resistors are favoured.

In the limiting faint light detection of astronomy we can have electron movements of the order of only, say, several per second along a high potential gradient. The notion of a continuous electric current through a conductor gives way to something more akin to minute charge displacement in an electrostatic context. In fact, the original approach to signal measurement was that displaced electrons were allowed to charge up a low capacitance for a given time interval, and the resulting voltage developed across the plates measured by an electrometer. This was later refined to the use of 'electrometer tubes' as a primary stage in the recording process.

Random movements of electrons in the measuring equipment at normal ambient temperatures can be associated with energies of order 0.1 eV. The signal should be able to generate a charge-voltage product (qV) at least comparable to this for distinct registration ($S/N \gtrsim 1$). Hence if f is the rate of energizing photon incidence at the cathode, we require $qV = (GQft)^2/nC \gtrsim 0.1$, where the capacitance C is in picofarads, and then n is the number of electrons in 10^{-12} C of charge ($= 6.241 \times 10^6$), while t represents the integration time. G denotes the gain of the tube expressed as anode electrons produced per released cathode electron. For a typical twelve-dynode photomultiplier G should be at least $\sim 3^{12}$, say 10^6, in value. Efficient registration at normal ambient temperatures thus implies a rate of incidence,

$$f \gtrsim \frac{\sqrt{6.2 \times 10^5 C}}{GQt}. \tag{5.3}$$

Charge measurement by conventional electrometers is stable for capacitors as small as tens of picofarads and for tens of seconds; so that the foregoing expression shows that the signal from a photomultiplier would rise above ambient thermal noise at even very low photon incidence ($\lesssim 10^{-3}$ Hz). But (5.3) also points up the important role of the photomultiplication G for signal measurement. In the absence of this avalanche enhancement ($G = 1$), only relatively bright stars could produce measurable signals with flux collectors of reasonable size (cf. Table 3.5). This restriction seriously affected the early days of stellar photoelectric photometry.

5.2.4 Noise in single-channel photometry

It is instructive first to consider the disturbances to signal registration which occur for a basic detector such as a photomultiplier.

There is a statistical noise due to the inherently non-steady photon influx. This is sometimes called shot noise, or the primary detection noise. A poissonian distribution is normal for the arrival of the primary photon stream from a source of constant emissive power. This gives rise to the familiar $\sqrt{n}$ concept for the S/N ratio of the primary signal ($\propto n$). This becomes the predominant noise source for higher incident fluxes.

The multiplicative effect of the secondary emission process in a photomultiplier should follow a similar pattern. The final form is quasi-poissonian, but with some usually significant departure, the full explanation for which is complex, but has been connected with non-uniformity of response over the dynode surfaces, or the fact that some electrons fail either to reach the next dynode or produce secondary electrons if they do arrive there. The multiplication noise can be characterized by a second moment parameter $\alpha = \overline{q^2}/\overline{q}^2$, where q is the observed output charge in each pulse. The spreading out of α has been represented by $\alpha = g(b+1)(1-k)/(g-1+k)$, where g is the gain at each electrode, k denotes the (very small) fraction of orginal photodetection pulses which fail to propagate, and b $(0 < b < 1)$ is an empirical quantity, typically ~ 0.2, related to unevenness of the multiplication.

These two inherent sources of noise, i.e. shot and multiplication, are supplemented by various other contributions, increasingly significant for fainter illumination. They are reducible by careful design and operation, but, in the case of 'giant' pulses due to cosmic ray incidence, for example, are difficult to eliminate entirely.

The most commonly encountered source of additional noise is 'dark current' — the background of low energy pulses present in the absence of cathode illumination. For sufficiently low illuminance this dark current can become comparable to the signal: a condition referred to as dark current limited detection. It may arise in astronomical contexts with a very dark sky, low scintillation and small entrance aperture, depending rather critically on ambient temperature.

The dark current is usually regarded as mainly thermionic emission from the cathode and perhaps some dynodes, but this is not the whole story. Dark current is indeed normally reduced with a reduction in ambient temperature, but sometimes low work function materials used for certain emissive surfaces can, at a given temperature, give rise to a lower dark current than that found from other tubes employing higher work function electrodes. Contamination of the near vacuum tube interiors by residual gas particles released from adsorption to components, or possible diffusion in of helium traces from the atmosphere, secondary cosmic ray effects, residual radioactivity or electroluminescence in the glass envelope, field emission from sharp corners on electrodes, or trapped charge from previous exposures to light have all been related to such non-thermionic contributions to the dark current. The cathode's photoemissivity is also reduced by lowering the temperature, so it is not obvious what the optimum working temperature will be for any particular tube — such data can usually be supplied by the manufacturer, however, from information determined during production.

The dark current proportion of detector noise is much less with a photodiode, because of both increased detective quantum efficiency (which may be at least 2–3 times greater in the visible, and much greater towards the infra-red), and the responsive surface being small and entirely illuminated. Though measurement noise is essentially much greater for photodiode than photomultiplier systems, as a result of the G factor difference, this does not necessarily cause the greatest loss of information in all photometric conditions. At a sufficiently bright level where inherent shot noise predominates, for example, the quantum advantage of photodiode detection can outweigh the then relatively negligible addition of measurement noise.

Voltages in the range 1000–2000 V between cathode and anode are typically required to operate a photomultiplier as a faint light detector. For any particular tube an S/N experiment can be performed, measuring dark current and signal from a faint source against applied voltage, to determine the optimum working voltage. Such experiments show that tubes have a quite sensitive response to voltage variation, due to multiplicative secondary emission. For example, according to the manufacturer's specifications for the EMI 9558 tube, which has been recommended for astronomical photometry, increasing the voltage from 1000 to 1200 V changes the sensitivity typically from 70 to 400 A lm^{-1}. The behaviour is approximately that of a power law with an exponent of about 8. In other words, if tube output currents for a constant source are to be kept constant to within 0.1%, the voltage must be regulated to within a factor 10^{-4}. This is rather a stringent limit, and calls for a high construction quality, keeping in mind that observatories are often located in remote sites, where it may be difficult to ensure a well-regulated basic supply. The strong recommendation for limiting drawn current (usually to a few milliamperes) in a unit capable of supplying several thousands of volts is also worth noting.

Tubes have a variety of operating idiosyncrasies, associated with 'warm-up' or 'fatigue' effects, dampness, the action of the background magnetic field or other effects. Some new designs have been particularly well suited to low-light-level pulse counting; though there is also an argument for staying with tube types already tested by extensive pioneering research. The astronomer using a photomultiplier soon becomes acquainted with the peculiarities of his or her tube, and develops a precautionary practice — e.g. minimizing applied voltage variation during an observing period — in a somewhat individual way.

Problems of instrumental instabilities are not necessarily too serious, depending on the nature of the observations. Differential photometry, for example, is quite stable against slow 'creep' effects. Fast photometry does not concern itself with variation on such timescales (hours). Instrumental calibration by reference to standard stars, on the other hand, really calls for the most stable of conditions, both externally and internal to the equipment.

The total noise of detection is then not just that of the photoelectric responder, since the signal has already been deteriorated by the atmosphere. The existence of static atmospheric extinction can be dealt with by well-known procedures of intercomparisons between different stars. But the atmosphere has also *dynamic* extinction properties on a wide range of timescales. At the high-frequency end these are covered by the term scintillation. This can be conveniently characterized as a 'modulation' of the signal, which is a representative proportion of it, m say, for a given night and integration period. Scintillation is a complex result of atmospheric inhomogeneities on the incoming wavefront, as sampled by the telescope's aperture; but, from the practical standpoint, can be treated as an empirical parameter adding mS into the noise.

The atmosphere also scatters light from all other sources, some of which are variable on short timescales. Even the steady component of this background light of the sky introduces a secondary noise, through its shot noise from the detector. On the best quality nights, when transparency is high, scintillation is low and adequate precautions have been taken to reduce inherent instrumental noise to a minimum, the background sky noise represents a final limiting for accurate measurement of weak astronomical sources — 'sky limited' detection.

5.2.5 Areal detectors and enhancers

Although the single-channel photometer has enjoyed a very productive run, particularly since the introduction of photomultiplier tubes in the middle years of the twentieth century, there are strong indications that we are moving to an era where detection with two-dimensional resolution is predominating. There are good reasons for this. Although the potential information yield becomes burdensome if one simply stores everything that an areal detector delivers, this is really an argument that the data logging has to be versatile enough to handle various observational requirements. In conventional stellar photometry one normally uses only a small subset of the information available. Efficient data handling then implies rejection of the rest, but what is retained, because of its effective *simultaneity*, is essentially more informative than single-channel, interrupted, photometry.

A number of different types of areal detector have been available for some time, starting, in principle, with the photographic procedures of the last century. Positive remarks can be made about photographs: they are permanent, relatively inexpensive and easy to store; they can easily be handled and inspected by eye. They are capable of very fine spatial resolution over a relatively very wide field, and continue an archival source, with a century

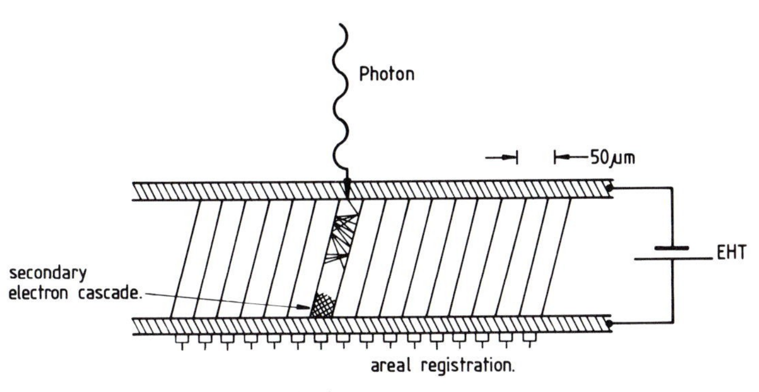

Fig. 5.9. A microchannel plate.

or more of accumulated observations, of inestimable value. Nevertheless, for accurate photometry their value is limited. They have a detective quantum efficiency an order of magnitude below photoelectric responders. This means relatively long exposures are required to produce measurable effects, during which time atmospheric transmissions can drift in a way which is hard to quantify precisely. The development process is itself non-linear, and imaging responses vary even over a given plate for various reasons. Photographic plates have had an important historic role in the development of astronomical observations, and may continue for certain special purposes, though this seems unlikely in specifically photometric researches.

The secondary emission cascade, which occurs in a photomultiplier, finds an interesting two-dimensional parallel in the 'microchannel plate' (Figure 5.9). In this device a bundle of semiconducting glass fibres is drawn out in an inclined or curved arrangement behind a plane which receives an incoming photon distribution. The input plane of fibre ends, where the image is first focused and initial photoelectrons produced, is normally in contact with cathode material, of somewhat lower work function than the channels themselves. Electrons liberated by the photon incidence then avalanche down the fibre, in response to an accelerating field of typically several kilovolts, and repeated incidences on the fibre wall. Gains in excess of 10^6 can be produced in this way. The anode array at the other end of the fibres may drive a two-dimensional photon-counting logger, or a phosphor screen upon which an intensified image is produced. The technical requirements of the microchannel plate, particularly with regard to channel evacuation and the application of high voltages, make for a somewhat delicate and expensive device.

Image intensifiers, in general, need not employ the secondary emission feature of the microchannel plate. They may simply utilize the energization

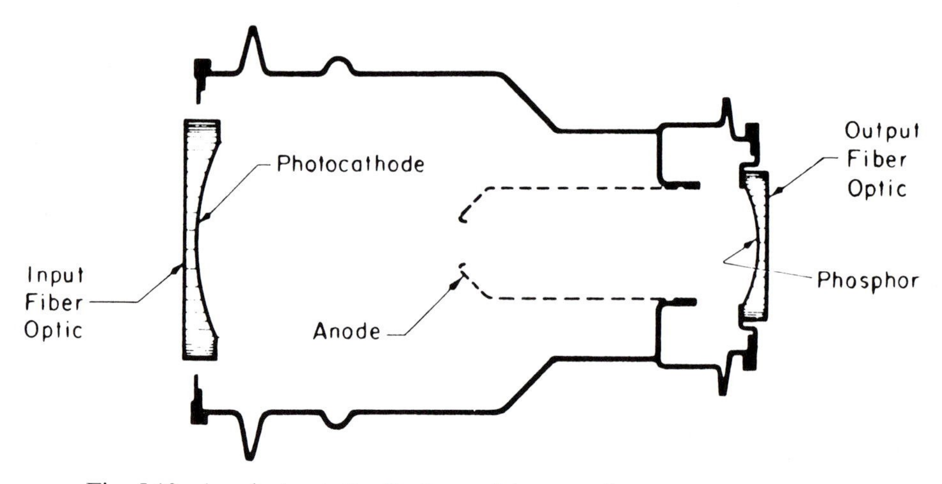

Fig. 5.10. An electrostatically focused image tube.

of photoelectrons by a strong electric field in a focused flight from cathode to anode (Figure 5.10). The accelerated electrons impinge upon a phosphor screen target where the intensified image is formed. Optical fibres may be used to reproduce this image, normally formed on a curved surface, into a conveniently flat format. Commercially available intensifiers can be made quite robust, and are able to be cascaded in several stages. Operating voltages are inevitably high, however, since this provides the basic source of intensification. This can give rise to operational awkwardness in typical dome observing conditions.

The development and availability of television type cameras raises interest in their application to astronomical work, where various related instrumentation has appeared — for example, the Image Photon Counting System (IPCS), developed at University College, London. A full discussion of this large technology is outside the scope of this book; there are a couple of points that can be made here, however.

Firstly, vacuum tube imaging detection necessitates two stages of electron information transfer, firstly in the image production on the target, and secondly in the sampling of the read-out beam. Although suitable cooling of the camera may allow noise fluctuations in the read-out beam to be reduced to acceptable levels, there is inevitably a limit to the extent to which strong local variations of intensity on the target can be faithfully reproduced in the statistics of the return beam. In short, precision in the read-out process appears inherently more difficult to acheive than in the direct charge relocation procedure of solid-state charge coupled devices.

Secondly, whilst sophisticated electronic cameras become increasingly within general reach, their basic purposes usually differ from those of astronomical photometry. Whilst it may be interesting to record a varying star on a video tape, for example, it is the *digitization* of the photometric data which is required for analysis. Such data may be available by suitable decoding of the relevant frames, but the process of removing information used to create a visual impression, which was not the main aim of the observation, introduces irrelevances.

5.2.6 Charge coupled devices (CCDs)

Charge coupled devices, or CCDs, are ultra-sensitive, semiconductor, 'chip'-based areal detectors which have been increasingly used as the detector element in astronomical instruments. Large observatories were developing their use in the mid-seventies. By the late eighties CCD usage had diffused through to quite small establishments and amateurs.

Each individual CCD detection site, can be regarded as a minute location where a doped semiconductor substrate (usually p doped silicon) is separated by an insulating layer (SiO_2) from an electrode (polysilicon). The electrode is charged in such a way that its electric field repels the semiconductor's majority of free carriers, holes (electrons) in the case of p type (n type) material. This produces a charge depletion region where any minority carriers — electrons (holes) — collect. Incident photons produce electron–hole pairs in the semiconductor substrate. The signal is then registered by the number of minority carriers trapped in the depletion region, and this should be directly proportional to the number of incident photons; at least up to some practical upper limit, where a saturation effect comes into force (related more to the rate of information transfer rather than inherent registration capacity). The quantum efficiency of electron–hole production is relatively high, say double or treble that of a photoemissive cathode in the visible region, and much higher in the red and near infra-red. CCDs are thus advantageous for low-light-level detection over an extended wavelength range.

CCDs are normally set up as imaging detectors. A grid of electrodes and suitably charged 'channel stops' is arranged on the chip so as to produce a two-dimensional array of depletion regions. Read-out of the accumulated charge in such regions after a suitable exposure is achieved by varying the potentials on the electrodes row by row, so that the charges in successive rows of detection sites are transferred to a serial output temporary register (Figure 5.11). Some noise may be introduced into the process at the information transfer stage depending on the rate of transfer, the ambient temperature,

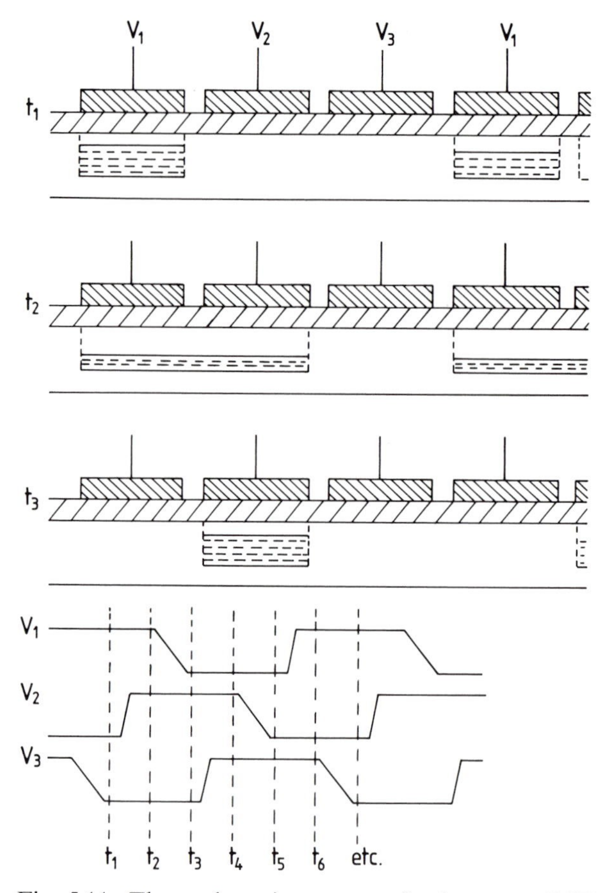

Fig. 5.11. Three phase image transferring on a CCD.

and certain more specialized aspects of the chip design. In astronomical contexts CCD detectors are often cooled to liquid nitrogen temperatures, so dark currents will be low and read-out efficiencies high even when data are transferred at kilobaud rates. The rows of an image are first transferred from the exposed to an unexposed storage area. This is a necessarily fast process, to avoid smearing in the still exposed region during transfer. The arrangement of the storage rows on a chip is a matter for design 'architecture'. Perhaps the simplest is the one shown in Figure 5.12 — one half of the chip is exposed, the other half masked with an opaque shutter.

Output of the exposed frame from the storage area to the data handling computer involves serial transfer via the register through a high-gain, on-chip (MOS) amplifier system, and this is a relatively slow procedure. Noise inevitably affects the charge measurement process, along the lines discussed in the lead up to (5.3). The relevant 'node' capacitance of the amplifier

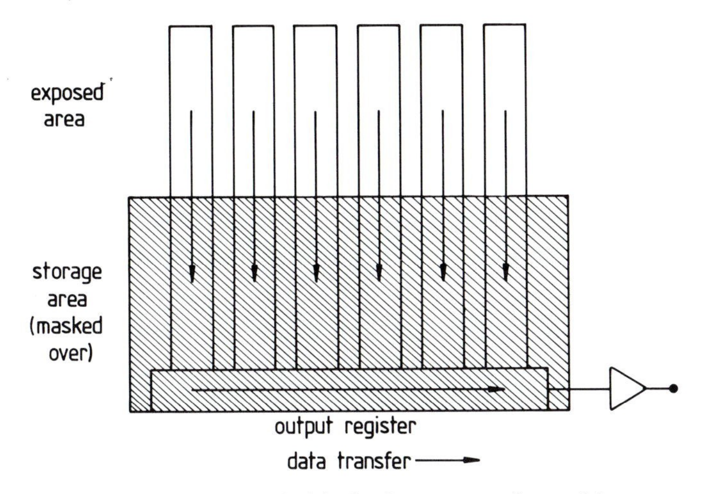

Fig. 5.12. Half covered chip in frame transfer architecture.

is typically only 0.1 pF, which, together with cooling, reduces electrometric noise ($\sqrt{kTC}$) to $\sim$ 100 electrons per pixel read-out. A way to recover the G advantage of (5.3) is to arrange an areal photomultiplicative device (e.g. a microchannel plate) in front of the CCD. Single primary photoelectrons can then be registered with negligible read-out noise.

From the chip output each pixel's accumulated charge is then reconstructed in a computer assigned array. At observer determined intervals, the accumulated digital information can be transferred to a screen, to enable direct visual inspection of how the image is built up. With suitable software the observer selects screened frames, or certain regions thereof, to be retained by a more permanent digital data logger, such as a magnetic tape or a disk. The final output of the process will thus allow more refined processing by computers at a later time.

The complexity of a CCD detection system implies the astronomical user is likely to want sufficient interaction with manufacturers to clarify technical requirements and capabilities. Thus, CCD chips intended for faint field imaging can be specially selected to be free from defects such as noisy pixels and bad columns. The wide spectral range of CCDs also requires some attention. Normal CCDs lack intrinsic blue sensitivity, because photons with wavelengths shorter than those of the yellow region do not penetrate as well into the the electron–hole pair-yielding depletion regions of the substrate. In order to increase blue and ultra-violet sensitivity the substrate may be thinned, but this is a delicate and relatively expensive operation. A cheaper alternative is to coat the surface to be exposed with a fluorescent dye, but since the lower energy photons are re-emitted in all directions there is then inevitably some loss of efficiency. In the red region, associated with the greater penetration of photons, there is also a potential relative loss of spatial resolution.

Operation of a CCD involves much more than obtaining a useful chip. The chip is potentially a very large data generator as can easily be seen from considering the information content of say a 1000 by 500 array, each pixel of which may have an event capacity of $\sim 10^4$. The whole may be recording exposures in a matter of seconds. It is not surprising, given this level of data generation, that a considerable proportion of time and effort in working with a CCD camera relates to the design of efficient image handling software.

A key task is the selection from the total amount of data generated that relevant to what is finally sought. With the photometry of variable stars, for example, one needs only to demarcate those pixel regions on the frame which contain stars of interest (i.e. the variable and its comparison(s)), sample some nearby background sky level, and store that data subset. On-line reductions, performed in an efficient system, allow the presentation of a preliminary form of light curve, more or less as it is observed.

The CCD-using astronomer can be helped by extensive technical experience deriving from more general applications, and commercially obtainable specialized equipment. Modern observatories will generally make available software facilities with special user-oriented computer command sequences, e.g. the Munich Image Data Analysis System (MIDAS) of the European Southern Observatory. Convenient as this may be, an increasing technological separation between the data acquisition system and its user carries some danger of discrepancies in unfamiliar or unseen parts of the processing creeping through to the end results unnoticed.

5.3 Conventional measurement methods

5.3.1 The DC method

Early measuring and recording techniques for photoelectric photometry, beyond simple electrometry, involved amplification of the photocurrent up to the point where it could be logged on a pen recorder. This is the analogue, or usually DC, procedure. An example of a schematic circuit which can perform the very large amplification required is shown in Figure 5.13. This circuit involves a commercially available operational amplifier arranged with a resistive negative feedback loop, in the manner nicely explained in Henden and Kaitchuck's (1982) book.

Let us consider the information transfer efficiency E from the detector through the DC amplifier. We regard the cathode, responding to a constant

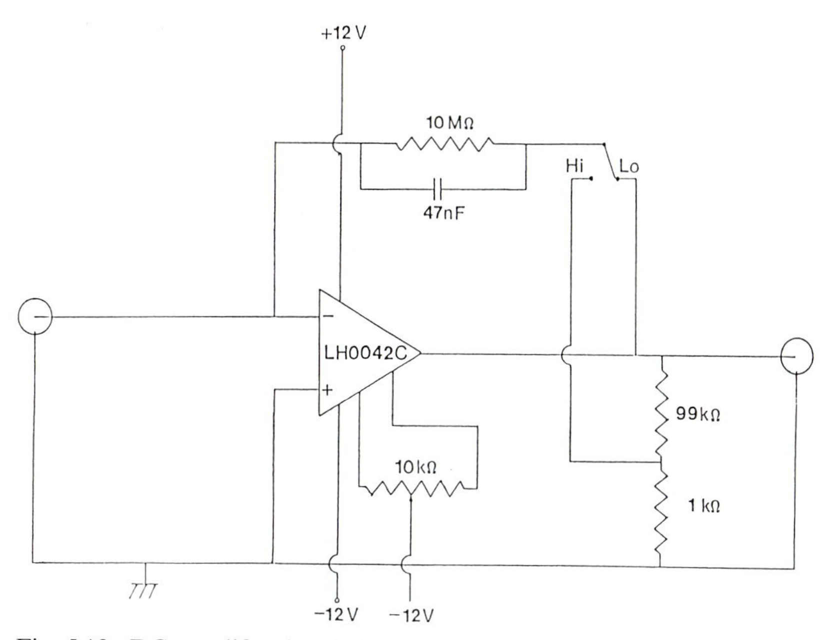

Fig. 5.13. DC amplification (schematic).

light input, as a current generator of a mean signal $\bar{I}$, onto which a noise component denoted $\sqrt{\langle i^2 \rangle} = \sqrt{\langle (I - \bar{I})^2 \rangle}$ is added. Independent additive noise sources a and b satisfy

$$\langle (a+b)^2 \rangle = \langle a^2 \rangle + \langle b^2 \rangle. \tag{5.4}$$

Now the ratio of detector output signal to input signal was specified to be G. Hence, if the thermal agitation of electrons in the amplifier's load resistor R generates an additional source of noise ('Johnson noise') i_R

$$\begin{aligned} E &= (S/N)^2_{out}/(S/N)^2_{in} \\ &= \frac{\langle i^2 \rangle}{\langle i^2 \rangle + \langle i_R^2/G^2 \rangle}. \end{aligned} \tag{5.5}$$

For the detector, the shot noise from photoelectrons of charge e, for a frequency bandwidth Δf, is given as

$$\sqrt{\langle i^2 \rangle} = \sqrt{2e\bar{I}\Delta f}. \tag{5.6}$$

The thermal noise in the resistor, in joules per coulomb, is similarly written as $\sqrt{\langle V_R^2 \rangle} = \sqrt{4kTR\Delta f}$, where, k is Boltzmann's contant, and T is the ambient temperature. Alternatively, we can put,

$$\langle i_R^2 \rangle = \frac{4kT\Delta f}{R}. \tag{5.7}$$

It is clear that,

$$E = \frac{\bar{I}R}{2kT/G^2e + \bar{I}R}. \tag{5.8}$$

The factor $2kT/e$ of the denominator has a numerical value $\sim$0.05 V at typical ambient temperatures ($\sim$300 K), which implies that the product of detector current and load resistance must be greater than $0.05/G^2$ to prevent serious information loss from this measurement noise. Again we see the importance of the photomultiplier's internal gain G in minimizing the effects of thermal noise.

The load resistance R couples with stray capacitance of the tube C_s to introduce a response time $\tau = RC_s$, so that if f is the mean frequency of arrival of electrons at the anode, we can rewrite (5.8) as

$$E = \frac{f\tau}{3.1 \times 10^5 C_s/G^2 + f\tau}, \tag{5.9}$$

where C_s is in picofarads, and the time unit is seconds. This equation gives a limit on available time resolution for a detector of given internal gain G and stray capacitance C_s. The stray capacitance of a photomultiplier tube assembly is typically of the order of a picofarad, so $f\tau \geq 10^{-6}$ to avoid serious signal degradation from amplification noise. This condition would be comfortably satisfied in any feasible application of DC amplification, because for the faintest signals, where the largest resistors would be used, τ would be large enough ($\gtrsim 10^{-3}$ s) that other more significant sources of noise would constrain f long before the low incidence ($\sim 10^{-3}$) corresponding to this Johnson noise limiting was reached. Bright sources ($f \sim 10^6$ in Table 3.5) will correspondingly admit very high time resolution.

The gain issue was underlying the comparison between photomultipliers ($G \sim 10^6$) and photodiodes ($G = 1$) in Section 5.2.4, though other factors are involved, apart from those previously mentioned. Firstly, improvements in component manufacturing have allowed stable resistances of very high values ($\gtrsim 10^9$ Ω) to become more easily available, with corresponding improvements to the ratio (5.8). Secondly, this ratio can also be reduced with effective cooling of components. Such points significantly bear on the choice of detector — photomultiplier or photodiode. The latter is constrained to DC recording, however, since individual photon events are, in principle, not registrable in the technique.

A common output medium for the DC method is the strip chart, or 'pen', recorder. This presents a visible, simple and permanent record of a photometric observing run. With a reliable strip chart one can take in, at a glance, the run of the data, the relative quality of the night, any special or spurious events, or commentary that the observer may have written alongside the pen's track. There are many types of pen recorder, and recent designs include more favourable features. Accurate linearity, constant known speed, and easy, smooth, continuous pen action are the most important of these. Their prices remain relatively high by current standards in electronics. As a final medium for data-logging, pen recorder charts have the disadvantage of requiring a human measuring process, which can be time consuming, as well as introducing inaccuracies or personal factors. On the other hand, they are a good idea as a parallel back-up medium, allowing spot-checks of a simultaneously running digitized recording, and giving some insurance, in case of the latter's failure during observations.

The input signal is originally not of analogue form, but consists of a sequence of individual pulses. There is an inherent loss of information in a measuring system whose inertia loses sense of the discrete nature of the initial data, but how serious this is depends on how faithfully the quantities sought, such as mean illuminance levels over given time and wavelength intervals, are delivered. In any case, the multiplicative component of tube noise inevitably adds into the direct current. This particular noise can be avoided if the rate of photoelectron avalanches from the anode (the PC method), rather than the net flow of charge, is measured. The DC approach generally faces problems connected with adequate component stability, failure to discriminate against non-signal contributions to the measured level, and recorder-imposed constraints on measurement, which hinder maximum information retrieval.

The advantages of automatic digital recording can still be achieved in a DC system via analogue to digital conversion. This has been effected with voltage or current to frequency converter (VFC or CFC) type circuits. In one arrangement (VFC) the anode signal is converted to a voltage which operates a voltage-controlled oscillator. In another (CFC) charge builds up until it triggers an output pulse of a definite size and shape (Figure 5.14). In this way, a digital information flow can be directly logged by computer. A simultaneously running hard copy can be maintained through the computer's printer.

5.3.2 *Pulse counting*

From the capacitance and gain values mentioned in the preceding section we find the voltage in a photoelectron pulse to be typically tens to a hundred

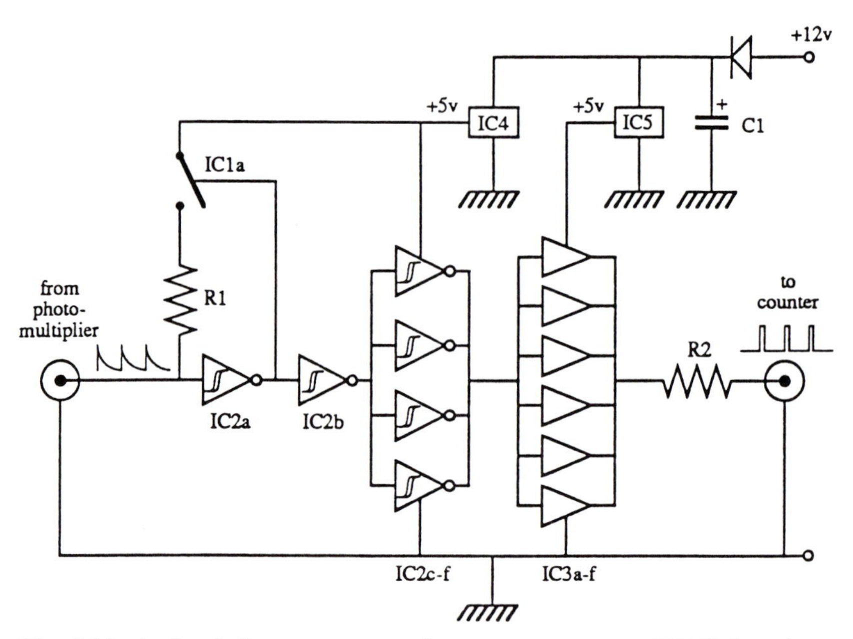

Fig. 5.14. A circuit for a current to frequency converter (CFC) based on a design of Rowe (1969). The components are: R1 = 27 kΩ, R2 = 56 Ω, C1 = 4.7 μF, IC1 = 4016 analogue switch, IC2 = 40106 hex Schmitt inverter, IC3 4050 hex driver buffer, IC4,5 = 78L05 power supply regulators.

or so millivolts. They arrive as a bunch of electrons spread over a time interval of the order of nanoseconds, depending significantly on the tube's type of construction. The pulses can vary in voltage considerably, both above and below a representative 50 mV, however, and are not all generated by the illumination. Light generated pulses tend to have distinctly higher voltages than thermionic ones, which can be intuitively understood in terms of different source electron energies. In fact, measured distributions show a considerably extended peak for very low pulse heights, associated with effects other than light detection ('stray pick-up'). An information advantage can therefore be achieved by electronic *discrimination* in favour of photoelectron pulses over background noise. This gain is in addition to the previously mentioned inherent removal of multiplicative tube noise in pulse counting.

The aim of discrimination is to maximize the photometric S/N ratio, and Figure 5.15 shows that with an appropriately assigned window on pulse heights progress can be achieved. In any case, it is necessary to cut out the very low pulse height tail to have manageable count rates. Actual discriminator settings are usually determined from experimental trials on individual tubes, since there can be appreciable variation in performance

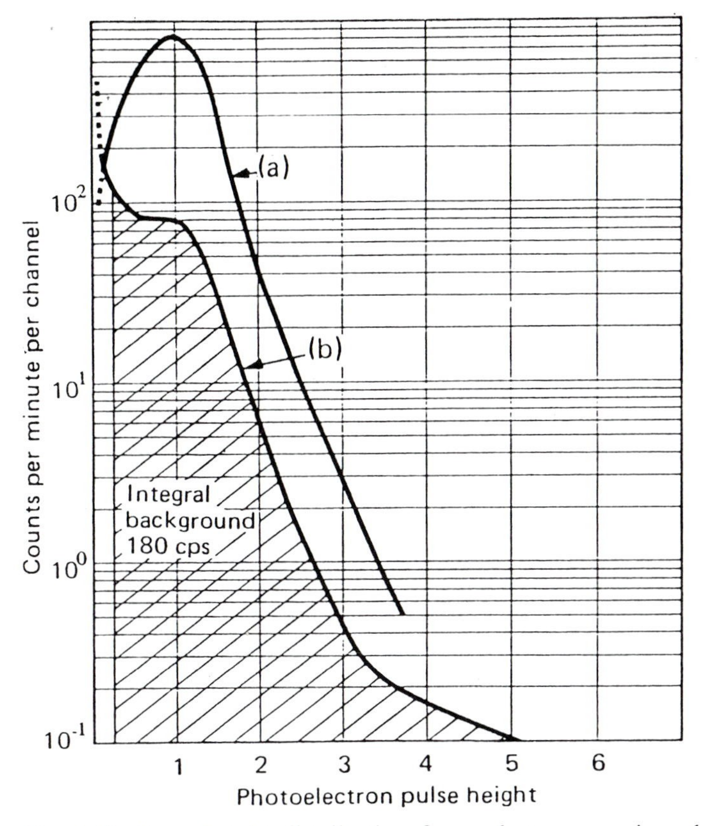

Fig. 5.15. Pulse height distribution for a photon counting photomultiplier, type EMI 9813 (from Wardle, 1984): (a) response to light source, (b) background.

from tube to tube. The optimum operating voltage is usually somewhat greater than for DC measurement — the constraint now being largely set by stability of the count rate at a given level of test illumination, before too high a voltage starts to introduce unwanted 'after pulses'.

The primary photon pulses from the tube, though measurable, are still inherently weak, and can compare with, for example, voltages which a few metres of cabling can receive from stray electromagnetic radiations of radio frequencies. Moreover, such cabling, required in an environment where recording electronics is physically separated from the components attached to the telescope, also reduces the peak and spreads out the voltage profile of the initial pulse. Hence, it is usually mandatory to introduce at least part of the pulse amplification in close proximity to the anode, often as a small box on the photometer head. The general arrangement of a PC system, as a block diagram layout, is shown in Figure 5.16.

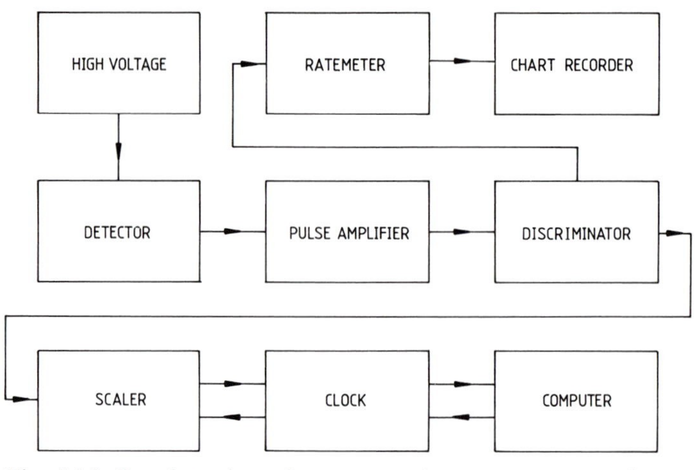

Fig. 5.16. Postdetection photon counting system: general arrangement.

The counting of individual pulses encounters time resolution (or 'bandwidth') limitation for bright sources. This can be appreciated in terms of 'pulse pile-up' — the definite chance that a pulse may indistinguishably merge with its neighbour when incidence becomes high enough. Measured count rates are thus systematically in error, to some extent. This will be slight when the true incidence frequency f is much less than the inverse time resolution ($1/\tau_d$), or amplifier bandwidth frequency f_b ($= 1/2\tau_d$), but it becomes progressively more significant as $f \to f_b$.

The amplifier sends to a counter pulses of a definite shape and size, typically $\sim$1 V, with as near a linear response to the input stream as conveniently possible. The counter thus needs a bandwidth at least as high as the amplifier, and usually it is an order of magnitude higher to ensure full resolution. In earlier times counting was performed by a separate 'scaler', sometimes with visible counting registers, from which numbers could be hand recorded. Nowadays, it is much more common to find such a unit on a special 'input/output card', inside a controlling and data-logging microcomputer.

The systematic underestimation of pulse incidence can be related to the net time resolution, or 'dead time' value τ_d, as follows. In unit time interval the number of incident pulses is f, while the number counted will be f_c. Each of these f_c events is associated with a dead time τ_d, so that the effective counting time is shortened by $f_c\tau_d$, and so

$$f_c = f(1 - f_c\tau_d) = f(1 - f_c/2f_b). \tag{5.10}$$

The distribution in electron arrival times at the anode can be as small as 10^{-9} s in modern tube designs with 'beam focusing'. Hence, the basic data-generator is potentially a high bandwidth (GHz) device. However, high measurement bandwidth f_b (low τ_d) brings with it more stringent requirements for well-screened, low-stray-impedence amplifier components, pushing up either likely costs or physical stability requirements. This need not be necessary for many astronomical programmes, where event incidence could be typically around several thousand hertz, as in, say, intermediate-band photometry of tenth mag stars with a $\lesssim$ 1 m class telescope (cf. Table 3.5). From (5.10) we require only $f_b \gtrsim 10^3 f_c$ to achieve 0.1% accuracy in standard photometric reductions (without a dead time correction), and in a differential photometry context even greater precision would be inferred. It implies that f_b ~10 MHz would be quite adequate for many practical situations.

For the most accurate work, and, in any case, for full awareness of instrumental parameters, it is desirable to have some knowledge of the operating dead time for any given photometer. This is sometimes arrived at from manufacturer's specifications, or by an error minimization technique on standard star data. It is preferable to have an independent empirical option for this, however, and a popular simple technique is to use two light emitting diodes (LEDs) mounted in such a way that all other light except theirs can be excluded from the photometer. Their light should be able to be maintained at an accurately constant level — perhaps by means of a battery source — and they should give enough light that some significant level of non-linearity will occur. Typically, their light output is equivalent to a pair of bright (naked eye) stars, and they are mounted at the top of the telescope with the cover tightly on.

We write f_1 for the count rate observed with the first LED on by itself, and f_2 for that with the second. When both LEDs are switched on together the count rate is f_3. In the absence of a dead time effect, we would have $f_3 = f_1 + f_2$, but in practice we should find that $f_3 = f_1 + f_2 - \epsilon$, where ϵ is a relatively small number.

Making use of (5.10), we have

$$\frac{f_3}{1 - f_3\tau_d} = \frac{f_1}{1 - f_1\tau_d} + \frac{f_2}{1 - f_2\tau_d}. \tag{5.11}$$

This can be solved for τ_d, to yield

$$\tau_d = \left[1 - \sqrt{\frac{(f_3 - f_1)(f_3 - f_2)}{f_1 f_2}}\right] / f_3, \tag{5.12}$$

from which, for small ϵ,

$$\tau_d \simeq \frac{\epsilon}{2\hat{f} f_3},$$

where $\hat{f}$ is the harmonic mean $(f_1 f_2)/(f_1 + f_2)$.

Any device (aperture stops, neutral density filters etc.) which allows a constant, reasonably bright light source to be divided into constant proportions may be used as an alternative to summing different LEDs — it is only required that f_1, f_2 and f_3 be accurately measurable and the true incidence rates exactly additive.

5.4 Bibliographical notes

The practical side of astronomical photometry was well covered, in a very readable style, in A. A. Henden and R. H. Kaitchuck's *Astronomical Photometry*, Van Nostrand Reinhold, New York, 1982. A comparable treatment is given in the recently revised (1988) *Photoelectric Photometry of Variable Stars* (Willmann-Bell) by D. S. Hall and R. M. Genet. Figure 5.3 comes from K. M. Nield's M Sc thesis, University of Waikato, 1987, 5.4 is from A. R. Sadik's Ph D thesis, University of Manchester, 1978, and 5.5 is courtesy of Optec Corporation. The idea of information transfer efficiency, underlying equations (5.4)–(5.9), was exposed in more detail in the article of A. Lallemande in *Astronomical Techniques* (ed. W. A. Hiltner) University Press, Chicago, 1962, p126, but it is also dealt with in Henden and Kaitchuck's text.

A number of books coming from the Fairborn Press, Mesa, Arizona, e.g. *The Photoelectric Photometry Handbook* (eds. D. R. and R. M. Genet) I, and II 1989; *Microcomputers in Astronomy* (eds. R. M. and K. A. Genet) I, 1983 and II 1984; and, more recently, the thought-provoking *Robotic Observatories* (ed. R. M. Genet and D. S. Hayes) 1989, are replete with instrumentation designs and possibilities. (Fairborn Press is now a division of AutoScope Corporation, an organization which produces automated astronomical photometric equipment.)

More technical details are to be found in texts such as *Detection and Spectrometry of Faint Light*, by J. Meaburn, Reidel, Dordrecht, 1976, and M. J. Eccles, M. E. Sim and K. P. Tritton's *Low Light Level Detectors in Astronomy*, Cambridge University Press, Cambridge, 1983, which have been sources for some of this chapter.

The section on detectors also draws on material provided, or cited, by manufacturers. For example, the electron tubes division of Thorne EMI have regularly produced detailed booklets on the properties of photomultiplier tubes, image intensifiers and the like. They cite special studies, such as R. Wardle's *Test Parameters and General Operating Rules for Photomultipliers*, Thorne EMI R/P 067, 1984, from where Figure 5.15 comes, or C. J. Oliver's paper in the same series, Thorne EMI R/P 066, which discusses part of Section 5.2.4 in detail.

Similarly, a wealth of new material on photoconductive detectors, especially CCD arrays, has appeared even as the present book was under preparation. CCD detectors were reviewed by C. D. Mackay in *Annu. Rev. Astron. Astrophys.*, **24**, 255, 1986, and discussed more recently in I. S. Mclean's *Electronic and Computer-aided Astronomy*, Ellis Horwood, Chichester, 1989. A readable backgrounder is J. Kristian and M. Blouke's article in *Sci. Am.*, **247**, 48, 1982, while a good selection of recent literature is given in W. Tobin's paper in the *Third New Zealand Conference on Photoelectric Photometry* (ed. E. Budding and J. Richard), published as a special issue of *South. Stars*, **34**, No 3, 1991.

A. A. Henden and R. H. Kaitchuck's book has good information on the recording techniques of Section 5.3. This includes the proper accounting for dead time, as discussed at the end of this section. C. Rowe's CFC circuit first appeared in *South. Stars* **23**, 63, 1969. Willmann-Bell also published (1989) software written by Kaitchuck and Henden and based on the 1982 text, and more recently (1991) more software, written by R. Berry, aimed at the amateur and small-scale user of CCD cameras for astronomical work.

A number of the diagrams used in this chapter (and the next) derive from students' theses. Thus versions of Figures 5.1, 2 and 4 appear in A. R. Sadik's PhD thesis (Univ. Manchester, 1978), 5.3 and 13 are in K. M. Nield's MSc thesis (Univ. Waikato, 1987), and Figure 5.14 is from the MSc thesis of M. C. Forbes (Univ. Wellington, 1990).

6

Procedures

This chapter is about the essential procedures for setting up and using a photometric system, and processing its data. It has two main sections, dealing with two basic calibration experiments. There follows a brief introduction to variable star photometry. The treatment and interpretation of data is concentrated on.

6.1 The standard stars experiment

The purpose of this is to calibrate a given 'local' photometric system to a 'standard' or reference system, based on detailed comparisons of published magnitude and colour values of standard stars, with corresponding measurements made with local equipment. The experiment is associated with the terms absolute, or all-sky photometry. To do it well normally requires very good, i.e. transparent and stable, sky conditions, but these nights are not so common at most observing locations. They are sometimes described as 'photometric nights', though certain kinds of high quality, differential, photometry have been carried out (notably with multi-channel photometers) in fairly cloudy conditions.

The choice and finding of particular standard stars is related to observing experience and particulars of the task. Specialist programmes are underway which continue to produce improved and more extended lists of standards in various photometric systems. But there are also certain well-accepted primaries (Chapter 3), which are bright (naked eye) and easy to find. A reasonable place for a preparatory start would be with the standard star lists tabulated in the *Astronomical Almanac*.†

Generally, the local system is taken to be set up in a similar way to the standard one, so that any systematic differences are regarded as small and

† See Section 6.4 for more source information.

linear. In what follows local magnitudes and colours are denoted by prime superfixes, standard values will be unprimed. A rearrangement of equation (3.23) will then yield, for a source of a given (constant) temperature, relations of the form:

$$m_{\lambda_1} - m'_{\lambda_1} = \epsilon C_{\lambda_1 \lambda_2} + \zeta_{\lambda_1}, \tag{6.1}$$

or

$$C_{\lambda_1 \lambda_2} = \mu C'_{\lambda_1 \lambda_2} + \zeta_{\lambda_1 \lambda_2}, \tag{6.2}$$

where it is easily deduced that

$$\epsilon \simeq \frac{1/\lambda_1 - 1/\lambda'_1}{1/\lambda_1 - 1/\lambda_2}, \tag{6.3}$$

or

$$\mu \simeq \frac{1/\lambda_1 - 1/\lambda_2}{1/\lambda'_1 - 1/\lambda'_2}, \tag{6.4}$$

where λ'_i denotes a mean wavelength of the filter-cathode combination actually used, while λ_i is the corresponding quantity for the standard system.

Measurements of standard stars do not immediately produce the appropriate magnitude values in the local system to substitute into the foregoing. These measurements are complicated by other effects, in which the role of the Earth's own atmosphere is most prominent. Concerning atmospheric extinction, a relationship of the form:

$$m_z = m_0 + k'X_z + k''X_zC + k'''X_z^2 + ... \tag{6.5}$$

was obtained in Section 4.1, where m_0 represents the required above-atmosphere magnitude, m_z denotes what is derived from actual measurements, C stands for standard colour, X_z the air mass corresponding to a star at zenith distance z in standard atmosphere thicknesses, and the quantities k', k'', k''' are extinction coefficients of various orders.

One approach is to use a set of observations of a number of standard stars, distributed evenly over the whole sky, to derive representative mean coefficients. In this way, all unknown quantities on the right hand sides of equations formed by combining (6.5) with (6.1) or (6.2) are determined in a single operation. We will see how this works out in practice by following through a particular example. In Table 6.1 a list of raw data points from standard star observations in the *UBV* system has been given.

The information in Table 6.1 is a typical basic data set, which can be used in a 'least squares' program to determine the coefficients in the foregoing equations. A few entries are missing, either because of unreliable or excessive

Table 6.1. *Data from DC photometry of standard stars.*

No.	BS	RA			Dec		Flt.	Time		Delfn.		Mlt.	Cat.
	No.	h	m	s	deg	min		h	m	Star	Sky		
1	100	0	25	22	-43	46.5	3	8	50.2	36.5	17.8	2	3.94
							2	8	51.1	72.3	17.9	2	0.17
							1	8	52.1	46.2	17.9	2	0.11
2	8848	23	16	27	-58	19.7	3	8	56.3	37.6	17.8	2	3.99
							2	8	57.7	60.9	17.9	2	0.40
							1	8	58.5	41.6	17.9	2	-0.02
3	9076	23	59	3	-65	40.3	3	9	2.1	29.3	17.8	2	4.50
							2	9	2.7	59.9	17.9	2	-0.08
							1	9	3.5	45.7	17.9	2	-0.28
4	9091	0	1	28	-29	48.9	3	9	9.8	32.3	17.9	1	5.01
							2	9	10.4	73.0	18.1	1	-0.15
							1	9	11.0	66.0	18.0	1	-0.55
5	8675	22	47	32	-51	24.4	3	9	16.5	48.2	17.8	2	3.49
							2	9	16.0		17.9	2	0.08
							1	9	15.6	66.8	17.9	2	
6	373	1	15	44	-2	35.4	3	9	21.7	32.8	17.8	0.5	5.41
							2	9	22.3	39.9	17.8	0.5	0.90
							1	9	23.2	36.3	18.0	0.2	0.44
7	531	1	48	45	-10	46.2	3	9	29.3	36.2	17.8	1	4.67
							2	9	30.5	62.3	17.9	1	0.33
							1	9	28.5	39.4	17.9	1	0.03
8	1030	3	23	54	8	58.2	3	9	36.2	49.3	17.8	1	3.60
							2	9	35.7	56.2	17.8	1	0.89
							1	9	37.0	34.7	18.0	0.5	0.61
9	811	2	43	19	-13	55.8	3	9	41.5	31.2	17.8	2	4.25
							2	9	42.5	65.8	17.9	2	-0.14
							1	9	40.6	51.1	17.9	2	-0.45
10	8969	23	39	5	5	32.1	3	9	48.0	32.5	17.8	2	4.13
							2	9	46.7	47.6	17.9	2	0.51
							1	9	47.3	32.7	17.9	2	0.00
11	8181	21	25	3	-65	26.6	3	9	51.6	45.3	17.8	1	4.22
							2	9	52.0	78.0	17.9	1	0.49
							1	9	52.7	52.3	17.9	1	-0.12
12	8353	21	52	54	-37	26.7	3	10	42.7	63.8	17.8	2	3.01
							2						-0.12
							1						
13	8431	22	7	24	-33	4.3	3	10	44.8	29.2	17.8	2	4.50
							2	10	45.4	54.0	17.9	2	0.05
							1	10	45.9	36.1	17.9	2	0.05
14	8551	22	27	0	4	36.6	3	10	50.0	44.8	17.8	0.5	4.79
							2	10	50.5	51.3	17.9	0.5	1.05
							1	10	49.7	28.2	18.0	0.2	0.89
15	8630	22	44	23	-81	28.3	3	10	58.5	48.5	17.8	2	4.15
							2	10	57.9	59.3	17.8	2	0.20
							1	10	57.2	37.4	17.9	2	0.11

readings, or inadequate source data. Additional information, e.g. the date of observation, is also needed to determine the air mass value X for each star at its time of observation. The filters, indicated by integers in the `Flt.` column of Table 6.1, require identification; e.g. 1 with ultra-violet, 2 with blue, and so on. Standard magnitudes and colours of the given stars, as listed in the *Astronomical Almanac*, are given under the heading `Cat.`, and correspond to V, $B - V$ and $U - B$ values, respectively.

The timing accuracy of one tenth of a minute translates to around a hundredth of a degree in angular measure, or $\lesssim 10^{-3}$ air masses at a zenith distance of $\leq 60°$, and so is adequate under normal circumstances. Similar considerations apply to the supplied sky positional coordinates.

The measurements given in this example were taken from a conventional DC driven chart recorder used at the Black Birch outstation of Carter Observatory (New Zealand) on a clear, stable night in 1983. The readings for the different stars, identified by their Bright Star Catalogue† (`BS`) numbers, are made with the recorder set to different gains to accommodate the various levels of output current. The deflections `Star`, `Sky` are therefore multiplied by the factors given in the `Mlt.` column before being used in the system calibration. Three digit representations indicate measuring accuracies of better than 1% by this means — approaching 0.1%, perhaps. This accords with the accuracy of the supplied catalogue data.

If we combine equation (6.1) with (6.5) we obtain

$$m_0 - m'_z = \epsilon C - k' X_z - k'' X_z C - k''' X_z^2 + \zeta \tag{6.6}$$

(an essentially similar right hand side, but with a colour difference on the left, results from combining the colour form (6.2) with (6.5)). We now assess the relative determinability of the various coefficients in (6.6) by the least squares method.

Consider the equation

$$z = ax + by + c \tag{6.7}$$

underlying a series of N actual measurements of the form

$$z_i - ax_i - by_i - c = e_i, \tag{6.8}$$

where e_i denotes some error, representing the combined effect of inaccuracies in the measurements of x_i, y_i and z_i.

† *Catalogue of Bright Stars* by D. Hoffleit and C. Jasehek, Yale University Observatory, New Haven, Conn., 1982.

Let us suppose that a and $c >> e_i$, but that $b \sim e_i$. The least squares method produces a set of 'normal equations' from the equations of condition (6.8), of the form:

$$\mathbf{X}\cdot\mathbf{a} = \mathbf{Z} \tag{6.9}$$

or, written out in full,

$$\begin{vmatrix} \sum x_i^2 & \sum x_i y_i & \sum x_i \\ \sum x_i y_i & \sum y_i^2 & \sum y_i \\ \sum x_i & \sum y_i & N \end{vmatrix} \begin{vmatrix} a \\ b \\ c \end{vmatrix} = \begin{vmatrix} \sum x_i z_i \\ \sum y_i z_i \\ \sum z_i \end{vmatrix}. \tag{6.10}$$

These equations can be inverted, to derive the coefficients vector $\mathbf{a}$ (i.e. $\mathbf{a} = \mathbf{X}^{-1}\cdot\mathbf{Z}$, where $\mathbf{X}^{-1}$ represents the inverse of $\mathbf{X}$). The equations for the coefficients a and b are of the form:

$$\left.\begin{aligned} a &= X_{11}^{-1}Z_1 + X_{12}^{-1}Z_2 + X_{13}^{-1}Z_3, \\ b &= X_{21}^{-1}Z_1 + X_{22}^{-1}Z_2 + X_{23}^{-1}Z_3. \end{aligned}\right\} \tag{6.11}$$

Equations (6.10) are well determined when the matrix $\mathbf{X}^{-1}$ is dominated by its central diagonal. Under such circumstances, we can write:

$$a \sim X_{11}^{-1}Z_1 \qquad \text{and} \qquad b \sim X_{22}^{-1}Z_2\ ,$$

where coefficients X_{11}^{-1} and X_{22}^{-1} are given by:

$$X_{11}^{-1} = [N\sum y_i^2 - (\sum y_i)^2]/D, \tag{6.12}$$

and

$$X_{22}^{-1} = [N\sum x_i^2 - (\sum x_i)^2]/D, \tag{6.13}$$

where,

$$\begin{aligned} D = {} & N\sum x_i^2 \sum y_i^2 + 2\sum x_i y_i \sum x_i \sum y_i - \\ & -\sum y_i^2(\sum x_i)^2 - \sum x_i^2(\sum y_i)^2 - N(\sum x_i y_i)^2. \end{aligned} \tag{6.14}$$

X_{11}^{-1} and X_{22}^{-1} also appear in the error estimates for the determined parameters, since:

$$\left.\begin{aligned} \Delta a &= \sqrt{X_{11}^{-1}e^2}, \\ \Delta b &= \sqrt{X_{22}^{-1}e^2}, \end{aligned}\right\} \tag{6.15}$$

where $e^2 = \sum e_i^2/(N-3)$, the variance of the observational errors e_i.

If we assume that the values of x_i and y_i are both evenly distributed over the same range, then it follows that $X_{11}^{-1} \sim X_{22}^{-1}$, from comparison of the two symmetric forms (6.12) and (6.13). Moreover, we lose no generality by setting the origin such that $\sum x_i$, and $\sum y_i = 0$, while, the x_i and y_i being independent, we should also find that $\sum x_i y_i \sim 0$. On this basis, we have

$$D \sim N \sum x_i^2 \sum y_i^2, \tag{6.16}$$

and

$$\left.\begin{aligned} \Delta a &\sim \sqrt{e^2 / \sum x_i^2}, \\ \Delta b &\sim \sqrt{e^2 / \sum y_i^2}, \end{aligned}\right\} \tag{6.17}$$

or writing $\sigma^2 = \sum x_i^2/(N-1) \sim \sum y_i^2/(N-1)$ for the adopted common variance of the observations of x and y,

$$\left.\begin{aligned} \Delta a &\sim e/\sqrt{N-1}\sigma, \\ \Delta b &\sim e/\sqrt{N-1}\sigma. \end{aligned}\right\} \tag{6.18}$$

Hence, as more observations are made, both a and b may be determined to a greater precision, though this depends on the leverage implied by the dispersion of the variables x_i and y_i, i.e. σ, as well as the inherent observational accuracy e. The point is that while a can be obtained to an accuracy which may well be a small percentage of its numerical value, that of b is comparable to the value of b itself, and so evaluating the coefficients as a single operation ignores the *relative* determinability of the sought quantities.

Referring back to equation (6.6), more leverage will normally come from the spread of catalogue supplied magnitudes, or even colours, than the air masses (a number usually varying between 1 and perhaps a little over 2 at most); so that the equations of condition are not really favourable for simultaneous determination of the extinction, especially not for *relatively* accurate determinations of the higher order extinction coefficients. The standard stars experiment thus becomes more directed towards calibration of the photometer, in which case it is advantageous to select the sample of standards from close to the zenith.

Other observations can be more specifically aimed at determination of the static extinction terms. If, instead of a set of standard stars of different colours distributed over the whole sky, we took just one reference star over a wide range of air masses, then the m_0 and ϵC terms in (6.6) would be effectively absorbed in the constant term. In a similar way, a pair of stars, closely separated on the sky, but of quite different colours, could be monitored for their differential magnitude at a range of air masses, and the second order extinction term thereby more easily prised out. Observations of this type occur quite naturally in the process of *differential* stellar photometry, described in more detail later. They are useful alternative sources of information on the extinction properties of the atmosphere.

In Figure 6.1 we indicate, by a flow chart, a computer program, which acts upon information such as that of Table 6.1 to determine values of the coefficients in equation (6.6) by the method of least squares.

We consider first the determination of the local-to-standard calibration coefficients for V of the UBV system, i.e. a determination of ϵ and ζ_V in (6.1). The information on extinction is taken to be separately available, and supplied as known coefficients. After some preliminary reductions on the raw input, and accounting for the extinction effects, catalogue minus local magnitude values are arrayed with corresponding standard $B-V$ data. This is applied directly in a linear least squares fitting to an equation of the form (6.1) with just two coefficients.

Relevant information (using the data of Table 6.1) is shown graphically in Figures 6.2 and 6.3. The least squares solution output follows as Table 6.2a. In the tabular list `X1` gives the $B - V$ value for each star. `Y` is measuring $V - V'$, with the zero constant (8.14) subtracted. If there were no dependence on colour these `Y` values would all randomly cluster around zero. Including a linear colour dependence in the residuals gives rise to the `YFIT` coloumn. Finally, the `DIF` column indicates the distribution of errors `Y - YFIT` in the fitting.

It will be noticed that stars 2 (BS 8848 = γ Tuc) and 6 (BS 373 = HD7672) in this match-up show unusually large errors. Two sets of catalogued values for V and $B-V$ on γ Tuc, consistent to within 0.01 mag, are given in the US Naval Observatory's (1970) *Catalogue*. A remark on HD7672 is, however, included to the effect that this star is suspected of variability. Whatever the explanation for such, not so infrequently met, discrepancies, a second run omitting these two stars produces a distinct drop in the errors, implying increase in overall precision.

Concerning the calculated quantities — less direct attention usually attaches to the zero constant ζ_V (= `A0`), which often involves various arbitrary scaling factors, though its variation from night to night, or on interchange of components, deserves note.[†] The value of ϵ (= `A1`) was derived as -0.16 ± 0.04 in the first fitting and -0.13 ± 0.02 with the omission of the two dubious data points. This quantity relates to the transmissive properties of the filter/cathode combination used, (Schott GG13 + GG14, and EMI 9813B

[†] For a photon counting system it is relatively direct to check the derived zero constant against the expected count rate at zero magnitude, using data such as that supplied in Table 3.5.

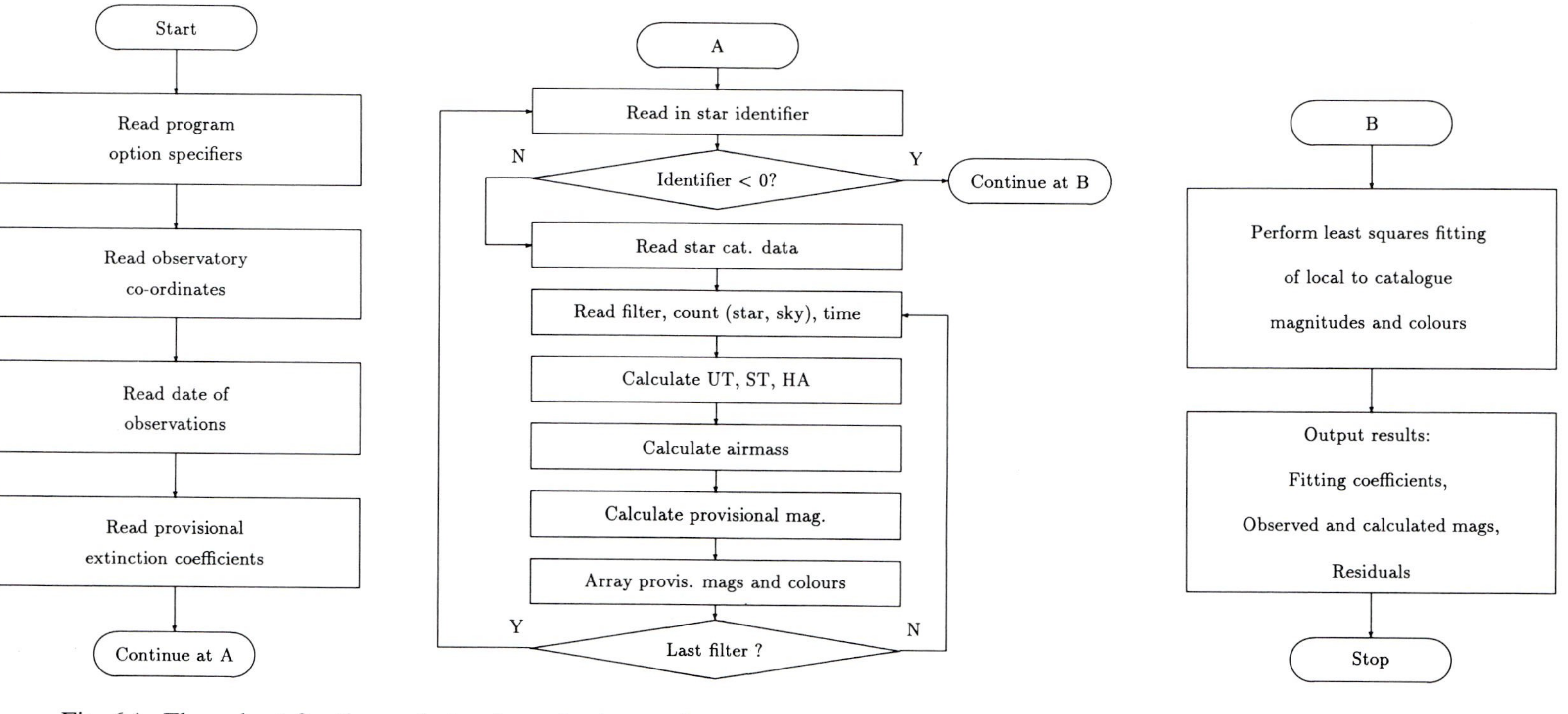

Fig. 6.1. Flow-chart for the analysis of standard stars data.

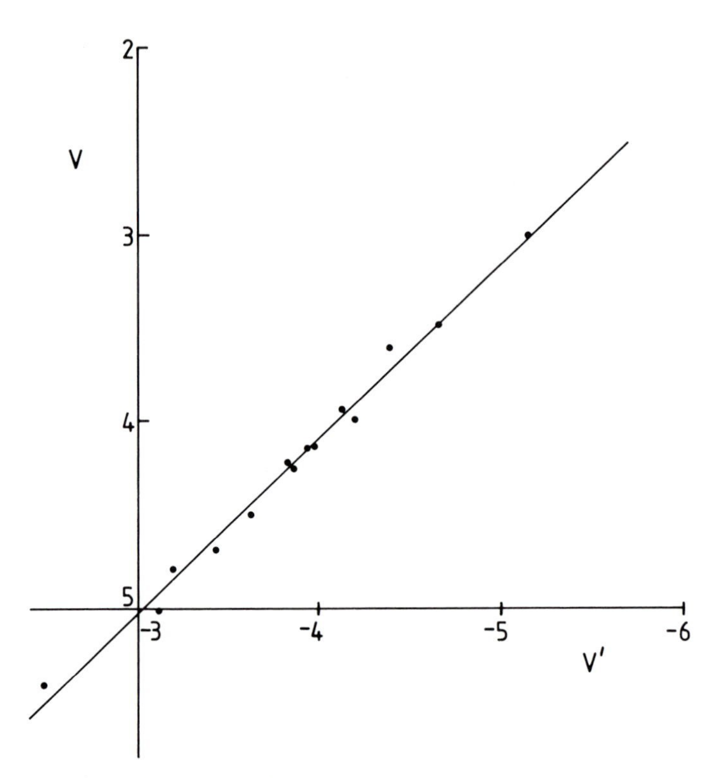

Fig. 6.2. Plot of catalogue V against measured V'.

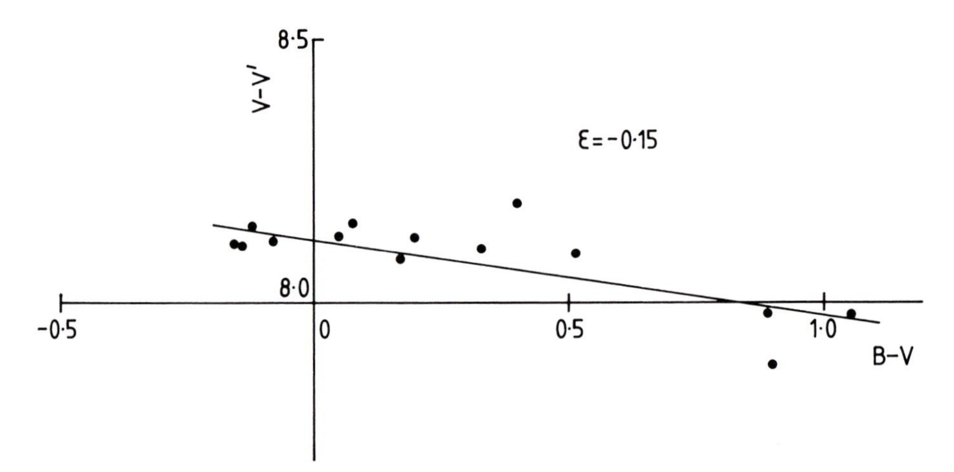

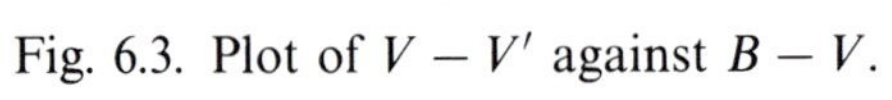

Fig. 6.3. Plot of $V - V'$ against $B - V$.

Table 6.2a. *Part of computer output for two-coefficient fitting to V data.*

STAR	X1	Y	YFIT	DIF	STAR	X1	Y	YFIT	DIF
1	.17	-.05	-.0267	-.0282	8	.89	-.15	-.1399	-.0078
2	.40	.06	-.0629	.1199	9	-.14	-.02	.0220	-.0426
3	-.08	-.01	.0126	-.0236	10	.51	-.04	-.0802	.0400
4	-.15	-.02	.0236	-.0463	11	.49	-.08	-.0770	-.0058
5	.08	.02	-.0126	.0301	12	-.12	.02	.0189	.0040
6	.90	-.25	-.1415	-.1077	13	.05	.00	-.0079	.0047
7	.33	-.03	-.0519	.0206	14	1.05	-.15	-.1651	.0128
					15	.20	.00	-.0314	.0298

```
THE COEFFICIENTS  A0, A1, A2, A3....
  .81361E+01     -.15719E+00
THEIR ERROR ESTIMATES
  .17211E-01      .35062E-01
```

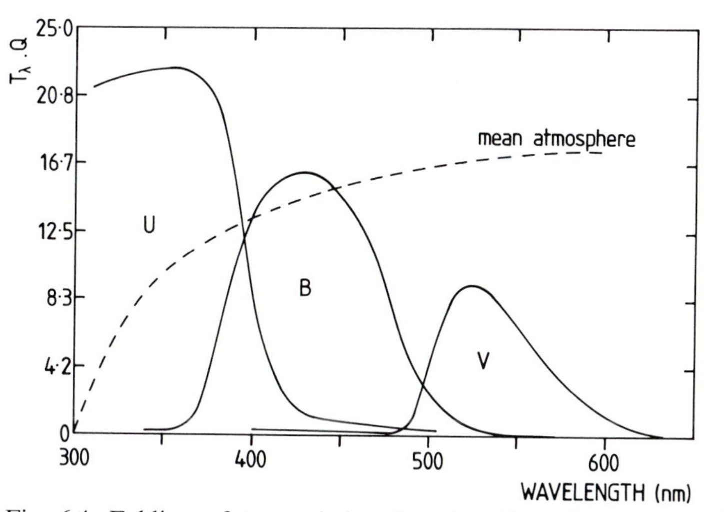

Fig. 6.4. Folding of transmission function T_λ and quantum efficiency Q, together with the mean atmospheric transmission.

(bialkali)) multiplied by the atmospheric transmission. This is indicated in Figure 6.4.

For the V' observations here reported, the mean wavelength is about 5330 Å, i.e. noticeably displaced to the blue of standard V (5500 Å in Chapter 3). Applying formula (6.3) we find $\epsilon = -0.143$, which lies between the two previously determined values, and within the given error estimates of either.

If we now extend the least squares fitting of equation (6.6) to include

Table 6.2b. *Three-coefficient fitting to V data, here listing input and main result.*

STAR	B-V	AIRMASS	V-V'	WEIGHT	STAR	B-V	AIRMASS	V-V'	WEIGHT
1	.17	1.04	7.87	1.00	7	-.14	1.48	7.82	1.00
2	-.08	1.10	7.90	1.00	8	.51	1.48	7.80	1.00
3	-.15	1.04	7.91	1.00	9	.49	1.17	7.82	1.00
4	.08	1.02	7.95	1.00	10	-.12	1.20	7.92	1.00
5	.33	1.36	7.83	1.00	11	.05	1.19	7.89	1.00
6	.89	3.21	7.35	1.00	12	1.05	1.84	7.62	1.00
					13	.20	1.33	7.87	1.00

```
THE COEFFICIENTS  A0, A1, A2, A3....
  .81517E+01    -.10700E+00    -.22059E+00
THEIR ERROR ESTIMATES
  .22798E-01     .27340E-01     .18160E-01
```

primary extinction, noting that there are a few points with air masses appreciably greater than unity, we find the value of ϵ little changed at –0.11, the zero constant slightly increased to 8.15, and an extinction coefficient k' of 0.22 (i.e. –A2 in Table 6.2b), which is not atypical of sky conditions at the observing location.

Insight into this result comes from Figure 6.5. The pattern we see is of a clustering of points at low zenith distance, together with a few outliers at larger air mass, which are disproportionately influential in fixing the slope. This situation typifies the results of standard star observations, but contrasts with differential photometry, where air masses are sampled in a much more uniform run.

The error estimates of Table 6.2b have increased quite significantly. Here we find an example of an important point, which will be frequently re-encountered. This concerns the accuracy, with which we can resolve a parameter, a say, in a given set $\mathbf{a}$ of m such parameters, associated with curve or surface fitting to data characterized by N equations of condition ($N > m$). This resolution will tend to deteriorate, as we increase the number of parameters m to be determined, while keeping N constant. It is relatively easy to demonstrate this principle in the two extending to three parameter example just encountered, and this is done in the following subsection.

Extending the parameter determination to the second order coefficient in (6.6) results in a further slight drop of residuals between calculated and observed points, as one might expect. However, parameter error estimates increase substantially. ϵ moves appreciably away from its independently predicted value, though both it and the second order extinction term k''_V

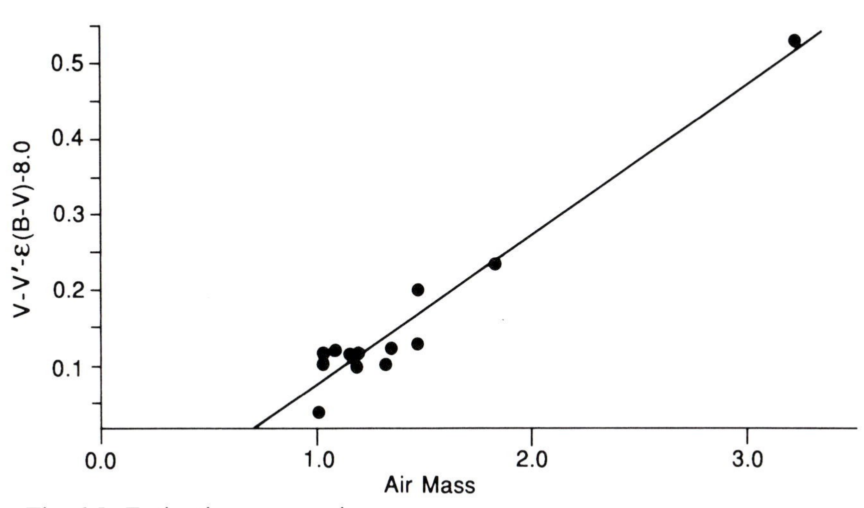

Fig. 6.5. Extinction versus air mass.

become less (in absolute value) than their errors. In fact, because of the relatively low range of air mass variation, the air mass colour product trends in a similar way to the colour alone, so the extra term introduced to the fitting correlates strongly with the colour coefficient.

Although the least squares fitting for three terms is still determinate, its solution distributes the information in parameter space too diffusely to do more than confirm our general expectations — e.g. a negative ϵ of about the right order of magnitude. It could be argued that accuracies would be increased by including more observation points in the determination, but in view of the $\sim \sqrt{N}$ division of the error estimates in (6.18), a large number of stars would need to be observed to improve the result substantially. It is better to determine the extinction parameters independently, and concentrate the present effort on the calibration coefficients ϵ, μ in (6.1), (6.2).

Turning to the colour equations, we can see one or two outlying points in the $B-V$ match of Figure 6.6.† The corresponding data and first-order fitting parameters in Table 6.2c show errors again coming from BS 8848 (star 2), and now noticeably BS 1030 (star 7). The value of μ (`A1`) at 1.12 ± 0.05 is in fair agreement with $\mu = 1.08$ calculated from formula (6.4). The value $\mu > 1$ can be understood by reference to Figure 6.4. While the V'-filter is centred somewhat blueward of the mean wavelength of V, the blue filter is in closer agreement with its standard specification ($\sim$ 4450 Å). Hence the difference $1/\lambda'_B - 1/\lambda'_V$ will be slightly smaller than $1/\lambda_B - 1/\lambda_V$, and so

† Stars originally numbered 5 and 12 in Table 6.1 are incomplete in the colour data, so are excluded from this fitting.

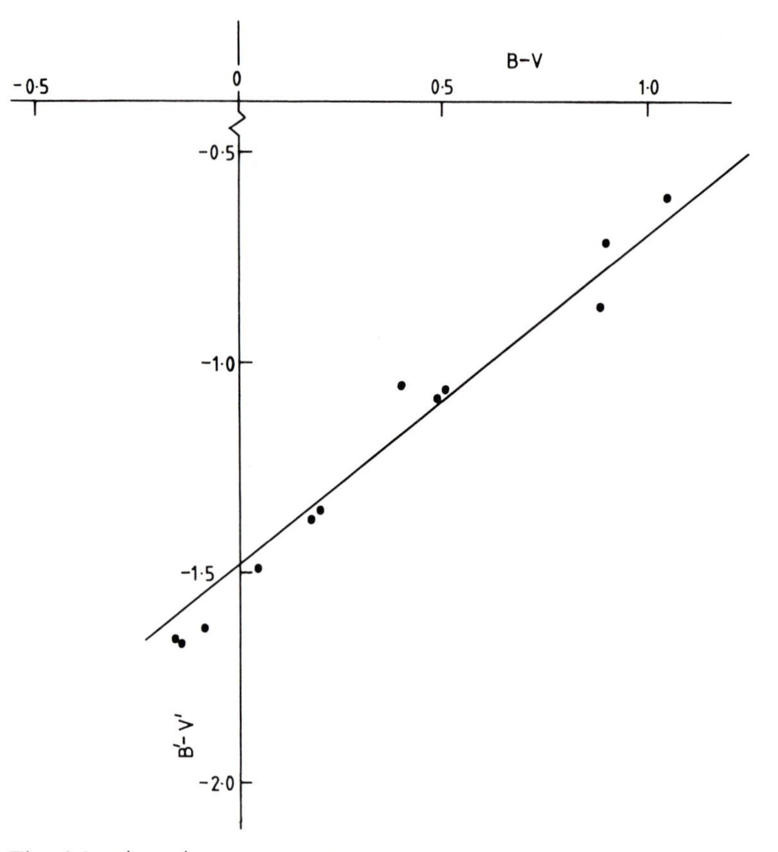

Fig. 6.6. $B' - V'$ versus $B - V$.

$\mu > 1$ according to (6.4). Table 6.2c gives $B' - V'$ (X1), which compares with $B - V$ when the zero term (A0) is subtracted off (Y). In order to make a good match, the $B' - V'$ values have to be rescaled (YFIT). The quality of the result is seen from the residuals list (DIF).

Repeating the fit with the exclusion of the troublesome BS 8848 and BS 1030 data, and a simultaneous determination of the first order extinction coefficient, gives $\mu = 1.01 \pm 0.02$ and $k'_{BV} = 0.06 \pm 0.05$. Although the result for μ appears formally more accurate, it is not what we expect; while the colour extinction coefficient implies transmission in B does not differ from that in V as much as it normally would (Figure 6.7). Another solution in which we adopt k'_{BV} to be 0.20 results in $\mu = 1.09 \pm 0.01$. Since both μ and k'_{BV} now accord better with reasonable expectation, and the equations are relatively poorly conditioned for the determination of k'_{BV}, this result is preferable.

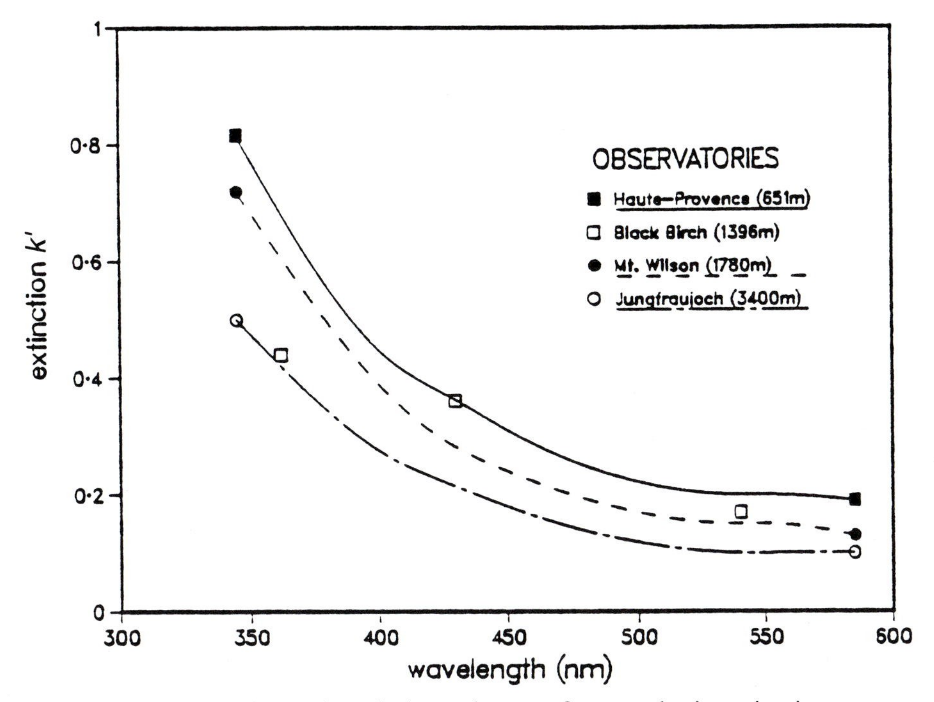

Fig. 6.7. Typical wavelength dependences of atmospheric extinction.

Table 6.2c. *Two-coefficient fitting to $B - V$ data.*

STAR	X1	Y	YFIT	DIF	STAR	X1	Y	YFIT	DIF
1	-1.37	-1.54	-1.5317	-.0112	7	-0.86	-0.82	-0.9664	.1435
2	-1.05	-1.31	-1.1766	-.1363	8	-1.68	-1.85	-1.8770	.0241
3	-1.63	-1.79	-1.8218	.0289	9	-1.06	-1.20	-1.1851	-.0178
4	-1.66	-1.86	-1.8590	-.0039	10	-1.08	-1.22	-1.2136	-.0093
5	-0.71	-0.81	-0.7941	-.0188	11	-1.49	-1.66	-1.6685	.0056
6	-1.23	-1.38	-1.3745	-.0084	12	-0.60	-0.66	-0.6721	.0092
					13	-1.35	-1.51	-1.5075	-.0054

```
THE COEFFICIENTS  A0, A1, A2, A3....
  .17129E+01      .11197E+01
THEIR ERROR ESTIMATES
  .62841E-01      .49881E-01
```

A similar processing of the $U - B$ data gives $\psi^{\dagger} = 1.14 \pm 0.04$ as a first result, as shown in Table 6.2d, and graphically in Figure 6.8. This differs from a calculated value, using (6.4), of $\psi = 0.955$; but, where the U filter is involved, there is a greater tolerance for disparity. This is due to both the relatively greater extent of noise in the ultra-violet at a given

† ψ is usually used in place of μ for $U-B$ colours.

Table 6.2d. *Two-coefficient fitting to U − B data.*

STAR	X1	Y	YFIT	DIF	STAR	X1	Y	YFIT	DIF
1	0.50	0.63	0.5692	.0651	7	1.03	1.13	1.1723	-.0380
2	0.44	0.50	0.4973	.0070	8	0.10	0.07	0.1116	-.0373
3	0.23	0.24	0.2583	-.0140	9	0.46	0.52	0.5236	.0007
4	-0.06	-0.03	-0.0696	.0438	10	0.37	0.40	0.4205	-.0162
5	0.91	0.96	1.0361	-.0718	11	0.50	0.57	0.5728	.0015
6	0.51	0.55	0.5807	-.0264	12	1.18	1.41	1.3344	.0799
					13	0.55	0.63	0.6286	.0057

```
THE COEFFICIENTS  A0, A1, A2, A3....
 -.52430E+00      .11354E+01
THEIR ERROR ESTIMATES
  .22582E-01      .36491E-01
```

signal strength, usually due to atmospheric interference (inhomogeneities have a greater relative effect), and also the role of the static atmosphere in determining the cut-off on the ultra-violet side. This can vary from night to night at a given site, so we should be prepared for nightly, or otherwise temporal, variations in ψ. The calculated ψ (0.955) comes from folding the combined filter-detector response with the continuous transmission function of the normal atmosphere, as tabulated by Allen (1973), which results in a mean wavelength of about 3600 Å. However, Allen's estimation of dust and aerosol haze corresponds to very good observing conditions, and this contribution, which was set at 20% of the model's net extinction at 3600 Å, can vary by an order of magnitude, depending on the weather.

Another difficulty affecting U observations can occur if a red-sensitive detector is used, arising from the red-leak of the filter, mentioned in Section 5.1. Under these circumstances, extra measurements are required to assess the scale of spurious additions to the signal, which will be more serious for cool stars.

The irregularities of U data have fed back into catalogued sources of information, so that, for instance, the US Naval Observatory's (1970) general compilation of UBV photometry shows a significantly greater scatter in $U - B$ values compared with corresponding V and $B - V$ for (presumably) constant sources. Proposals have been made on how the $U - B$ indices can be straightened out, but for the most part, users of the UBV system appear ready to live with some level of inaccuracy when traded off against the system's broadband light grasp for faint objects, and its general familiarity.

A fitting in which the effect of atmospheric extinction was taken to

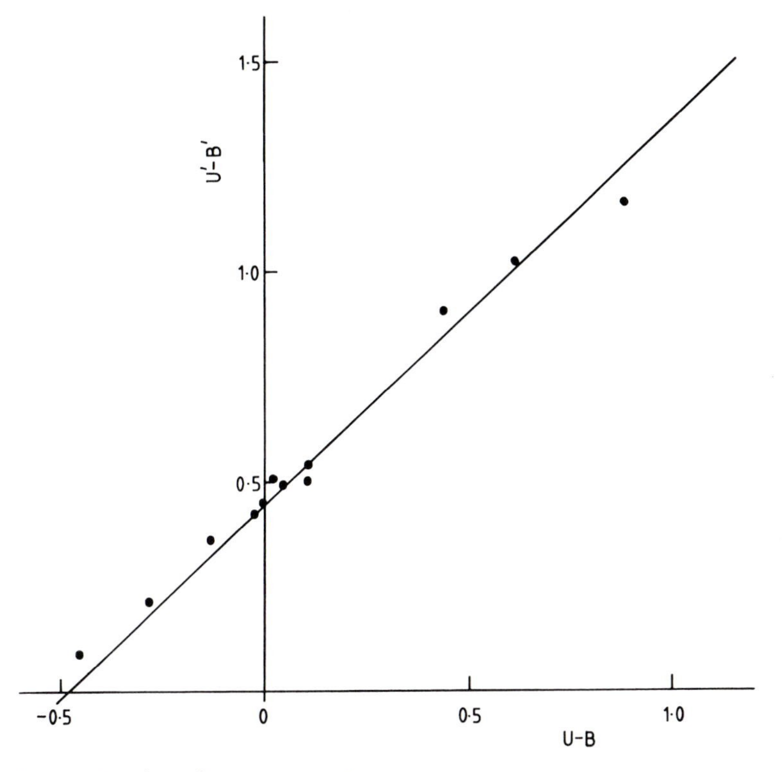

Fig. 6.8. $U' - B'$ versus $U - B$.

be unknown (rather than assigned a typical value) produced the result $\psi = 0.87 \pm 0.06$, and $k'_{UB} = 0.10 \pm 0.05$, but without significant improvement in the overall goodness of fit. In fact, the first value (1.14) is closer to the mean of a dozen roughly contemporaneous determinations, which, as a rule, show $\psi \gtrsim 1$ for the Black Birch photometer.

We have thus followed through the processing of observations of standard stars in detail. The calibration of other sytems follows along essentially similar lines, whether broad-, intermediate- or narrow-band filters are used. The inclusion of higher terms in (6.6) would not normally be needed in other than broadband work (cf. Chapter 3), however, this remark should be qualified, since, due to known difficulties of definition, the *UBV* system is itself not a particularly accurate one. More recently introduced photometric systems carry inherently higher precision. The use of higher order terms in the expansion (6.6) has then to be examined against available accuracies of standards before one can say where truncations should be made. In Section 4.1.1 we found that these higher order terms (in k'', k'''), are of order $W_2/\lambda_0^2 W_0$. If the standard system is aiming at 0.001 mag accuracies, therefore, this bandwidth related ratio has to be less than 10^{-3} to justify neglect of the second order terms (cf. Table 4.3).

6.1.1 *Parameter determinacy and set size — a simple case*

Let us write the normal matrix **X** which determines **a**, as in (6.9), in the two-parameter case simply as

$$\mathbf{X} \equiv \begin{vmatrix} x & u \\ u & y \end{vmatrix}.$$

The error variance $\Delta a^2 = A^2 e_2^2$, where A^2 is given by (6.15), using the inverse of X, by

$$A^2 = \frac{y}{xy - u^2}, \tag{6.19}$$

and e_2^2 is the variance of the residuals in this two-parameter fit. Note that the real determinacy of the normal equations implies that **X** is a positive definite, symmetric matrix, which means, in effect, that $x > 0, y > 0$ and $xy > u^2$.

Now when we introduce a third, still determinable, parameter into the derivation, then **X** becomes **X**$'$, still positive definite, where,

$$\mathbf{X}' \equiv \begin{vmatrix} x & u & v \\ u & y & w \\ v & w & z \end{vmatrix}$$

and so,

$$\Delta a^2 \rightarrow A'^2 e_3^2 = \frac{(yz - w^2)e_3^2}{(xy - u^2)z - (xw - uv)w - (yv - uw)v}. \tag{6.20}$$

From (6.20) we can write, after a little rearrangement,

$$A'^2 = A^2 \left(\frac{1 - w^2/yz}{1 - H/zX} \right), \tag{6.21}$$

where H satisfies $X' = zX - H > 0$, and X, X' denote the determinants of **X**, **X**$'$, so that $H = xw^2 + yv^2 - 2uvw > 0$, since $(\sqrt{x}w - \sqrt{y}v)^2 > 0$, and $xy > u^2$. Note also that $0 < w^2/yz < 1$, so that (6.21) is of the form,

$$A'^2 = A^2 \left(\frac{1 - \delta_2}{1 - \delta_3} \right), \tag{6.22}$$

where $0 < \delta_2, \delta_3 < 1$.

Thus, $A'^2 \geq A^2$ if $\delta_3 \geq \delta_2$, i.e. if $H/X \geq w^2/y$. But, multiplying out H and X, we find,

$$yH - w^2X = (yv - uw)^2 \geq 0, \tag{6.23}$$

so the required inequality is established. The result can be generalized to the $m \to m+1$ extension in a similar, but more protracted, manner. Hence, the error factor A never decreases with the addition of new parameters. Whether the standard deviation error Δa increases or not depends on whether the net improvement in the overall fitting overrides the deterioration in Δa. This tends not to occur, in curve-fittings we encounter, after the specification of a few major determinables of the model.

6.2 Differential photometry

The second procedure set out in this chapter is that in which an object of particular interest is compared with some reference source or sources. This often leads on to the production of light curves. The need to know about the reliability of equipment and its calibration by accepted standards, though, makes performance of the first procedure a practical prerequisite.

The difference in brightness to be measured is usually that of one star from another at more or less the same time. It can refer to one and the same star, however, as with narrowband spectral intercomparison, or in fast photometry of short-term fluctuations. Slow drifts of background sky brightness or transparency are then neglected against the scale of fast intrinsic variation. This idea carries over to more normal timescales, if background variation can be neglected in comparison to that investigated. In practice, the issue boils down to the *relative* frequency with which the object is compared with another star, which depends on both ambient conditions at the time and location of observation, and the nature of the object.

Near simultaneous comparison of one star with another similar one located close by in the sky implies a relaxation of the stringency on atmospheric conditions, compared with standard stars' photometry. If exactly simultaneous comparison is possible, i.e. using a more-than-single channel, or areal, photometer, then useful data can be gathered through quite hazy or variable sky conditions. Stability through at least a few stellar observation cycle times (typically several minutes each) would be the minimum requirement for conventional single-channel photometry.

Traditional visual observations involved a sequence of reference stars. This developed, with instrumental photometry, into the selection and adoption of one main comparison star for a particular object. This star is, at some favourable time, compared with a standard sequence, so that its magnitude can be standardized. Its magnitude is also checked for constancy — usually against another reference 'check' star. Since any variability of the comparison would be inherently independent of the check, under normal circumstances, constancy of the comparison/check ratio is a sufficient guarantee that both stars are constant within some prescribed limit of precision. Any variability of this ratio (corrected for non-intrinsic effects) would imply extending the comparison process to more stars, and replacing whatever original reference was found to vary with a suitable alternative. The choice of a main comparison usually results from a compromise between various kinds of suitable nearness to the object, i.e. nearness in the sky, in magnitude, and in colour. Such nearness reduces the role of the correction terms in (6.6), when that equation is cast into a differential form. The observational procedure is, of course, normally preceded by catalogue or background literature searches on stars of interest, when information on comparison stars often comes to light.

Non-intrinsic effects include the role of 'sky' variation. Along with repeated measurements of object, comparisons and check(s), it is standard practice also to measure the background radiation from the immediate vicinity of the programme stars through the same photometer entrance aperture. This background light is subsequently subtracted out. In its simplest form then, differential stellar photometry is the repeated measurement of light from two or three stars and the nearby sky.

A normal starting point, in making preparations, is location of the object on a finding chart. In the case of variable stars, the *General Catalogue of Variable Stars (GCVS)* provides a reference for such a chart for each variable.† This is tantamount to a definition of the object in some cases, as other forms of identification (i.e. coordinate specification) are more prone to surreptitious error, particularly with faint or little-known objects. A number of alternative sources exist from which objects can be located apart from the *GCVS*, but, in any case, identification of the field requires pattern-recognition. This operation is eased with the aid of suitably scaled maps covering the right magnitude range. A back-up technique, useful with fainter objects, involves starting from some 'unmistakable' bright star and proceeding through a sequence of fainter ones until the required object is centred. Computers are being increasingly applied to image location: in the first stage by displaying representations of relevant extracts from stored catalogues as field overlays; with more sophistication an image matching algorithm can be executed which essentially guarantees identification.

The position of the object also, of course, comes into planning the ap-

† The fourth edition (ed. P. N. Kholopov) published by 'Nauka' Publishing House, Moscow, is in three volumes (1985a,b; 1987).

Table 6.3. *A typical segment of raw data produced by the data-logger program.*

```
 0               00:00:00    0   0
NAME- murray some haze on the horizon
DATE D-M-Y= 21-10-1987
FILTERS :
 2 = U
 3 = B
 4 = V
 5 = HB WIDE
 6 = HB NARROW
 0 = END FILTERS
INTEGRATION PERIODS OF EACH FILTER :
 2 = 10 SECS
 3 = 10 SECS
 4 = 10 SECS
 5 = 20 SECS
 6 = 30 SECS
OBJECT 1 IS R Arae
OBJECT 2 IS COMP
OBJECT 3 IS CHECK
OBJECT 9 IS SKY
COUNTS              TIME     FILT OBJ
 28              19:58:24    2   9
 27              19:58:34    2   9
 26              19:58:44    2   9
 38              19:58:56    3   9
 38              19:59:06    3   9
 37              19:59:16    3   9
 7               19:59:26    4   9
 7               19:59:36    4   9
 6               19:59:46    4   9
 12              20:00:08    5   9
 11              20:00:28    5   9
 11              20:00:48    5   9
 5               20:01:20    6   9
 4               20:01:50    6   9
 4               20:02:20    6   9
 906             20:08:06    2   2
 957             20:08:16    2   2
 952             20:08:26    2   2
 1606            20:08:36    3   2
                     :
                     :
 0               00:00:00    0   0
END 01:46:54
```

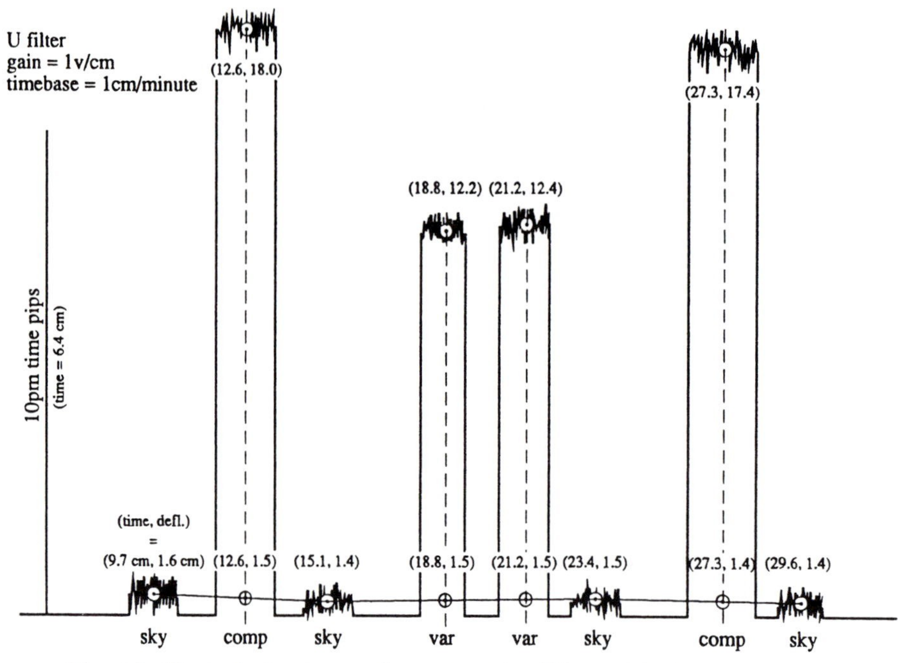

Fig. 6.9. Typical chart recorder output in differential photometry.

propriate times for observing. This is often a fairly elementary matter, but sometimes special attention is required if there are particular events, e.g. relatively rare eclipses to be covered, or bright times of the lunation to be avoided. Useful, in this context, are those small 'spherical astronomy' computer programs,[‡] which can, for example, arrange for plots of altitudes for the object corresponding to particular times of night on particular dates, or give background on the lunar phase, or twilight duration.

Table 6.3 shows the beginning of a typical raw data file produced by a microcomputer carrying out the data-logging and control functions indicated in Figure 5.1 for differential photometry of the variable star R Arae.[†] The photometer is operating as a single-channel DC-CFC instrument (Section 5.3.1) taking all its targets sequentially. Apart from some header type information, which specifies extra particulars on the observations to be collected, the bulk of the data file consists of sets of four basic pieces of information, i.e. count, time (when the count was registered), filter used, and target (object, comparison, check or sky).

[‡] See e.g. P. Duffet-Smith's *Astronomy with your Personal Computer*, 1990, Cambridge University Press, Cambridge.

[†] This, and subsequent similar data on R Arae in this section, have been freely quoted from the MSc Thesis of Murray C. Forbes of the Victoria University of Wellington (1990).

The reduction program which operates on this information carries out essentially the same task as that one done by hand in the measurement of pen recorder charts (Figure 6.9). The fundamental arithmetic task is to put

$$\Delta m' = -2.5 \log \left(\frac{\text{object} - \text{sky}}{\text{comparison} - \text{sky}} \right) \tag{6.24}$$

for the differential magnitude $\Delta m'$. This equation has the magnitude difference (object – comparison) positive when the object is fainter than the comparison. Light-curve data are sometimes presented in the opposite sense, i.e. negative $\Delta m'$ values when the object is fainter.

The program is now required to determine the appropriate set of numbers for object, comparison and sky to substitute into (6.24), and what time, and filter identifier, to affix to the set. Having determined initial, raw data $\Delta m'$ values, the program corrects them for differential extinction and local to standard differences.

This is conveniently carried out as a two-stage operation. We first write, using (6.6), to sufficient accuracy,

$$\Delta m'_0 = \Delta m' - k'_m \Delta X_z - k''_m \bar{X}_z \Delta C', \tag{6.25}$$

to determine the differential magnitude outside the atmosphere $\Delta m'_0$. Here ΔX_z is positive if there is a greater air mass to the object than the comparison, keeping the same convention for Δm as (6.24). $\bar{X}_z$ denotes the mean of object and comparison air masses. With successive filter wheel positionings corresponding data sets are processed, so that the extinction-corrected local colours $\Delta C'_0$ are derived in turn. Corrections to the standard system are then applied as

$$\Delta C_0 = \mu \Delta C'_0, \tag{6.26}$$

or

$$\Delta m_0 = \Delta m'_0 + \epsilon \Delta C'_0. \tag{6.27}$$

The data file presented in Table 6.3, after some preliminaries, starts with data on the sky and comparison star before moving to the object. In this way, the output file is formed as a set of object $\Delta m'$ values, which have had appropriate sky and comparison numbers *interpolated* to the times of object readings. In reliable photometric conditions relatively less time need be spent monitoring and checking the comparison and sky. Their apparent variations should be small and slowly drifting. The object can then be concentrated on, and supporting data related to it.

The controlling microcomputer performs preliminary calculations of $\Delta m'$,

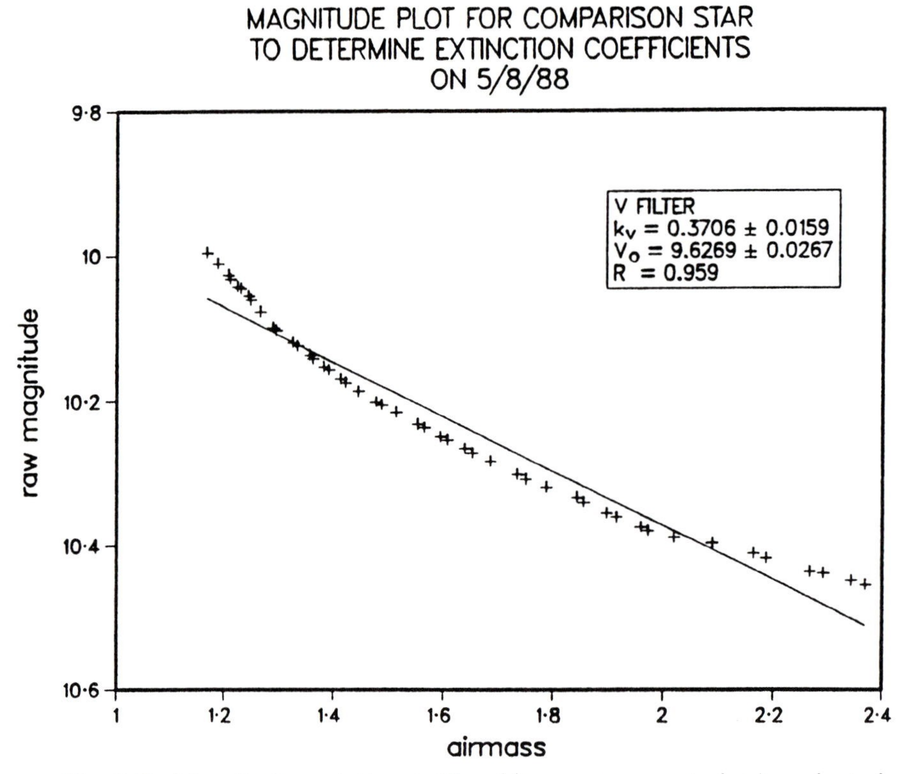

Fig. 6.10. Magnitude variation (with arbitrary zero constraints) against air mass.

and can graphically display the results on the screen, enabling on-line judgement of the progress of the observations. Another convenient real time calculation is that of a 'quality factor' — a number which indicates how the spread of numbers in a set of counts compares with the Poisson distribution expected for shot noise limited statistics. For example, the first three U readings of the comparison star (`FILT` = 2, `OBJ` = 2) in Table 6.3 have mean 938.3 and sample standard deviation 28.1. The quality parameter, variance/mean, has an acceptable value of 0.84 in this case. The computer program which processed this particular data sounds a 'beep' to draw attention to an unreliable datum when this ratio exceeds 3. Such quality measures, considered alongside acceptable S/N ratios, strongly bear on the worth of the data set as a whole.

In Figure 6.10 the apparent magnitude of a comparison star, with an arbitrary zero point, is plotted against air mass. Note the relatively even distribution of points over a range of air masses, as compared with that shown in Figure 6.5. Although there is a quasi-linear increase of magnitude

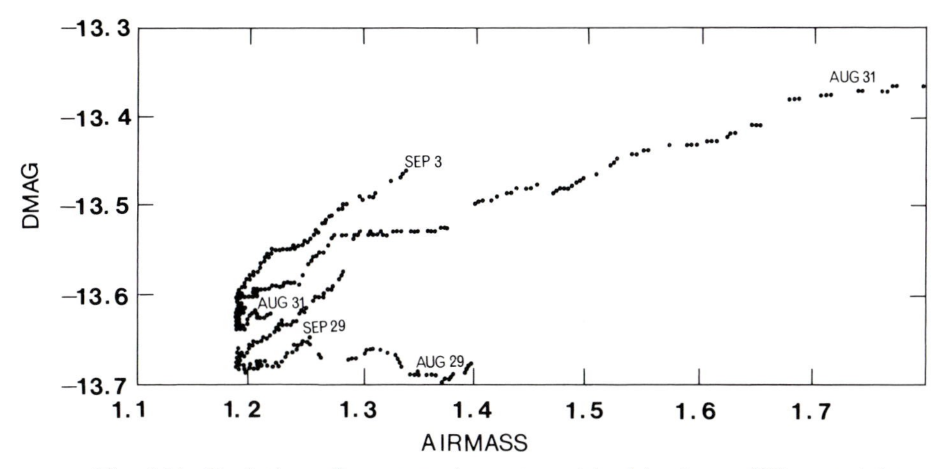

Fig. 6.11. Variation of a comparison star with altitude on different nights.

with air mass, a distinct curvature is evident in the trend. Curvatures in comparison star magnitude versus air mass plots are also visible in Figure 6.11, where a number different nights' data have been superposed. They indicate characteristic transparency waves, reflecting the inherently dynamic behaviour of the real terrestrial atmosphere. Although linear approximation would be an oversimplified representation of such a trend as a whole, it can be adequate to determine the *differential* extinction at normal sampling rates. Even the main static term in the atmospheric extinction typically only introduces corrections in the third significant digit in $\Delta m'$. Taking into account time-dependent properties of the transparency would, in typical sampling conditions, i.e. comparison readings at $\sim$ 10–15 minute intervals, cause alterations to the derived $\Delta m'$s of much less than the shot noise fluctuation from point to point, and so would be lost in subsequent smoothing.

The program would also usually reduce time measures to an appropriate scale for variable star work, such as Heliocentric Julian Date (or its Barycentric equivalent), and probably also the phase of variation, in the case of a periodic variable object. In order to determine the heliocentric (or similar) correction, applied to remove that light-time effect in observations of stars, arising from the orbital motion of the Earth about the Sun, we refer again to the rotation-matrix method of Chapter 4. The task is to find the x-coordinate of the centre of motion, which for convenience we denote as the Sun, when the x-axis points to the object. Starting with a coordinate system whose x-axis points to the Sun, we transfer to the object by the following sequence of rotations:

$$\mathbf{x}' = \mathbf{R}_y(-\delta)\cdot\mathbf{R}_z(\alpha)\cdot\mathbf{R}_x(-\epsilon)\cdot\mathbf{R}_z(-\theta)\cdot\mathbf{x}, \tag{6.28}$$

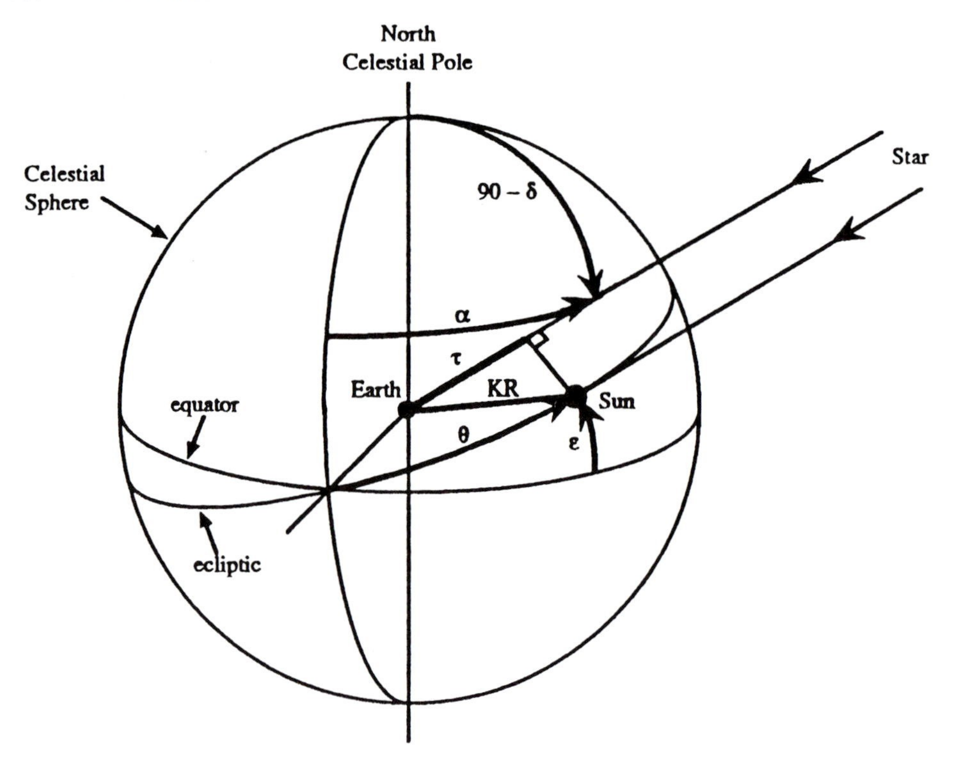

Fig. 6.12. Geometry for the heliocentric correction.

where the object has equatorial coordinates α, δ, while the Sun lies at longitude θ along the ecliptic plane, which is inclined at angle ϵ to the equator (Figure 6.12). The vector for the Sun's coordinates in the starting system is simply (1,0,0). Multiplying out from the right, therefore, it is easy to derive the required x-coordinate as

$$x_{\odot} = \cos\delta(\cos\alpha\cos\theta + \sin\alpha\cos\epsilon\sin\theta) + \sin\delta\sin\epsilon\sin\theta, \qquad (6.29)$$

which when multiplied by the light-travel time for the Earth–Sun distance (499 s or 0.005775 d) gives the quantity to be subtracted from the observed to specify the heliocentric time.

6.2.1 Typical photometric comparison: an example

We start with the small section of data file presented in Table 6.4a. The data give V' filter measurements of a variable star (R Arae), its comparison (BS 6114) and background sky, measured with a CFC converter attached to the standard type single-channel photometer of Carter Observatory.

This segment of data is reworked into the form of Table 6.4b, where the individual groups of three measures have been averaged, and their standard deviations calculated. The time has been changed from local to Universal Time measure. The sky and comparison readings on either side of that of the variable are now interpolated to the time of the variable's measurement (Table 6.4c). The same process is applied to the B' data. The operation to be performed is essentially simple, but programming details can become cumbersome when written to cover a variety of possible data sequences.

Using representative extinction coefficients measured from contemporaneous data the following relevant quantities are now assembled: $\Delta V' = 1.042 \pm 0.007$; $\Delta B' = 1.085 \pm 0.009$; $k'_V = 0.371 \pm 0.016$; $k''_V = 0.00 \pm 0.01$; $X_z(var) = 2.6971$; $X_z(comp) = 2.6932$. Substituting into (6.25), we find

$$\begin{aligned} \Delta V'_0 &= (1.042 \pm 0.007) - (0.001478 \pm 0.000064) - (0.00000 \pm 0.00035) \\ &= 1.041 \pm 0.008, \end{aligned} \quad (6.30)$$

where the quoted uncertainties are standard deviations. The relatively small effect of differential extinction, even with the fairly large air masses involved, is clear. This shows that very accurate extinction coefficients are usually not so necessary in differential photometry, though it can often provide them.

With adopted values $\epsilon = -0.039 \pm 0.068$ and $\mu = 0.992 \pm 0.016$ for relevant transformation coefficients, we proceed to find:

$$\begin{aligned} \Delta V &= 1.039 \pm 0.008 \\ \Delta(B - V) &= 0.042 \pm 0.013. \end{aligned}$$

Table 6.4c gives the time of observation as 10.9003 h UT, and the date (Table 6.4a) is 21/10/87. Using positional data from the *Astronomical Almanac* for 1987, and the *General Catalogue of Variable Stars* the heliocentric correction works out from (6.29) as –0.0031 d (light from R Arae would reach the Sun before the Earth on the date in question), giving the Heliocentric Julian Date of the observation as 244 7089.9511. With the epoch of primary minimum at (HJD) 242 5818.028 and period 4.42509d, we find some 4807.1165 revolutions have occurred since the initial epoch, i.e. the orbital phase at the time of observation is 0.1165 (41.9°).

The computer program may perform scores of such reductions on the data of a typical photometric run. The resulting points form a corresponding portion of the variable's (differential) light curve when plotted.

The comparison star (BS 6114) has been included with standard star observations on various nights in order to put its apparent magnitude and colours on the standard scale. Representative values for V and $B - V$ are

Table 6.4. *Differential photometry data processing: (a) raw data*

```
DATE D-M-Y= 21-10-1987
FILTERS :
 4 = V
INTEGRATION PERIODS OF EACH FILTER :
 4 = 10 SECS
OBJECT 1 IS R Arae
OBJECT 2 IS COMP
OBJECT 9 IS SKY
COUNTS          TIME      FILT OBJ
                 :
                 :
                 :
 450          22:35:46     4   2
 441          22:35:56     4   2
 445          22:36:06     4   2
 3            22:40:14     4   9
 2            22:40:24     4   9
 2            22:40:34     4   9
 169          22:53:56     4   1
 170          22:54:06     4   1
 170          22:54:16     4   1
 435          23:04:28     4   2
 439          23:04:38     4   2
 433          23:04:48     4   2
 3            23:08:58     4   9
 2            23:09:08     4   9
 2            23:09:18     4   9
                 :
                 :
                 :
```

5.64 and 0.01, respectively. Hence, the standard V magnitude and $B - V$ colour of R Arae at the given time are 6.68 and 0.05. Note that these quantities cannot be given with as high a precision as the magnitude and colour differences.

6.3 Light curves of variable stars

Observing variable stars is one of the main pursuits of stellar photometry. In a strict sense, all stars necessarily vary in the rate at which they emit

Table 6.4 (cont.) *(b) partially reduced, (c) comparison and sky interpolated to times of variable readings.*

object	mean UT (hours)	mean count rate (counts/sec)	std dev count rate (counts/sec)
comparison	10.5975	44.53	0.45
sky	10.6719	0.23	0.06
variable	10.9003	16.97	0.06
comparison	11.0758	43.57	0.30
sky	11.1508	0.23	0.06

object	mean UT (hours)	mean count rate (counts/sec)	std dev count rate (counts/sec)
comparison	10.9003	43.92	0.25
sky	10.9003	0.23	0.04
variable	10.9003	16.97	0.06

radiation, but the term variable star usually refers to something distinctly measurable, observed over measurable timescales.

There is a large class of variables whose range of variation is, say, half a magnitude or greater, over a period of order hours, days or weeks, with mean brightness more than that of twelfth magnitude. These are popular and relatively easy targets for observation. The retrieval of a differential light curve forms a particular objective, and if the pattern of light variation is strictly repetitive, over a manageable period, as with 'textbook' eclipsing binary systems, this becomes a well-defined task.

The light curve notion itself is really a convenient fiction for reconciling observational data with the graphical form of a continuous mathematical function. This function may be derived from the formal representation of a physical model, but observations are almost invariably a non-uniform time series of data points: a discontinuous set, which seldom, if ever, repeats itself exactly with successive cycles. Such data, when plotted out, however, generally show that some continuous light curve can be constructed to which the measurements approximate, to within a specifiable observational error. If this constructed curve is that of a model-dependent theoretical function, the way is opened to interpretating the data in terms of parameters characterizing the model. This idea is developed in subsequent chapters.

The light curve thus forms a useful basis for analysis, particularly as such analysis can usually separate out differential effects from the mean reference level of light. Physical quantities of interest are thence derived from the differential data alone. If the mean light level can be accurately fixed, it will yield additional information, of course.

6.4 Bibliographical notes

The classic reference for much of the subject matter of this chapter is the paper of R. H. Hardie in *Astronomical Techniques* (ed. W. A. Hiltner) University Press, Chicago, p178, 1962. The procedures discussed there were elaborated on in A. A. Henden and R. H. Kaitchuk's *Astronomical Photometry*, Van Nostrand Reinhold, New York, 1982, C. Kitchin's *Astrophysical Techniques*, Adam Hilger, Bristol, 1984 and W. A. Cooper and E. N. Walker's *Getting the Measure of the Stars*, Adam Hilger, Bristol, 1988. More recently Henden and Kaitchuck produced their *Astronomical Photometry Software for the IBM-PC*, Willmann-Bell, Richmond, Va., 1989, complete with three computer diskettes, enabling the user to apply such procedures directly to data.

In fact, the impact of computers on data processing is a major change that has occurred since the time of Hardie's paper. This development continues everywhere, and indications about coding included with the present chapter should be regarded as illustrative only. Programmers are continually introducing more efficient and 'user-friendly' algorithms, in purpose-oriented computer languages, to deal with data reduction and analysis of the kind presented here. Examples of such techniques have been presented in the *I.A.P.P.P. Commun.* and the Fairborn Press literature mentioned in the bibliography section of the preceding chapter, or in articles such as R. Elston and M. Zeilik's in *Publ. Astron. Soc. Pac.*, **94**, 729, 1982. Detailed considerations for approaching 0.001 mag accuracies in calibrating a photometric system were provided in A. T. Young *et al.*'s paper in the proceedings of the IAU Joint Commission Meeting on *Automated Telescopes for Photometry and Imaging*, Buenos Aires, 1991 (ed. S. Adelmann) to be published by the Astronomical Society of the Pacific.

With regard to calculations required in preparing for observations, besides the book by P. Duffet-Smith, mentioned in the text, the similar one by J. Meeus *Astronomical Formulae for Calculators*, Willmann-Bell, Richmond, Va., 1988, could be mentioned. A number of such books were reviewed by E. Bergmann-Terrell in *South. Stars*, **34**, 59, 1991.

Data showing the variations of extinction coefficients at an astronomical observatory over extended periods of time appear from time to time distributed throughout the literature, e.g. W. J. Schuster in *Rev. Mex. Astron. Astrofis.*, **5**, 149, 1982; T. S. Yoon and S. W. Lee *J. Korean Astron. Soc.*, **15**,

59, 1982. The subject was reviewed by J. Stock in *Vistas Astron.*, **11**, 127, 1969.

Source data on the astronomical constants used in the sidereal time and heliocentric correction calculation are reported in the annually produced *Astronomical Almanac*, which also lists standard stars in the *UBVRI* and *uvbyβ* systems, as well as a great deal of other circumstantial information. The *Astronomical Almanac* is published jointly by the US Government Printing Office, Washington and Her Majesty's Stationery Office, London. Much of its information is also available in computer diskette form.

Standard star lists continue to be upgraded. Of the more basic references for the *UBV* system we can quote H. L. Johnson and W. W. Morgan, *Astrophys. J.*, **117**, 313, 1953; H. L. Johnson and D. L. Harris III, *Astrophys. J.*, **120**, 196, 1954; B. Iriarte *et al.*, *Sky Telesc.*, **30**, 25, 1965, though a number of variables are included in this list, so some care is required in selecting good standards from it. In that connection a valuable source book is the work of V. M. Blanco, S. Demers, G. G. Douglass and M. P. Fitzgerald in *Publ. US Nav. Obs.*, **21**, 1970 (second printing), which compiles most, if not all, published *UBV* measures of stars up to that date. In the Southern Hemisphere A. J. W. Cousins continues to improve the accuracy of his standard star lists in a series of publications in the *Mon. Not. R. Astron. Soc. South Africa* — **47**, 16, 1988 (with J. D. Laing) provides a more recent example dealing with the Magellanic Clouds. The *E-region* standards (A. W. J. Cousins and R. H. Stoy, *R. Obs Greenwich – Cape Bull.* **49**, 1962) are also popular with southern observers. Perhaps more accessible are the two articles in *Mem. R. Astron. Soc.*, **77**, 223, 1973, and **81**, 25, 1976. Alternative lists are from A. U. Landolt in *Astron. J.*, **88**, 439 and 853, 1983, and J. A. Graham *Publ. Astron. Soc. Pac.*, **98**, 244, 1982. The SIMBAD database facility should also be referred to for extensive reference coverage on particular stars.

7

Basic Light Curve Analysis

7.1 Light curve analysis — general outline

In this chapter we approach some classic problems of photometric analysis. We start with the light curves of eclipsing binary stars. These reduce, in their simplest form, to regular patterns of variation which can be understood by reference to simple models of stars in a simple geometrical arrangement. Estimates for key parameters can be directly found from inspection of the salient features of a light curve. This is a useful preliminary to more detailed analysis. The basic issue underlying this and subsequent chapters, however, concerns the setting out of a general method for parameter value estimation. This represents a subsection of the field of optimization analysis, or the optimal curve-fitting problem.

We are given a set of N discrete observations $l_o(t_i)$ $(i = 1, ...N)$ in a *data space*, which have a probabilistic relationship to an underlying real variation, dependent in a single-valued way on time t. This real variation, whatever its form, is approximated by a fitting function, $l_c(a_j, t)$, say, which is, formally, some function of the independent variable t, and a set of n parameters a_j $(j = 1, ...n)$. A subset m of these parameters we regard as determinable from the data.

The object is to transfer the $l_o(t_i)$ information from data space to the a_j information in *parameter space*. Different sets of parameters map different versions of the fitting function, overlapping different data values, so that the net probability of such versions varies in response to the chosen parameter set. An *optimal* solution to the curve-fitting problem is provided by a set of m parameters which maximizes this probability. It can be shown that if data ordinates are normally distributed about their mean at each t_i this maximization is equivalent to the minimization of χ^2, formed by

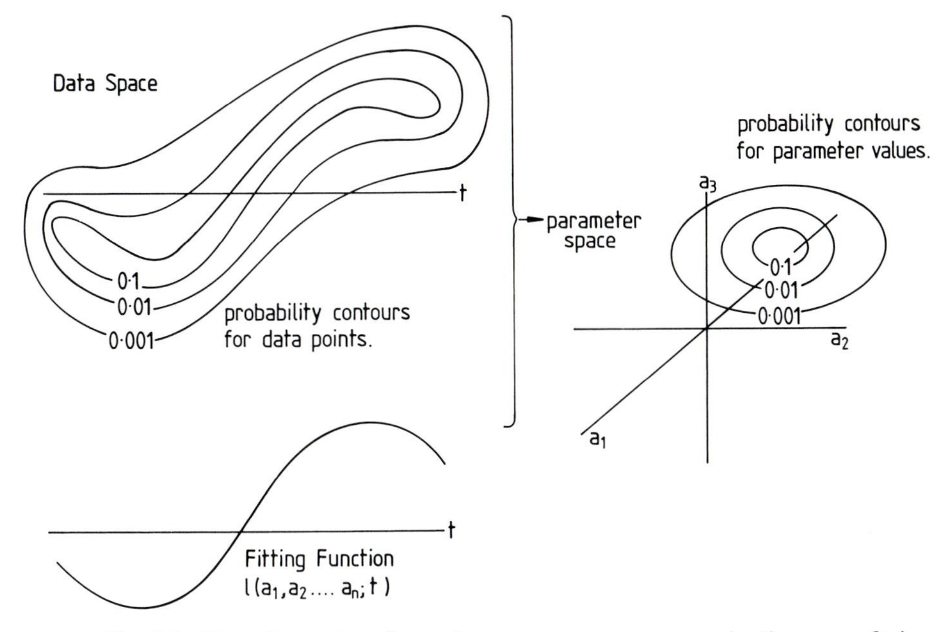

Fig. 7.1. Transformation from data to parameter space in the curve-fitting problem.

$$\chi^2(a_j) = \sum_{i=1}^{N} [l_{oi} - l_c(a_j, t_i)]^2/\sigma_i^2, \tag{7.1}$$

where σ_i^2 denotes the variance of the distribution at t_i.

The full number of parameters (n) depends on the adopted form of the fitting function, but the parameters are generally not free to take any real value, and are sometimes constrained so that one parameter is a definite function of some other(s). Some of them are usually fixed in value from separate evidence than that of the data sample $l_o(t_i)$ ($i = 1, ...N$).

The observational data are finite in extent, implying a limitation on their *information content*, or, equivalently, the number m of independent parameter values in the set **a** to be specified by the curve-fitting solution. Ideally, the transformation would be from closed contours of constant probability in the data space to closed (ellipsoidal) contours in m-dimensional parameter space (Figure 7.1), but as m increases from unity a certain number will be reached m_{max}, beyond which this no longer holds good, and there is no longer a unique optimizing parameter set. As m is increased uncertainty factors Δa_j (simply related to the expected errors of the determinations of the a_j — cf. Section 6.1.1) increase progressively, until, at some point just beyond m_{max} (the 'information limit'), such factors, for one or more a_j, formally pass through a singularity where the solution becomes indeterminate, or non-unique. In other words, a finite data sample can only yield a finite amount of information.

There are three fairly distinct aspects to the curve-fitting problem: 1) specification of the initial data set in an appropriate form, $l_o(t_i)$; 2) specification of a physically reasonable and mathematically tractable fitting function, $l(a_j, t)$; and 3) the optimization of the match of l_o to l_c. The procedure is relatively clear for standard eclipsing binary data, with important yields in astrophysical information.

7.2 Eclipsing binaries — basic facts

Eclipsing binaries are essentially a subset of 'close binary systems' — pairs of stars whose separation is typically an order of magnitude greater than a component mean radius. There are some known cases where the companions are relatively well separated (e.g. α Coronae Borealis), though the likelihood of observing eclipses in binary systems will clearly drop with increasing displacement of the pair.

It can also be easily seen that component radii will not, in general, be extremely dissimilar, because if one star were very much larger than the other (by an order of magnitude or more, say) then the system's light would tend to be dominated by that larger star. The eclipse of the small star by the large one (an 'occultation'), or the passage of the small star in front of the large disk (a 'transit') would then have only a slight effect, evading easy discovery, unless the small star were relatively very bright, per unit area of its surface, at the wavelength of observation.

Eclipsing binaries are thus pairs of stars showing a regular cycle of variation in apparent brightness, due to their orbital motions lying in a plane inclined at a small angle to the line of sight. The term orbital inclination means the angle between the axis of the orbit and the line of sight; so that for eclipsing binaries, this angle is usually not far from 90°, and, in any case, can be shown to be greater than about 58°. There is then a succession of alternating photometric minima, transit followed by occultation, and so on. The eclipses are described as complete when the outline of the smaller disk projects entirely within that of the greater one: a total eclipse if an occultation, or an annular one if a transit. If an eclipse is not complete it is partial.

During the course of an orbital cycle of an eclipsing pair there are two photometric minima; however, one of these can involve such a low loss of light as to be hardly noticeable. The deeper eclipse minimum, caused by the eclipse of the star which has greater brightness per unit area (associated with a higher temperature photosphere) is usually called the 'primary' minimum, and the star eclipsed at this minimum referred to as the primary star. Similarly, the shallower minimum is known as the 'secondary' one and corresponds to the eclipse of the secondary star.

An interesting atlas of light curves of eclipsing binary systems was produced some years ago by Fracastoro, including examples that show distinct effects of orbital eccentricity. In the great majority of cases, however, orbits are effectively circular, and there are physical reasons for this. Hence, circular orbits represent a practical and convenient simplification for common analysis. Some elementary deductions can be made at once on this basis: thus the durations of both eclipses are the same. The same photospheric areas are in eclipse at corresponding phases through either minimum of the light curve, and so the ratio of light lost at such phases yields a corresponding ratio of surface flux averaged over these eclipsed areas, thus approximately measuring the surface brightness ratio of the two photospheres.

The matter of which star is called the primary can be occasionally confusing. In spectroscopic contexts, the primary would normally be regarded as the component putting out more light altogether — not necessarily the same thing as the star of greatest brightness per unit area — so the spectroscopic primary can differ from the photometric one. Also, when stellar evolution of a binary is considered, the term primary usually refers to the originally more massive star, though it need not always remain so. There is very compelling evidence that stars lose appreciable mass during the course of their evolution, at a rate that depends on the initial mass of the star. In addition to this general effect, mass may be *transferred* between the components in certain scenarios of 'interactive' binary evolution. The originally more massive primary may then become the secondary star.

Concerning derivable information on eclipsing binaries, it is necessary to combine observational data — photometric, spectroscopic and/or other types — to maximize what can be stated; but light curves do provide a reasonably self-contained branch of the problem.

There have been, conventionally, three basic categories in the empirical classification of light curve types: the 'Algol' (EA) type, the 'β Lyrae' (EB) type, and the 'W Ursa Majoris' (EW) type (Figure 7.2). The conventional prototype names are, at least in the first two cases, somewhat inappropriate, and simple letter designations (i.e. EA, etc.) are preferable, and will be used in what follows.

The EA type light curves have roughly constant light levels (to within, $\sim$0.1 mag) outside of the eclipses, which are thus clearly marked, or at least

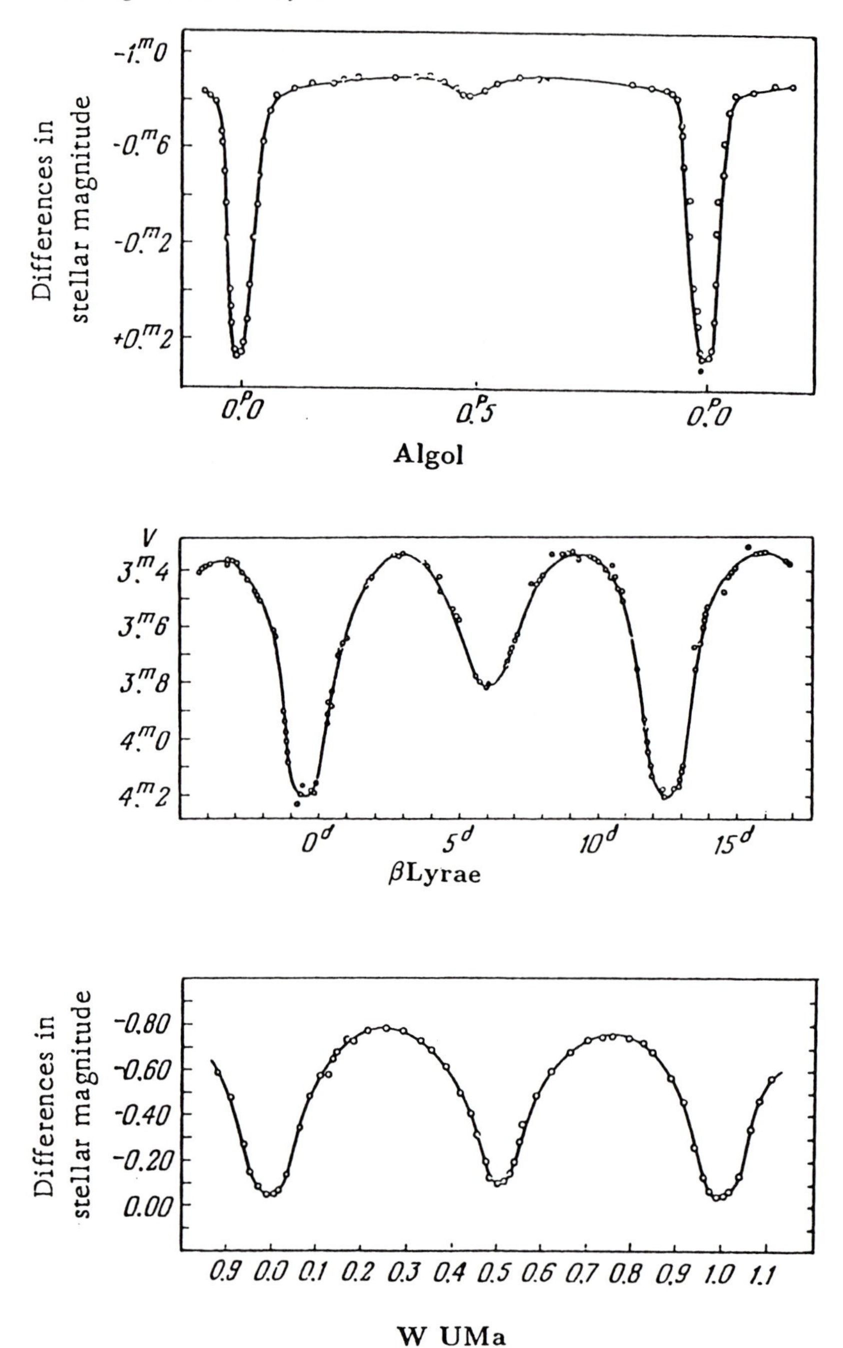

Fig. 7.2. Photometric classification of eclipsing binary light curves.

the primary minimum is. It is possible to subdivide this light curve type further, and the designations EAD and EAS have been used for this. The EAD light curves, for orbital periods of a few days, usually result from the mutual eclipses of a pair of not-too-close normal Main Sequence stars. (Eclipsing systems composed of pairs of normal giants are relatively rare, and involve much longer periods.) Both minima can be discerned, and their depths have a simple relationship to each other if the spectral types (and therefore, inferred approximate surface effective temperatures) are known. The depth of the primary minimum cannot exceed little more than one magnitude at optical wavelengths, in consequence of the Main-Sequence-like properties of the stars.

The EAS type light curve is exemplified by Algol itself, but perhaps more obviously by such completely eclipsing Algol systems as U Cep or U Sge. The primary eclipse is deep because of the relatively large size but cool photosphere of the 'subgiant' secondary. The relatively low mass of this star can be inferred from the low scale of proximity-induced tidal distortion of the bright primary, so that the light curve shows little of the interminimum curvature (at least at optical wavelengths) associated with such effects. An EAS light curve is distinguished, then, by a deep primary minimum (greater than 1 mag) and the secondary minimum depth being so shallow as to stand apart clearly from the trend of secondary to primary depth ratios predictable for Main Sequence pairs. Note that it is not *necessary* for an Algol-like system to exhibit an EAS type light curve, even if an EAS light curve were always produced by an Algol-like pair of stars.

The EB type light curve shows a continuous pattern of variability including the rounded interminimum regions associated with more pronounced tidal interactions. It is more frequently associated with earlier type binaries, and generally closer relative separations. There are some early type binaries with EB type light curves with absolute separations which are quite large compared, for example, with the radius of the Sun: β Lyrae itself is an example. These systems, whose evolved character can be inferred from their relatively long, and often slowly varying, orbital periods, are physically different from the simpler situation of a pair of close, early type, Main Sequence binaries.

With EW type light curves the proximity-induced interminimum rounding reaches such a scale as to merge imperceptibly into the eclipse minima, which can thus no longer be clearly empirically separated out. It is a fact, which is not yet perfectly understood, that the EW binaries also show more or less equal depths of primary and secondary minima. Hence, a light curve showing very pronounced out-of-eclipse rounding, but with minima depths differing by more than one or two tenths of a magnitude, should be assigned an EB classification. In fact, such a light curve is very rarely encountered, EB binaries being preferentially found among systems of somewhat longer period and earlier type than the EW systems. Genuine W UMa type binaries all appear to have periods appreciably less than one day in duration. They are also predominantly made up of late type pairs, by far the largest proportion, by spectral class, being G type primaries and secondaries.

Thus, EAD and 'normal' (short-period) EB binaries can be usually clearly interpreted in terms of unevolved, Main-Sequence-like pairs. When we move to the EAS, long-period EB and EW binaries, however, we encounter a range of properties of varying awkwardness for explanation, which have sometimes been termed paradoxes. Perhaps the most well known is the 'Algol paradox' of the EAS binaries. This can be readily sensed from the photometry, as indicated above. The deep primary minimum would imply a more evolutionarily advanced, and therefore, presumably, more massive companion, in contradiction to the low mass ratio indicated by the absence of tidal proximity effects. A second Algol paradox has been considered in recent years, connected with the difficulty which standard interactive binary evolution theory has in getting the Algols into their observed 'gentle' configurations, without drastic 'run-away mass-loss' or 'spiral-in' effects.

A 'β Lyrae paradox' exists for the longer period EB stars. In order to avoid extremely large masses in the explanations of spectroscopic data on these binaries, it is necessary to suppose that the more massive components are actually *not seen*, but buried in a 'thick disk'.

Finally, there is a 'W UMa paradox' for the EW binaries. This is associated with these stars being so close as to make 'contact' with each other's surfaces. A star in such a situation would have its outer boundary conditions fixed by dynamical constraints imposed by the contact configuration. But stars of given mass and composition fix their own outer boundaries, by the Vogt–Russell theorem of classical stellar structure. The ratio of masses to radii fixed by the contact configuration is, in general, incompatible with that of pairs of single stars of the same composition, hence such binaries are expected to have an inherent instability. Yet the W UMa type stars are observed in such high relative numbers as to show that any such instability does not have the significant effects expected.

Although the foregoing classification of light curve types is simple and empirical, it does point to what gives rise to them. Some alternative approaches to classification are designed to emphasize this underlying physics more strongly; but this would also imply a greater degree of prior analysis (and hypothesis) on the data.

A well-known scheme is that of the detached, semidetached and contact systems, based on the relationship of the components to the so-called Roche critical surface; limit or lobe (cf. Figure 7.3). If a component having expanded up to this surface from below were to expand still further, a dynamical instability (entailing mass-loss) would ensue. Detached pairs can be largely identified, in this way, with the EAD type light curves, and also some shorter period EB binaries. EAS type light curves are similarly closely associated with semidetached systems, though the latter need not always produce a light curve free from significant interminimum rounding if the evolved secondary happens to be of earlier than usual type — u Her is a frequently cited example of this. As suggested before, EW light curves and contact binaries are fairly well identified with each other, though it is possible to produce an EW type light curve from a very close, but not actually in-contact, pair. A more widespread term nowadays for the contact pairs is 'common envelope' binary. The photospheric surface is considered to lie between the inner and outer critical Roche surfaces. This picture, though not without some theoretical difficulties, offers advantages in accounting for the EW type light variation, as well as other properties of these stars.

Apart from these more basic classifications of the eclipsing binaries certain other groups have been identified. One such group, the RS CVn binaries, has attracted much attention in recent years. Though not all of these binaries undergo eclipses, as RS CVn itself, most of the orginally identified members of the group came from studies of eclipsing binaries with peculiarly distorted light curves. The RS CVn stars show a generalized or enhanced range of phenomena parallelling the magnetodynamic activity of the Sun. Dwarf novae are another group now regarded as binary systems, and interpreted through interactive binary evolution. Some of them eclipse, and the resulting light variation can provide very useful information on the geometry of the configuration. In a more extreme physical condition are X-ray binaries, believed, like dwarf novae, to have resulted from interactive binary evolution involving initially relatively massive stars. For some of these systems eclipses are observed for the X-ray radiation itself, while others show very enhanced scales of proximity effects in the optical domain. Qualitatively similar effects have recently been recognized in 'symbiotic stars'; while other kinds of stellar peculiarity, e.g. old novae, or the 'barium star' syndrome, are increasingly related to the effects of binarity.

Although about 4000 eclipsing binaries are known at the present time, curiously, hardly any close *eclipsing* binary has been found to be a member of a globular cluster, though they have been seen in the galactic field of our nearest neighbour galaxies. In principle, eclipsing binaries offer a direct

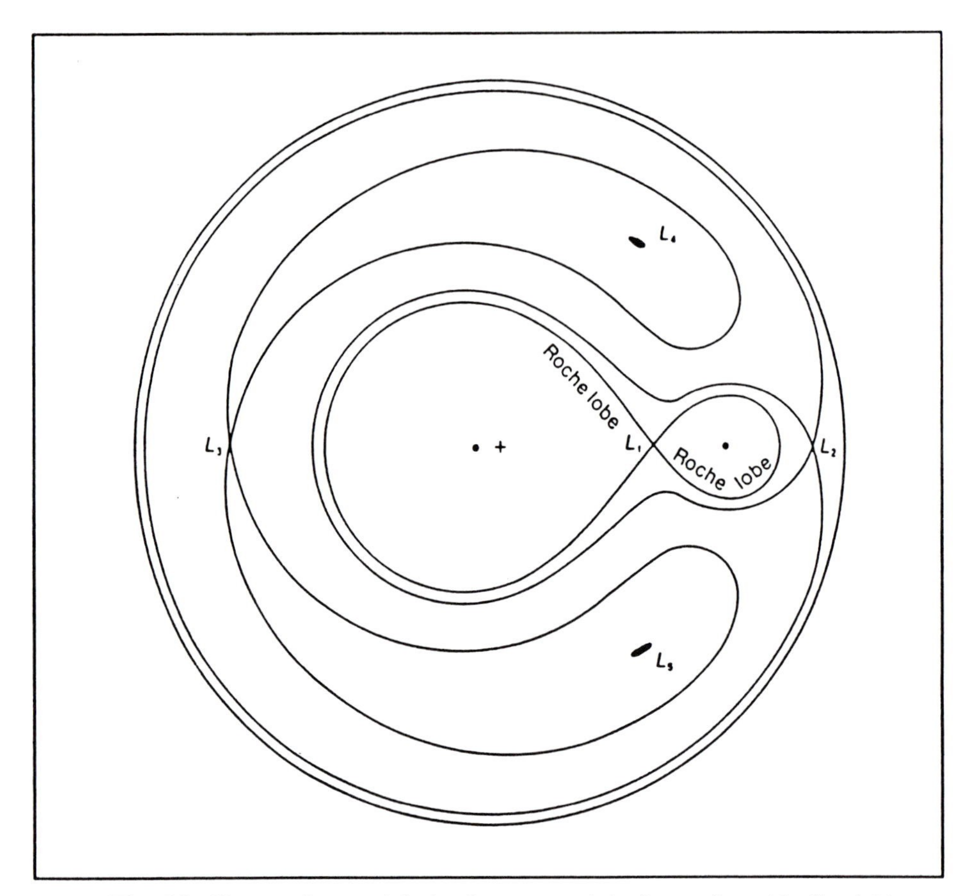

Fig. 7.3. Key equipotentials in the equatorial plane of an idealized binary star.

means for retrieving quantitative (empirical) information on stars over very large tracts of space.

7.3 Hand solution of light curves

The geometrical implications of eclipses (only) affecting a pair of stars, assumed spherical in form and moving in circular orbits for present convenience, can be gathered from the simple representation in Figure 7.4. We first define some commonly used symbols and terminology, thus: U — 'unit of light', i.e. a normalized reference level for the ordinate scale. Illuminance levels, in arbitrary units, are divided by a representative out-of-eclipse value, for numerical convenience. Light levels can then be rescaled by adjusting

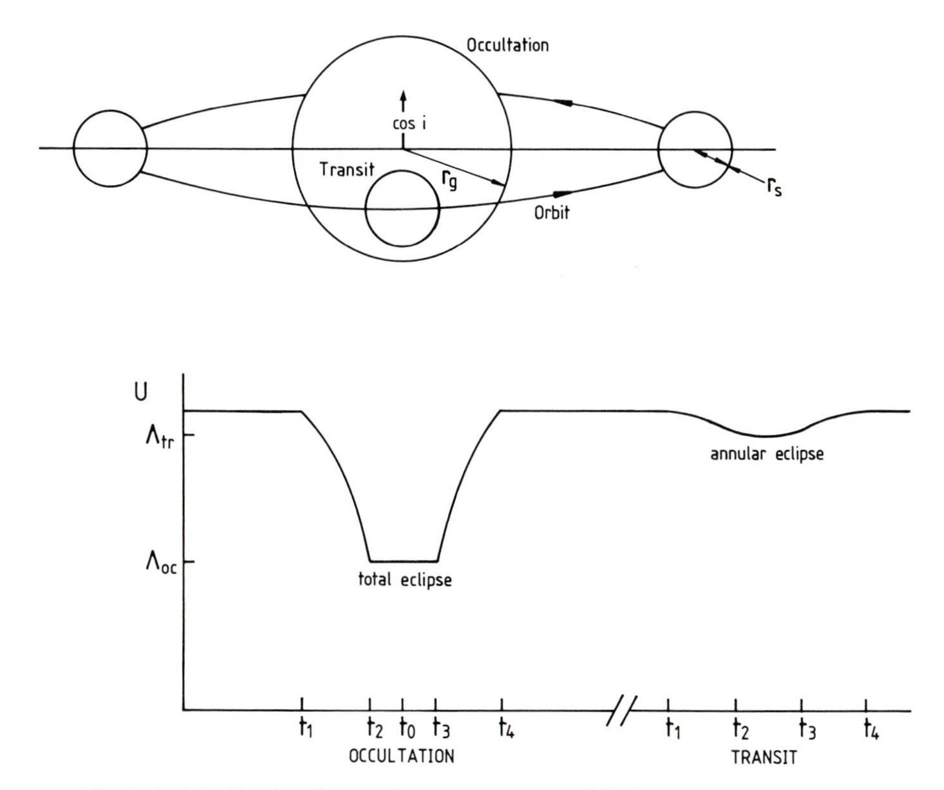

Fig. 7.4. Idealized eclipsing binary system and light curve.

U as part of the optimization procedure; L — luminosity of one of the components, normally expressed in the final outcome of the analysis, as a fraction of U; r — radius of a component in units of the mean separation of the centres of the two stars; i — orbital inclination, i.e. the angle between the line of sight and the axis of the orbit; θ — orbital phase angle. This is usually reckoned from the phase of primary mid-minimum, but allowance is made for a zero phase correction $\Delta\theta_0$, which is added to the values of θ to improve the curve-fit.

We write Λ to denote light levels regarded as constant for some range of phases, and may use suffixes for Λ, L or r, thus: tr or oc for that star which is in transit or occultation; g or s for greater or smaller size; 1 or 2 for the star being eclipsed or eclipsing. A flat region at the bottom of one of the minima would correspond to the total eclipse of the smaller star. This level is then denoted Λ_{oc}. If only two stars contribute to the observed light and its variation, we have:

$$\left.\begin{array}{lcl}\Lambda(\text{between minima}) & = & U, \\ \Lambda_{oc}/U & = & L_g, \\ (U-\Lambda_{oc})/U & = & L_s.\end{array}\right\} \quad (7.2)$$

Certain quantities relevant to the abscissal scale are: mean period P; times of beginning and end of eclipses ('first and fourth contacts') t_1, t_4; times of beginning and end of totality t_{2oc}, t_{3oc}; times of beginning and end of annular phase t_{2tr}, t_{3tr}.

Consider now a right-handed rectangular coordinate frame (x, y, z), such that the z-axis coincides with the line of sight, and the y-axis lies in the plane formed by the z-axis and the axis of orbital revolution passing through the centre of the primary star — the plane whose intersection with the orbit marks the zero point of the phase θ. The unit of distance is the orbit's semidiameter. We then find, for the coordinates of the secondary star:

$$\begin{array}{lcl} x & = & \sin\theta, \\ y & = & -\cos\theta\cos i. \end{array}$$

By squaring and adding we obtain a basic equation for the the separation of the two star centres projected onto the tangent plane of the sky, conventionally denoted δ, i.e.:

$$\delta^2 = \cos^2 i + \sin^2\theta \sin^2 i. \quad (7.3)$$

The orbital phase is calculated from the primary mid-eclipse t_0, at time t, as $\theta = 2\pi(t - t_0)/P$, where the epoch t_0 satisfies:

$$t_0 = (t_1 + t_2 + t_3 + t_4)/4, \quad (7.4)$$

though a more elaborate method would be used in practice for this determination. The 'folding paper' method, in which observations during a minimum are plotted out on a transparent sheet, which is then folded so that the two halves of the minimum curve are superposed, has been a common simple approach. In any case, the underlying idea is to average times from the two sides of the minima.

P is deduced from determined times of minima, and can be expressed with considerable accuracy when observations have covered many light cycles. Thus, if we determine t_0 to an accuracy Δt, then P (if constant) is known to an accuracy $\Delta t/N$; where N is the number of complete cycles of the light variation covered. For many of the well-known eclipsing binaries N runs into several thousands or tens of thousands, which would justify the quoting of P values sometimes up to eight decimal places in days, as seen in published data. Some systems do not have a sensibly constant period, however, and the specification needs to be qualified by an awareness of this.

The orbit size is so small in comparison with the distance of the observer that perspective effects are negligible. Regarding the two stars as disks in the plane of the sky, in which the orbit projects as an ellipse (Figure 7.4), then shows

$$\cos i \leq r_g - r_s \tag{7.5}$$

to be the condition that total-annular eclipses occur. If complete eclipses occur, and we compare corresponding points within the total and annular limits, the area eclipsed is that of the disk of the smaller star. If the surface mean intensity of this star is F_s, we can write:

$$\frac{1 - \Lambda_{oc}}{1 - \Lambda_{tr}} = \frac{F_s}{\{F_g\}}, \tag{7.6}$$

where $\{F_g\}$ represents a mean surface brightness of the greater star, averaged over the area of the smaller star's disk.

Neglecting, for the moment, the variation of intensity over the larger disk (limb darkening), we deduce that if the primary minimum corresponds to a transit then the greater star is intrinsically brighter (and *vice versa* for an occultation). Transit primary minima normally occur for pairs of stars on the Main Sequence. We find, fairly directly, that

$$(U - \Lambda_{tr})/U = L_g(r_s/r_g)^2\Psi, \tag{7.7}$$

where Ψ is generally quite close to unity (insensitively depending on the limb-darkening effect). Since $\Lambda_{oc}/U = L_g$, if we put $k = r_s/r_g$ we will have,

$$k \simeq \sqrt{(U - \Lambda_{tr})/\Lambda_{oc}}. \tag{7.8}$$

Hence, the light levels at the completion of the two eclipses, if available, give an approximate value for the ratio of radii of the two stars.

In the idealized light curve of Figure 7.4 at the external tangencies (t_1, t_4) the projected separation of centres δ_1, is

$$\delta_1 = r_g(1 + k), \tag{7.9}$$

and at the internal tangencies (t_2, t_3) the corresponding separation δ_2 is

$$\delta_2 = r_g(1 - k). \tag{7.10}$$

Taking (7.8), (7.9) and (7.10), and using (7.3), we can approximately evaluate the three geometric unknowns: r_g, r_s and i. The two luminosity quantities, U and L_g (L_s is constrained by (7.2) in this model), together with the epoch of primary mid-minimum t_0, make up the list of basic parameters to six. Taking things a step or two further, we encounter a coefficient of limb darkening u_g as an argument of the function Ψ in (7.7), which influences the curvature at the bottom of the transit minimum. There is a corresponding parameter u_s for the smaller star, though its effects are less discernible in the occultation ingress and egress. In any case, eight parameters correspond to a reasonably complete specification for the simple 'spherical model' light curve under consideration.

This analysis is essentially equivalent to that first discussed long ago by Henry Norris Russell, one of the great factfinders of stellar properties in the first half of the twentieth century. Of course, there are various shortcomings in this deliberately simplified approach, but it is instructive to test things out on an actual example.

For this G. E. Kron's photoelectric data for the fifth magnitude eclipsing binary YZ Cas, obtained on the 36-inch refractor of the Lick Observatory in the late 1930s provide a classic case. Although Kron's photometer did not have the same low noise detective efficiency as those of more recent type, the exemplary quality of his results is probably due to: i) the star is relatively quite bright (a naked eye object), ii) the telescope aperture was quite large for the magnitude of object being monitored, and iii) Kron was persistently careful in his observations of minima, of which he observed a few dozen between 1937 and 1940. The object is a textbook example of a pair of Main Sequence stars in a circular orbit, whose plane lies very close to the line of sight. Proximity effects are very slight, and there are no significant variations to the light curve apart from the eclipses themselves. Kron averaged out small, randomly distributed errors in the large quantity of individual data points, which were reduced to some 27 points in each minimum of his blue light curve; and 22 and 20 in the primary and secondary minima, respectively, of the red filter observations.

Reading from Kron's (1942) red (6700 Å) light curve, we have

$$\begin{aligned} U &= 1.000, \\ \Lambda_{tr}(\text{inner contacts}) &= 0.745, \\ \Lambda_{tr}(\text{centre}) &= 0.733, \\ \Lambda_{oc} &= 0.898. \end{aligned}$$

Taking the average of the two transit levels as representative to substitute into (7.8), we find $k = 0.539$. The internal and external tangencies occur at phases 3.5° and 12.3°, respectively. Using equation (7.3) with (7.9) and (7.10), and the foregoing value of k, we derive the three geometric elements, as follows:

$$r_g = 0.139,$$
$$r_s = 0.075,$$
$$i = 88.9°.$$

Although obtained in such a simple way, these results compare quite tolerably with those of much more elaborate analysis. Such methods have a place, then, in providing initial estimates for more detailed procedures.

If the light curve showed only partial eclipses, i.e. no identifiable internal tangency, a starting point is still possible using the equality condition of (7.7). This would give an upper limit for i and lower limits for the component radii.

A procedural development was already made in the days of Russell by the introduction of a function α, giving the fraction of a star's light lost during eclipse. α depends on the projected separation of centres δ, the ratio of the radii k, and some parameter (or parameters), determining the limb darkening over the disk of the star being eclipsed. As we have seen, if the eclipses are complete, it is possible to determine directly from the light levels during annular and total phases approximate values for both fractional luminosities L, and the ratio of radii k. With an appropriate value of L_1, the light loss of a star being eclipsed could be listed as a set of empirical α values for observations of a given minimum.

The classical method depended on having tables of calculated α values for a comprehensive range of values of k and a conveniently normalized measure of projected separation, p say, over a range of plausible values of the limb-darkening coefficient u. These α tables were inverted to provide separate sets of tables of p for given α, k and u. Hence, with u set, and determined k and α values, the corresponding p values were looked up. Equation (7.3), for varying values of the phase during eclipses, would then provide a set of equations of condition for the radius of the eclipsed star (used in scaling the δs to the normalized p values), and the orbital inclination i.

This approach provided a reasonably manageable method of analysing eclipsing binaries in the precomputer era. Nowadays, the use of tables seems very dated. Apart from that, there were certain other restrictive assumptions, glossed over here in order to provide an overview, which a modern analyst would want to reach beyond.

7.4 Computer-based light curve analysis

The use of equation (7.3) for certain special points on the light curve formed the traditional approach developed by Russell, Merrill and others. Such methods have been sometimes described as 'indirect', so as to distinguish a 'direct' procedure, concentrating on the equation of the light curve. This direct approach turns out to be well suited to computers, and has the advantage of using all available data. It exemplifies the curve-fitting problem of Section 7.1.

Firstly, however, it is required to present the data in the right initial format. Greater efficiency obtains if the number of data points can be reduced without a significant loss of information. This was encountered with Kron's data for YZ Cas, where the term normal points was met. The construction of a program to 'bin'[†] data, is relatively commonplace, with ready-made software widely available. The question of bin size arises here, but the optimum arrangement is difficult to specify strictly. Some regions of a light curve may be intrinsically more informative than others for the quantities sought. In order to avoid information loss, any error in the assumption of linear variation of light over the small interval whose data are averaged should be less than the error of the mean in that interval. Hence, if δl is the real first order (linear) variation of light for a small change of phase $\delta\phi$, while Δ denotes the accuracy with which the variation can be specified, we require the second order variation $\delta(\delta l(\delta\phi)) < \Delta l(\delta\phi)$. (Much greater intrinsic accuracy attaches to the phase determination than the light level.) For most of the eclipsing binary light curves normally encountered we can expect $\delta(\delta l) \lesssim 10^{-3}\delta\phi_d^2$, for a phase bin $\delta\phi_d$ degrees.

Failure of the condition is more likely near the tangencies, where sudden changes of slope occur. We have already seen that such points can be especially informative, so that bin size should be kept as small as conveniently possible in their vicinity. The foregoing estimate suggests that 1° intervals would be adequate for eclipse minima binning when normal point accuracy (Δl) is about 1%.

The suitability of the chosen bin sizes can be checked *a posteriori* from any significant variation of parameter values with an increase in the number of used data points. This would be a useful test once an acceptable parameter set has been found for a given data set. There are some light curves which show sudden rapid variations, flash effects, or fluctuations. Clearly, in

[†] Subdivide the range of abscissae into suitably sized intervals and perform standard statistical tasks, such as determining the mean and standard deviation within each interval.

these, and similar, situations the normal points idea would have to be used carefully, if at all.

Formulae and tables for α values were originally provided by Russell and Merrill at Princeton, and Tsesevich in the Soviet Union. In order to combine the transit and occultation forms of the older sources, however, we can write $k = r_2/r_1$ regardless of which star is larger, so allowing $0 < k < \infty$. The separation argument $d = \delta/r_1$, i.e. normalized to the radius of the eclipsed star, is also convenient, so that $\alpha = \alpha(u_1, k, d)$ is the fractional light loss of the eclipsed component in this notation. We can then write

$$l_c(\phi) = U - L_1\alpha(u_1, k, d), \tag{7.11}$$

as the basic equation of the light curve in the spherical model. The quantities u, L, k and d are different for primary (pr) or secondary (se) minimum region, though the latter three have very simple interconnections, i.e. $L_{1,se} = U - L_{1,pr}$, $k_{se} = 1/k_{pr}$, $d_{se} = d/k_{pr}$. Equation (7.11) can be substituted into (7.1) for each observation at phase ϕ_i and compared with the corresponding observed l_{oi}.

For a linear law of limb darkening, in which the intensity I at a direction θ to the outward normal is of the form $I(\theta) = I_0(1 - u + u\cos\theta)$, it can be easily shown that

$$\alpha = \frac{3(1-u)}{3-u}\alpha_0^0 + \frac{3u}{3-u}\alpha_1^0, \tag{7.12}$$

where integrals of the type α_n^m have appeared. These are defined as

$$\pi\alpha_n^m = \iint_A x^m z^n dx dy, \tag{7.13}$$

where A denotes the eclipsed area (cf. (4.32)). The α function for a spherical star, with limb darkening expressed as a series in powers of $\cos\theta$ ($\equiv$ z-coordinate of the corresponding surface point), being eclipsed by a spherical companion, is then given by a corresponding series in terms of α_n^0 integrals.

The more generalized integrals α_n^m appear when departures from sphericity are included. Recursion formulae are available which enable the whole range of α_n^m integrals to be evaluated, provided a few fundamental integrals are given. This range of integrals can be computer generated and applied to light curves which show proximity effects.

α INTEGRALS — The two basic α integrals — α_0^0 and α_1^0 — take the following forms:

(i) Annular case $d \leq 1 - k$, $k < 1$,

$$\alpha_0^0 = k^2, \tag{7.14}$$

$$\alpha_1^0 = \frac{2}{3}(\Lambda(\kappa_a, \beta) - \\ - \frac{1}{3\pi\sqrt{1-(d-k)^2}}\left\{[d^2(d^2 - 2dk - 5 + 8k^2) + \right. \\ + dk(8 - 14k^2) + (1-k^2)(4-7k^2)]E(\kappa_a) - \\ \left. - [d^4 - (5+2k^2)d^2 + 6d - 2 + k^2(1+k^2)]F(\kappa_a)\right\}\Big), \tag{7.15}$$

where we require complete elliptic integrals E and F, with modulus κ_a given by $\kappa_a = \sqrt{4dk/[1-(d-k)^2]}$. $\Lambda(\kappa, \beta)$ is Heumann's Λ-function, given as

$$\Lambda(\kappa, \beta) = \frac{2}{\pi}\left(E(\kappa)F(\kappa', \beta) + F(\kappa)E(\kappa', \beta) - F(\kappa)F(\kappa', \beta)\right),$$

where the incomplete elliptic integrals E and F, with modulus $\kappa' = \sqrt{1-\kappa^2}$, have argument β given by

$$\beta = \arcsin\left(\sqrt{\frac{1+k-d}{1+k+d}}\right).$$

Note the special case for $d = 1-k$

$$\alpha_1^0 = \frac{4}{3\pi}\left[\arcsin\sqrt{k} - \frac{1}{3}(1+2k)(3-4k)\sqrt{k(1-k)}\right],$$

(reducing simply to $\alpha_1^0 = \frac{4}{9}$, when $d = \frac{1}{4}$). Another special case occurs if $d = k$, when we have, $\kappa_a = 2k$, and then,

$$\alpha_1^0 = \frac{1}{3} + \frac{2}{9\pi}\left[(8k^2-4)E(\kappa_a) - (1-4k^2)F(\kappa_a)\right].$$

Finally, if $d = 0$ we can write, simply,

$$\alpha_1^0 = \frac{2}{3}\left[1-(1-k^2)^{3/2}\right].$$

(ii) Partial case $1+k > d > 1-k$,

$$\alpha_0^0 = \frac{1}{\pi}\left[\arccos(s) + k^2\arccos(\mu) - d\sqrt{(1-s^2)}\right], \tag{7.16}$$

$$\alpha_1^0 = \frac{2}{3}(1 - \Lambda(\kappa_p, \xi) + \\ + \frac{1}{3\pi\sqrt{dk}}\left\{2(d^2 - 4 + 7k^2)dkE(\kappa_p) - \right. \\ \left. -[d^3k + 5d^2k^2 - d(3+4k-7k^3) + 3(1-k^2)^2]F(\kappa_p)\right\}\Big). \tag{7.17}$$

In these expressions for the partial case we have introduced intermediate quantities $s = (1 + d^2 - k^2)/2d$, $\mu = (d - s)/k$. Again we encounter elliptic integrals, with modulus $\kappa_p = 1/\kappa_a = \sqrt{[1-(d-k)^2]/4dk}$, and, for Heumann's Λ-function, argument $\xi = \arcsin\left[\sqrt{2d/(1+k+d)}\right]$. Another special case occurs if $d = k$, i.e. $\kappa_p = 1/2k$,

$$\alpha_1^0 = \frac{1}{3} + \frac{2}{9\pi k}\left\{(16k^2 - 8)k^2 E(\kappa_p) - [k^2(16k^2 - 10) + \frac{3}{2}]F(\kappa_p)\right\}.$$

(iii) Total case $d \leq k - 1$, $k > 1$,

$$\alpha_0^0 = 1, \tag{7.18}$$

$$\alpha_1^0 = \frac{2}{3}. \tag{7.19}$$

Standard algorithms for the elliptic integrals E and F, in both complete and incomplete forms are relatively easy to find.†

GOODNESS OF FIT : THE χ^2 STATISTIC — In the present, and subsequent, curve-fitting problems we use the χ^2 statistic (7.1) to measure the 'goodness of fit'. This is not the only possibility, but it is reasonably simple to determine, and has received a good deal of attention in statistical literature, so that much useful supporting information is readily available.

χ^2 gives an appropriately weighted sum of squares of residuals between observational data l_{oi} and the theoretical fitting function $l_c(a_j, t_i)$. The variance σ_i^2, in principle, may vary from one point to another, or from one region of a light curve to another; but, in practice, a nominally average value is often inserted, at least initially. This is generally found from separate accuracy determinations for the observations (e.g. the scatter in the comparison – check star data).

The object now becomes the minimization of χ^2, which decreases as the goodness of fit increases. If a true minimum of χ^2 can be found, then 'in a χ^2 sense' the corresponding set of parameter values is optimal. The technique of χ^2 minimization is essentially a programming matter. The details need not concern us unduly — nowadays, fast and reliable optimization routines are widely available from software suppliers. A couple of general points about χ^2 minimization in the given context can be observed.

Firstly, it can be appreciated inituitively that a rapid curvature of the curve formed by the intersection of the χ^2 hypersurface, in the vicinity of its minimum, with the plane corresponding to any particular unknown,

† e.g. Hofsommer, D.J. and Van der Riet R.P. *Numer. Math.*, **5**, 291, 1963.

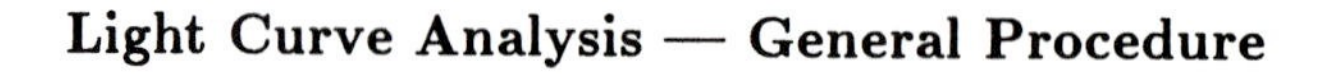

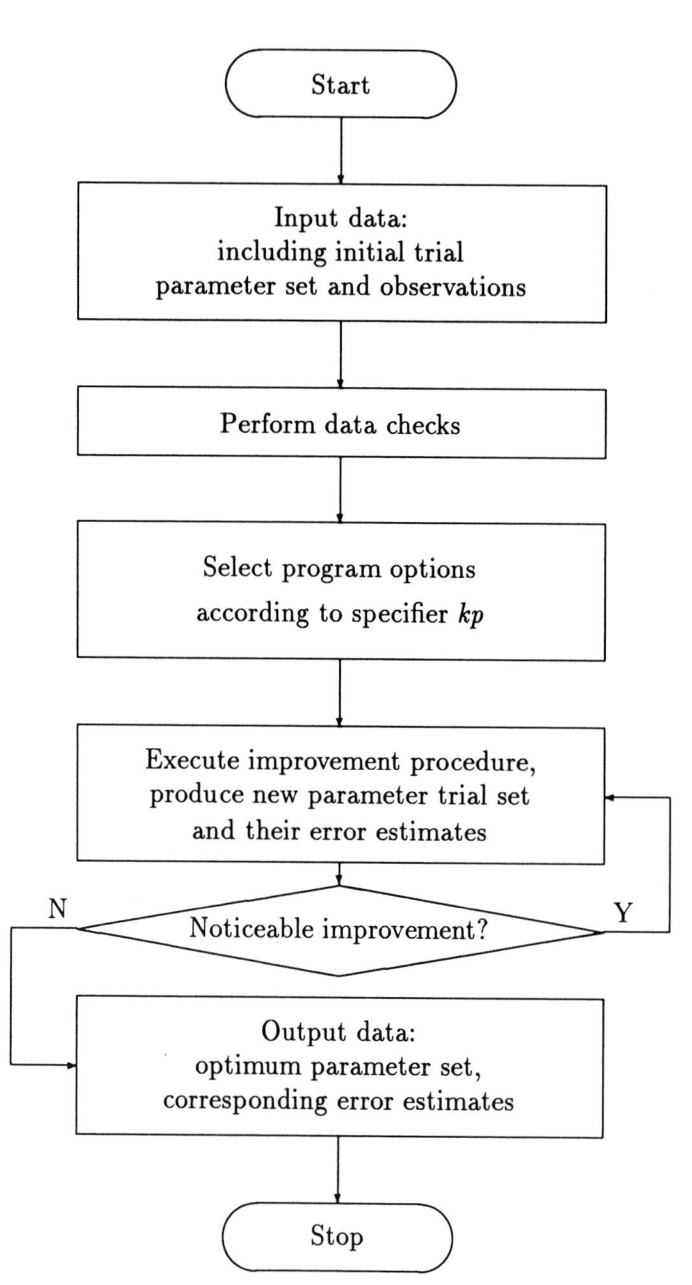

Fig. 7.5. Flowcharts showing the main features of light curve analysis programs.

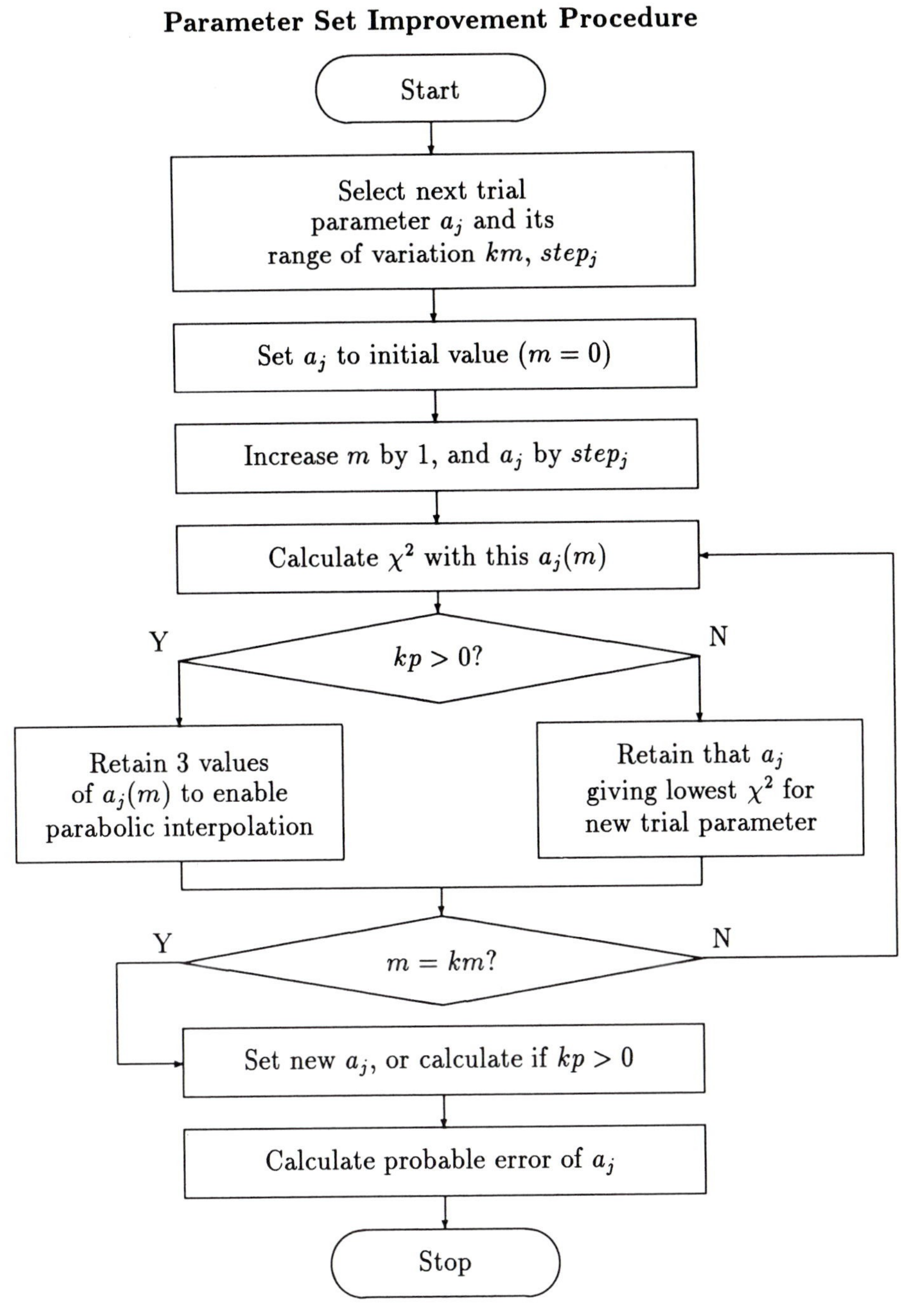

Figure 7.5 (cont.)

locates the value of that unknown with high relative accuracy. Conversely, a shallow curvature means that the parameter in question does not have a large influence on the goodness of fit, and cannot, therefore, be extracted with high precision. It can be shown that the increase of χ^2 by unity in any particular a_j, χ^2 plane corresponds to an uncorrelated assessment of the error of the parameter a_j as the variance Δa_j^2.

Secondly, the overall confidence in the solution may be judged by the resulting optimized χ^2 values. Thus, for example, Kron's 6700 Å light curve of YZ Cas has a total of 42 normal points. If all eight parameters of the model were to be determined it would mean 42 – 8 = 34 'degrees of freedom' (ν) to the data. The obtained minimum χ^2 value divided by ν — the 'reduced χ^2' — may be compared with its probability in the $\chi^2(\nu)/\nu$ distribution, as given in tables such as those of Pearson and Hartley (1954). According to the evaluated probability, interpretations are made about the estimated observational errors, or the adequacy of the underlying model. Summarizing program structures for fitting light curves by χ^2 minimization are indicated by means of the accompanying flow charts (Figure 7.5). The program block where χ^2 is evaluated makes repeated calls to the fitting function at each observed phase value, using each trial set of parameters. The fitting function is contained in a subprogram nested within that for the evaluation of χ^2, and in turn it may contain many subprograms to work out required expressions or integrals.

Calculation of χ^2 is called from within the improvement subprogram (Figure 7.5) carrying out the optimization strategy, i.e. searching, by an organized method, for the set of parameters which locates the minimum χ^2. This optimization subprogram is itself called within an outer program concerned with first reading input information, and then, after the optimization run, presenting output results.

One simple procedure which can be used is that of 'parabolic interpolation'. The idea is based on approximating $\chi^2(a_j)$ to a parabola in the immediate vicinity of its minimum. This allows an estimate for the optimum value of a_j, a_{jmin}, say, to be determined from three separate values of $\chi^2(a_j)$. Thus,

$$a_{jmin} = a_{j3} - \frac{b_j[\frac{1}{2} + \chi^2(a_{j3}) - \chi^2(a_{j2})]}{[\chi^2(a_{j1}) + \chi^2(a_{j3}) - 2\chi^2(a_{j2})]}, \tag{7.20}$$

where b_j is the step size used to increment a_j, i.e. $b_j = a_{j2} - a_{j1} = a_{j3} - a_{j2}$.

A formal error estimate of the determination Δa_j may be similarly calculated at the same time, thus

$$\Delta a_{jmin} = b_j \sqrt{\frac{2}{\chi^2(a_{j1}) + \chi^2(a_{j3}) - 2\chi^2(a_{j2})}}. \quad (7.21)$$

This error formula implies that the value of a_{jmin} is uninfluenced by current values of the other parameters of the set a_k, $k \neq j$. This is normally not the case — but for a reasonably determinate problem the errors derived from (7.21) are of the same order as those calculated by the more complicated procedure which takes such interaction effects into account.

With a provisional estimate of a_{jmin}, one can follow on similarly with the other parameters. The whole calculation proceeds iteratively until a desired smallness of χ^2, and its variation with each iteration, has been achieved; or a specified computation time exhausted. It should be noted that this termination procedure is a practical part of the program, but does not guarantee that *the* minimum χ^2 — the optimum solution — has been located, or even that a single optimum exists (the 'uniqueness problem').

This program outline is very simple and, of course, more refined techniques are possible. As optimization problems go, though, the present example is relatively direct (not many unknowns to be evaluated), so even step-by-step parabolic interpolation usually furnishes well-fitting parameter sets without excessive demands on computer time. Provided only several parameter values are sought, more elaborate procedures will usually confirm that an acceptable approximation to a true minimum of χ^2 is located in this way.

Application to YZ Cas — Computer-based χ^2 minimization has been used to provide an optimal parameter set for the same data on YZ Cas, already discussed. The results of eight parameter optimization fits to the two light curves are given in Table 7.1 and shown graphically in Figures 7.6 and 7.7. The closeness of the radii, inclination and fractional luminosities to the previous simple derivations can be seen.

From the relative smallness of the provisional error estimates it follows that the χ^2 hypersurface sweeps down relatively sharply onto most of the solution parameters, at least when treated in isolation from each other. This must reflect on the high accuracy of the observational data (~0.001 mag). The fitting to the primary minimum in blue is slightly better than the corresponding red light curve, but the net goodness of fit measures for both pairs of minima are quite comparable.

There is a noticeable difference between the derived values of k for the light curve in the two colours, which is greater than the given error estimate. This result was also obtained by Kron, though he did not lay much weight on it. The possibility of some real difference in measured size at the two wavelengths must be immeasurably slight if the two stars' photospheres are

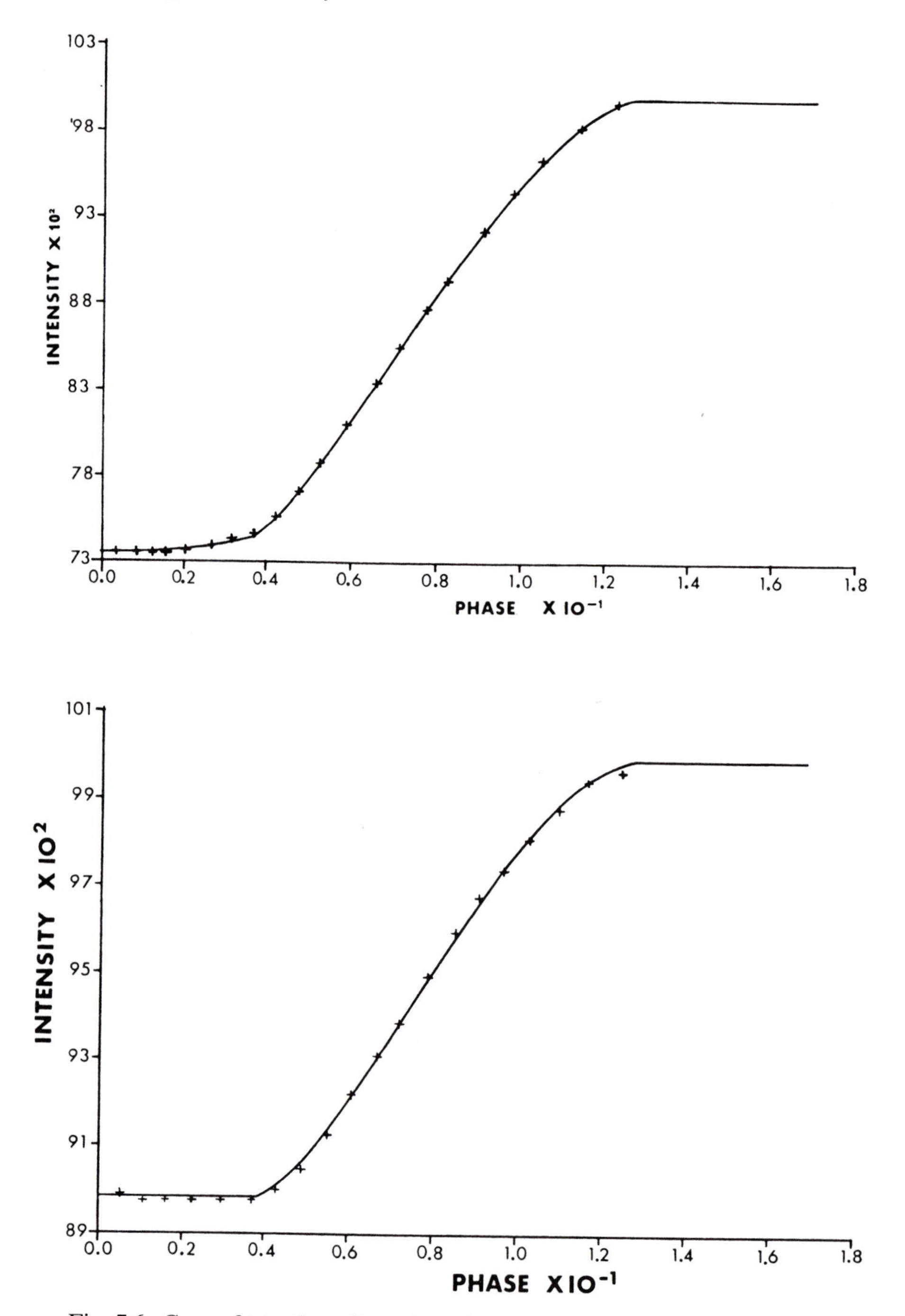

Fig. 7.6. Curve-fit to the eclipse data for YZ Cas (red light).

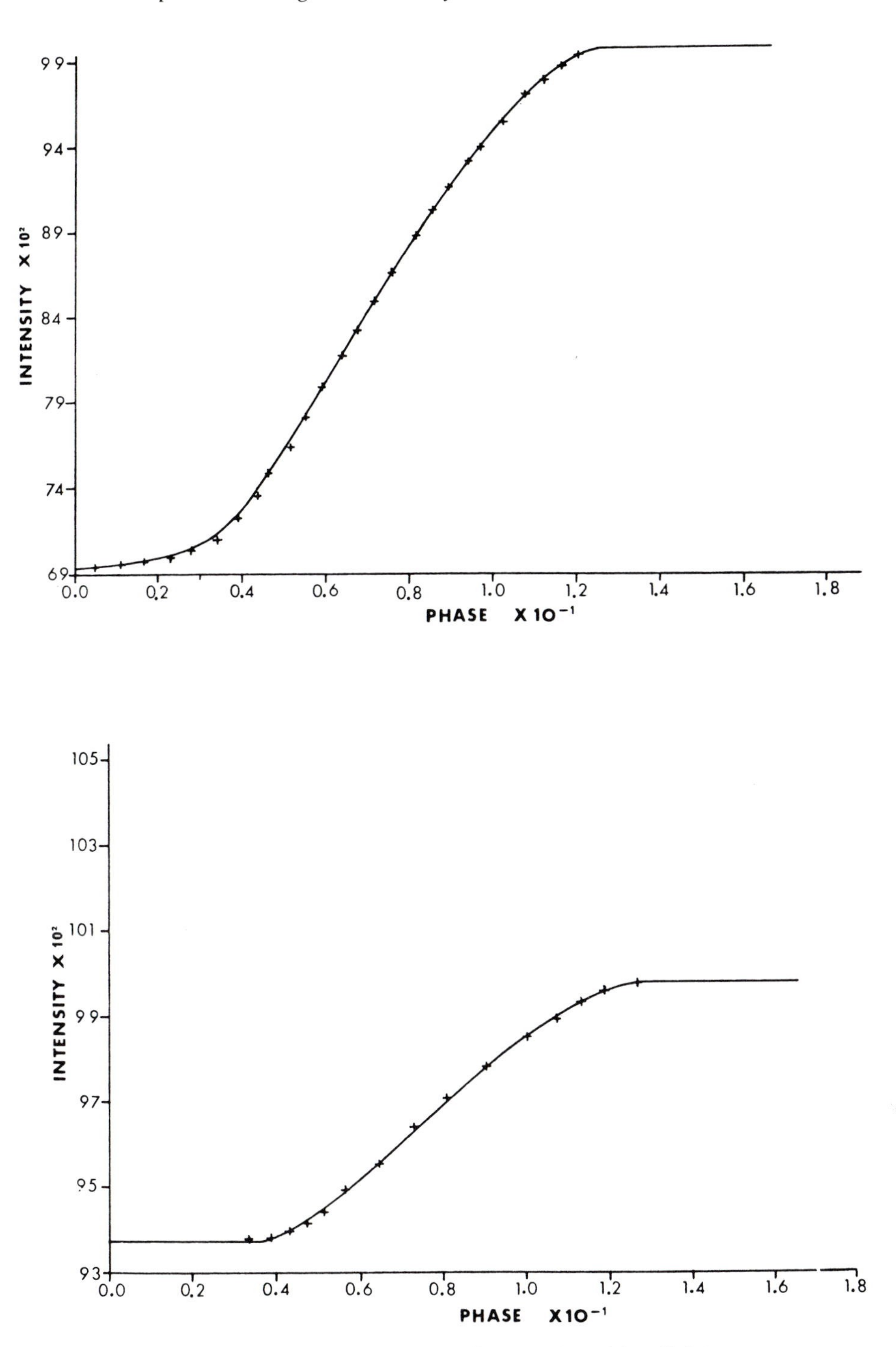

Fig. 7.7. Curve-fit to the eclipse data for YZ Cas (blue light).

Table 7.1. *Optimal parameter sets characterizing YZ Cas.*

Parameter	$\lambda = 670$ nm			$\lambda = 450$ nm		
	Kron's value	Opt. value	Prov. err. estimate	Kron's value	Opt. value	Prov. err. estimate
U	1	0.9988	0.0003	1	0.9983	0.0003
L_g	0.8974	0.8979	0.0009	0.9375	0.9371	0.0003
L_s	0.1026	0.1009	0.0009	0.0625	0.0611	0.0004
r_g	0.1443	0.1448	0.0002	0.1420	0.1448	0.0002
r_s	0.0758	0.0758	0.0002	0.0766	0.0773	0.0003
i	88.17	88.62	0.04	87.95	88.37	0.04
$\Delta\theta_0$	0	0.147	0.007	0	0.153	0.006
u_g	0.329	0.282	0.006	0.491	0.506	0.002
u_s	0.48	0.54	0.08	0.45	0.39	0.14
χ^2	145.2	17.1		188.3	24.9	
σ	0.0013			0.0013		
v	27			35		

anything like those of normal stars. The inclination values also appear rather more different than their quoted errors — about the same for both light curves — would suggest.

Such discrepancies question the simple use of (7.21) for error estimates. In fact, when the *correlated* errors are properly worked out, we find (Section 6.1.1), that the estimates grow with an increase in the number of sought parameters. This is the effect which limits the number of independent parameters which can be determined. Well-fitting parameter sets may be directly derivable from the illustrated procedure, but then care is required not to mistake accuracy of the curve-fit with accuracy of our knowledge of parameter values. Thus, the derived limb-darkening coefficients, except that of the smaller star at 4500 Å, are in reasonable accord with recent theoretical predictions for the types of stars in question (Main Sequence A0 and F5) when the effects of interdependence on error estimation are taken into account, but the errors quoted in Table 7.1 must be far too low.

7.5 Bibliographical notes

The generalities of the initial section recall much from P. R. Bevington's popular *Data Reduction and Error Analysis for the Physical Sciences*, McGraw-Hill Book Co., New York, 1969, though there are a number of more recent texts, in wide distribution, dealing with optimization techniques and curve-fitting problems (e.g. D. R. Adby and M. A. H. Dempster's *Introduction to Optimization Techniques*, Chapman and Hall, London, 1974). Practical computational methods are included in the *Numerical Recipes* of W. H. Press *et al.*, Cambridge University Press, Cambridge, 1986. I. J. D. Craig and J. C. Brown's *Inverse Problems in Astronomy*, Adam Hilger, Bristol and Boston, 1986, and the *Konechno parametricheskie obratnyie zadachi astrofiziki*, (Russian), Izd. Moskov. Univ., Moscow, of A. V. Goncharski, S. Y. Romanov, and A. M. Cherepashchuk, 1991, discuss a general range of data to parameter conversion issues.

With regard to eclipsing binary light curves, whose discussion starts in Section 7.2, there are broad reviews of the subject, in the context of variable stars, in C. Hoffmeister, G. Richter and W. Wenzel's *Variable Stars*, Springer Verlag, Berlin, 1984; and V. P. Tsesevich's *Eclipsing Variable Stars*, John Wiley and Sons, Chichester, 1973, from which Figure 7.2 derives. A. H. Batten's *Binary and Multiple Stars*, Pergamon Press, Oxford, 1973, gives a nice background to physical interpretations, as does J. Sahade and F. B. Wood's *Interacting Binary Stars*, Pergamon Press, Oxford, 1978, which contains a very full bibliography up to the year of publication. A comprehensive set of physical reviews appeared also under the title *Interacting Binary Stars*, ed. J. E. Pringle and R. A. Wade, Cambridge University Press, Cambridge, 1985.

The atlas referred to is that of M. G. Fracastoro, *An Atlas of Light Curves of Eclipsing Binaries*, Oss. Astron. Torino, Turin, 1972. Extensive bibliographic information on over 3500 eclipsing binaries was given in the *Finding List for Observers of Interacting Binary Stars* of F. B. Wood, J. P. Oliver, D. R. Florkowski and R. H. Koch, published by the University of Pennsylvania Press, Philadelphia, *Astron. Series*, **12**, 1980 (fifth edition). A very large amount of factual information on these stars has also been compiled by M. A. Svechnikov, of the Ural State University, whose most recent *Catalogue* (1990, with E. F. Kuznetsova) contains, at least approximate, parameter sets for almost all known examples. Apart from that, Dr Svechnikov also produced a comprehensive review *Catalogue of the Orbital Elements, Masses and Luminosities of Eclipsing Binary Systems* (in Russian) in 1986, published by the University of Irkutsk. Figure 7.3 was adapted from a diagram presented by R. L. Bowers and T. Deeming in their *Astrophysics I*, Jones and Bartlett, Boston, 1984 (their Figure 17.5).

Sections of Z. Kopal's (1959) *Close Binary Systems*, Chapman and Hall, London, 1959, have been updated and rewritten, firstly in *Language of the Stars*, Reidel, Dordrecht, 1979, and more recently in *Mathematical Theory of Stellar Eclipses*, Kluwer Academic Publ., Dordrecht, 1990; though the 1959 book contains most, if not all, of the basic formulae utilized in Section 7.3, as well as full reviews of the classical approaches of H. N. Russell *Astrophys. J.*, **35**, 315, 1912; and H. N. Russell and J. E. Merrill *Contrib. Princeton*

Univ. Obs., No. 26, 1952. J. B. Irwin's article in *Astronomical Techniques* (see Chapter 2 bibliography) p584 is still useful, showing the broad effects of main parameter variation on light curve morphology, for instance.

Section 7.4 is largely drawn from the author's 1973 analysis of YZ Cas in *Astrophys. Space Sci.*, **22**, 87. Kron's original data on YZ Cas were given in *Lick Obs. Bull.* No. 499, 1939, and *Astrophys. J.*, **96**, 173, 1942.

8

Close Binary Systems

In this chapter we proceed to more general effects in close binary light curves, with a wider sample of data sets.

8.1 Orbital eccentricity

Eccentricity affects eclipsing binary light curves in three main ways: (i) Displacement of the secondary mid-minimum, D, with respect to the halfway point between successive primary mid-minima. Let D be counted positive when the secondary occurs *after* the halfway phase. (ii) Unequal duration of the minima. Let S represent the ratio of secondary to primary minimum duration. (iii) Asymmetry of the shapes of the minima. Inspection of the light curve for these effects, particularly the first two, will allow approximate determinations of two additional parameters introduced by orbital eccentricity.

Whilst the meaning of orbital eccentricity e is clear, alternative formulae are possible for the orientation ϖ of the major axis of the elliptic orbit with respect to the line of sight. The relation:

$$\varpi = \phi + 90^\circ - \nu \tag{8.1}$$

is here taken as definitive, where ϖ is called the longitude of periastron and ϕ and ν both measure the orbital position angle of the secondary with respect to the primary star; ν from the periastron position, i.e. ν corresponds to the 'true anomaly'; and ϕ from the 'inferior conjunction' (zero point of the phase scale — see Figure 8.1). For most normal eclipsing binaries, the line of sight inclines at a low angle to the orbital plane, leading to the relationships between ϖ, D and S given in Table 8.1 (strictly valid only when the line of sight lies in the orbital plane).

Table 8.1. *Displacement D and relative duration S of the secondary minimum in an eccentric eclipsing binary system in relation to periastron longitude ϖ.*

	$D > 0$	$D < 0$
$S - 1 > 0$	$0 < \varpi < \frac{1}{2}\pi$	$\frac{1}{2}\pi < \varpi < \pi$
$S - 1 < 0$	$\frac{3}{2}\pi < \varpi < 2\pi$	$\pi < \varpi < \frac{3}{2}\pi$

The signs of D and $S - 1$ in Table 8.1 point to $S - 1$ being related to $e \sin \varpi$ and D to $e \cos \varpi$. Such relations are not difficult to formulate and have been sometimes used to obtain values of e and ϖ. Computers enable us to dispense with simplifying linearizations, however, and allow precise calculation of the effects of eccentricity at a general inclination, even if such effects cannot be made completely explicit.

In the case of circular motion d $(= \delta/r_1)$ is given, from (7.3), by

$$d = \frac{1}{r_1}\sqrt{\sin^2 i \sin^2 \phi + \cos^2 i}. \tag{8.2}$$

Note that in the circular case the argument ϕ in (8.2) is the same as the orbital phase θ in (7.3). Orbital eccentricity affects the formula for d in two ways: (i) r_1 has to be multiplied by a scaling factor, which is easily shown to be $(1 + e \cos v)/(1 - e^2)$ from the theory of elliptic motion,† and (ii) $\sin^2 \phi$ is replaced by $\cos^2(v + \varpi)$ using (8.1). So instead of (8.2) we write

$$d = \frac{(1 - e^2)}{r_1(1 + e \cos v)}\sqrt{\sin^2 i \cos^2(v + \varpi) + \cos^2 i}. \tag{8.3}$$

The true anomaly v is related to light curve's phase θ, which increases linearly with time, by a rather complicated function, involving eccentricity e and 'mean anomaly' $M = \theta + M_0$, where M_0 is the mean anomaly at primary mid-minimum. M, like v, is measured from the direction of periastron to the secondary, as projected from the centre of the primary. We thus write,

$$v = v(e, M) = v(e, \theta + M_0). \tag{8.4}$$

It is convenient, at least internal to the optimizing code, to use M_0, and not ϖ, together with the eccentricity e, for the two additional parameters characterizing an eccentric eclipsing binary system. For the usually quoted

† See e.g. W.M. Smart's *Textbook on Spherical Astronomy*, Cambridge University Press, Cambridge, 1977.

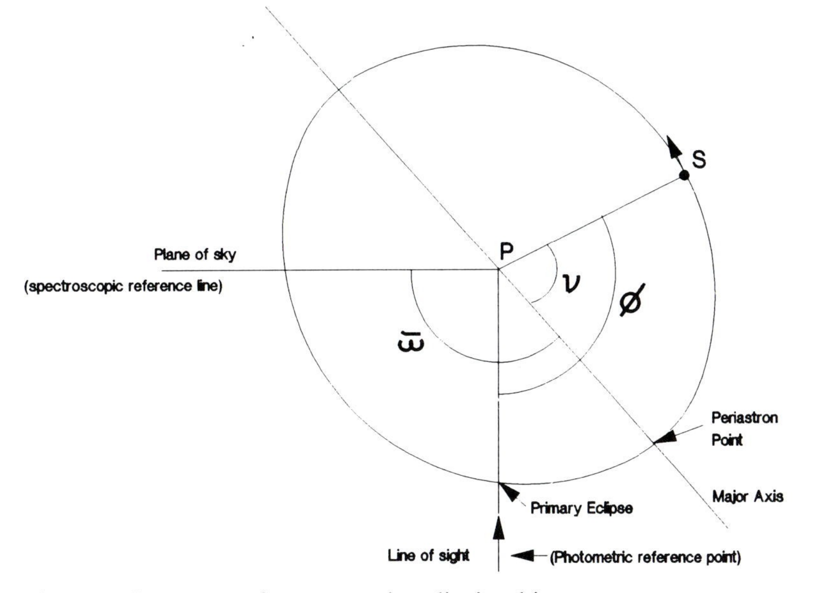

Fig. 8.1. Geometry of an eccentric eclipsing binary.

quantity ϖ, from (8.1) it follows that

$$\varpi = 90^\circ - v(e, M_0), \tag{8.5}$$

since $\phi = 0$ when $\theta = 0$, at the point where $M = M_0$.

Equations (8.4) and (8.5), which involve only the two parameters e and M_0, are substituted into (8.3) to determine the quantity d, used in the evaluations of the light variation (7.11). A preliminary estimate for M_0 is deduced, given a quoted value of ϖ, from (8.5), as $M_0 \approx 90^\circ - \varpi$. The equation which relates true to mean anomaly, (8.4), may be computed by standard means, such as the iterative use of Kepler's equation. As a check on calculations, note that

$$\mid v(e, \theta_s + M_0) - v(e, M_0) \mid \approx 180^\circ, \tag{8.6}$$

where θ_s is the observed phase of the secondary minimum. Figure 8.2 shows the application of such procedures to a real, eccentric close binary system.

8.2 Proximity effects

It can be shown that when two stars are sufficiently close to each other that the sum of their two mean radii, expressed as fractions of the separation of the two mass centres, becomes greater than about 0.75, the pull from sphericity becomes so great as to draw the two photospheric surfaces into

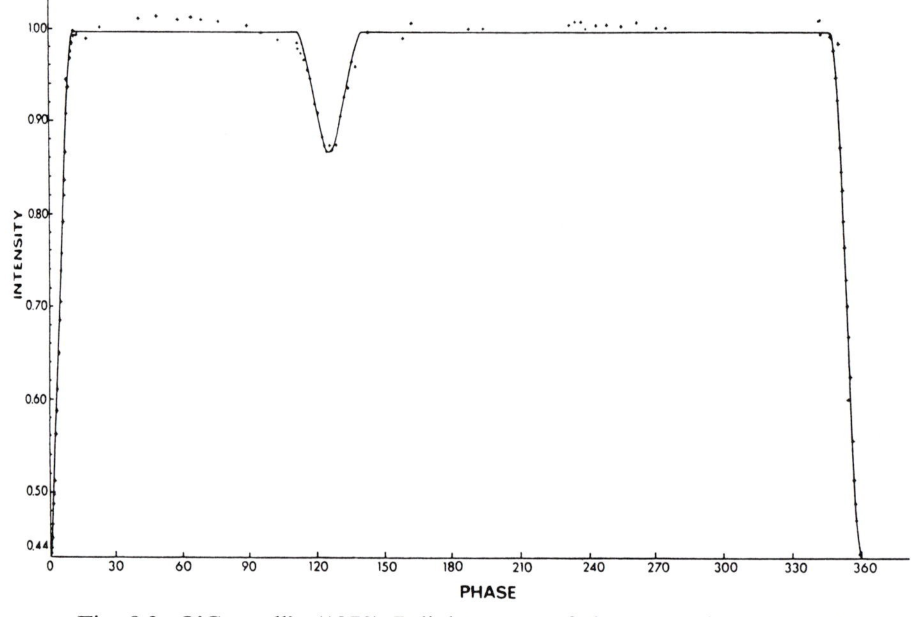

Fig. 8.2. O'Connell's (1970) B light curve of the eccentric eclipsing binary system V477 Cygni.

contact. The majority of observed close binaries, however, are composed of separate bodies, whose departures from spheres — 'ellipticity effect' — together with a radiative interaction — 'reflection effect' — introduce the photometric proximity effects. These often can be conveniently dealt with by a linearized approach, in which they are regarded as relatively small and additive. The light variation is modelled in a fairly direct way, so long as we include only the separable basic contributions of the whole star on its companion; i.e. drop the higher order effects of the tide on one star affecting that of the other. Such higher order interactions start with terms comparable to the square of the principle distortions, and at most would generally be of the order of a few per cent. The underlying theoretical description (e.g. the 'Roche' model) may have physical inadequacies at this level of accuracy.

The basic idea for the linearized approach can be expressed as follows:

$$\text{Observed light curve} \equiv \text{'spherical model' light curve} + \text{'distortions'}.$$

The spherical model light curve is that of the eight-parameter model described in Chapter 7, which is readily extended by another three extra parameters: two (e and M_0) to cover the case of elliptic motion, and one more in case of a separate 'third light' (L_3) contribution. In this case the constraint $U = L_1 + L_2$ is relaxed, L_2 is treated as a separate parameter, and then $L_3 = U - (L_1 + L_2)$ (> 0).

The variation of light due to proximity effects for each star is written in the form

$$\Delta\mathscr{L}_i = \{\Delta\mathscr{L} - \Delta^{(\alpha)}\mathscr{L}\}_{ell}(\tau_i, q_i; a_j)L_i + \{\Delta\mathscr{L} - \Delta^{(\alpha)}\mathscr{L}\}_{ref}(E_i; a_j)L_{3-i}$$
$$i = 1,2;\ j = 1,11 \tag{8.7}$$

(where the mass ratio $q \equiv q_1 = m_2/m_1 = 1/q_2$). The expression $\{\Delta\mathscr{L} - \Delta^{(\alpha)}\mathscr{L}\}$ represents an appropriate integration, with the eclipsed part of the distortion $\Delta^{(\alpha)}\mathscr{L}$, zero for all phases at which the star is uneclipsed, subtracted out during eclipses. The paramater dependence of these integrations is indicated by the contents of the regular parentheses. In addition to the eleven parameters a_j used in the generalized spherical model, five new ones appear — namely: q, $\tau_{1,2}$, and $E_{1,2}$ — making up to 16 the final number of independently specifiable parameters for this 'standard eclipsing binary' model. Since in most cases the orbit is essentially circular, only 14 of them usually require specification.

The full treatment of proximity effects, even in the linearized form here discussed, would take us beyond the scope of the present text, but relevant references are given in the later bibliography. Nowadays many investigators are prepared to accept and apply already written and reasonably tested software that others have made available, especially where it is possible to check details of the coding against source formulae. Programs such as Wood's WINK, and the Wilson and Devinney (W-D) numerical light curve synthesizer are often cited in this context. The program FIT, documented in some detail by Zeilik *et al.*,† derives from programming the explicit form of (8.7), and is more particularly relevant to the details of the following discussion.

Concerning the five new parameters, they could be treated as empirically determinable unknowns. However, in general, observations are not precise enough to resolve each of the parameters independently, particularly when correlation effects are present, i.e. photometric effects resulting from variation of a certain parameter can be closely simulated by suitable variations of others. The analytic form of the fitting function suggests that theory can provide guidelines, or actual values, for some of the pertinent quantities. Also, separate evidence is often available, such as from multi-wavelength photometry, enabling component temperatures to be assessed; or from spectroscopic or polarimetric data. Spectroscopy often yields a value for the mass ratio q, while polarimetric data may enable an independent estimate of the inclination i.

† *Operator's Manual for Light Curve and Spot Fitting Programs*, M. Zeilik, E. Budding, M. Rhodes, D. Cox, Internal Report of Institute for Astrophysics, University of New Mexico, Albuquerque, 1989.

The 'gravity-brightening' coefficients, τ_i, come from classical theory of a star's outer envelope. If the flux through these layers depends only on the temperature gradient (the 'diffusion' approximation), and the latter is constant along equipotential surfaces (the 'Clairaut' stability criterion), then the flux will be inversely proportional to the relative separation of these surfaces. But gravity itself has such an inverse proportionality. Hence, we deduce that the flux is proportional to gravity ('von Zeipel' law). It is well known that for a physically distorted star this line of reasoning contains a paradox, in that it leads to a situation where the temperature at, say, the pole of a rotating star ought to be the same as that at the equator (on the same equipotential), though the heating flux in these two directions is different (i.e. more heating at the pole). More detailed arguments suggest that a restorative circulation sets in to bring down the horizontal temperature differential, while the difference in gravity-dependent flux remains in place. In practice, the von Zeipel approximation appears to lead to a reasonable accord with observations, at least for those stars whose envelopes are dominated by a flux essentially propagated by radiative transport. For convective heat transport through the subsurface layers the position is less clear, both theoretically and observationally.

The reflection coefficients E_i can be derived by assuming that the incident flux from the companion leads to a local increase of temperature, such that the bolometric emission at the new temperature is equal to the sum of the original plus received fluxes. The local increase of flux at a particular wavelength is then approximated by multiplying the temperature derivative of the black body expression (3.8) by the calculated local temperature increase. This is called the black body approximation, and is usually adequate to represent the observed scale of radiative interactions in normal close binaries to available accuracies. It becomes inaccurate when there are very large differences in the components' temperatures, e.g. the optical light curves of X-ray binaries containing cool 'red' and hot subdwarf components.

An alternative, and even simpler, scaling factor (i.e. unity) occurs for the 'pure scattering' approximation, where only the incident flux integration is involved in calculating the reflected light. Anisotropy of the reradiation, i.e. limb darkening of the reflected contribution, has usually been neglected in eclipsing binary light curve contexts, though sometimes an additional, quasi-empirical albedo factor is introduced to scale a derived E_i value toward something more in line with observed data.

From the scale of manifest effects, it appears that *primary* gravity brightening τ_1 and *secondary* reflection E_2 would usually be better determined than the corresponding quantities τ_2 and E_1. How well observations support theory for these parameters has been the subject of many discussions. For trial purposes, the diffusion law model for gravity brightening (i.e. flux $\propto$ temperature gradient, and temperature constant on gravitation equipotentials) in the black body approximation may be followed, while a similar black body simplification is supported, in many cases, for the reflection coefficients. The appropriate formulae at wavelength λ, and primary and secondary unperturbed surface temperatures T_1 and T_2 then reduce to:

$$\tau_{1,2} = \frac{c_2}{4\lambda T_{1,2}[1 - \exp(-c_2/\lambda T_{1,2})]}, \tag{8.8}$$

and

$$E_{1,2} = \tau_{1,2}\left(\frac{T_{2,1}}{T_{1,2}}\right)^4 \frac{\exp(c_2/\lambda T_{2,1}) - 1}{\exp(c_2/\lambda T_{1,2}) - 1}. \tag{8.9}$$

A significant computing economy is achieved by the use of analytic formulae to evaluate the light distortions. Six basic integrals are required: namely, α_0^0, α_1^0, and four related, simpler expressions, usually denoted as $I^0_{-1,i}$, $i = 0 \ldots 3$; the rest are determined by recursion formulae. The six involve only elementary algebraic expressions or elliptic integrals. The latter can be evaluated at a comparable speed to the former ($\sim 10^{-5}$ s), by the use of fast and accurate modern algorithms. A typical ten-iteration five-parameter optimization with $\sim$100 data points, involving numerical evaluation of the curvature Hessian at optimum, its inversion and the calculation of its eigenvalues and eigenvectors can be achieved in less than an hour of computing time on a 'AT' type computer equipped with a fast arithmetic co-processor. A comparable amount of time would be used in just one single 5000 mesh-point numerically integrated light curve.

8.3 A 16-parameter curve fitter

Once the fitting function for the standard eclipsing binary model has been constructed it may be set in a suitable programming environment where a set of optimal parameters are derived, in an essentially parallel way to that discussed for the spherical model and Kron's light curves of YZ Cas. A paramount question is how many parameter values can we determine. If this is not the full sixteen, then we need to supply the underived ones, from reasonable theory or otherwise.

If one attempts to derive all sixteen parameters of this model from a typical eclipsing binary light curve one finds it generally not so difficult to produce a very good fit, provided one starts with a set of parameters near enough to a deep minimum of χ^2. If this is not the case, then, as the improvement sequence iterates, one frequently sees the set wandering off into a realm of parameter space where values become physically unmeaningful; perhaps very large limb-darkening coefficients, or negative luminosities. Such effects can happen even with a near minimum χ^2 starting set — the solution proceeds down a 'level valley' in the χ^2-parameters manifold, with a slight change in one parameter being matched at each step by slight changes in another, or a group of others, χ^2 being marginally decreased as the correlated parameters drift along, away from the domain of physical feasibility. The corresponding curve-fit may well look good, but the derived values can contain some embarrassing surprises.

The real test, for this context, comes from inverting the 'curvature Hessian' of the n parameters a_i — the square symmetric matrix formed by the elements $\partial^2\chi^2/\partial a_i \partial a_j (i, j = 1, ...n)$ — in the vicinity of the optimum, to obtain the corresponding correlated error matrix. A good explanation of this procedure is given in Bevington's (1969) book.† Almost inevitably, with a sixteen by sixteen matrix (apart from its time consuming derivation) using typical photometric data sets for standard close binaries, we would encounter negative elements on the leading diagonal of the error matrix, implying that the minimum is not properly defined; i.e. the contours of χ^2 in the parameter space do not close down to a point. It is not a strict minimum, because in its vicinity there is a level region, where, to available accuracies, slight changes of one group of parameters, which would increase χ^2, can be matched by an appropriate combination of others bringing it back down to numerically the same value again.

A guide to the number of independent parameters, which a given data set of N points can determine, comes from considering what the number of coefficients n would be if the data set was optimally modelled by a set of orthogonal functions, e.g. a Fourier series. The unrealistic wobbles of the fitting function with insignificant decreases in χ^2 which come when too many terms are included in such a fit are well known in empirical curve-fitting contexts. Since the coefficients of the orthogonal functions, when determined from uniformly distributed data sets across the full range of phases (0 to 1) are all independent of each other, the number n required to minimize $\chi^2/(N-n)$ specifies, in some sense, the information content of the data.

In the general curve-fitting problem the coefficients of the set of orthogonal functions have a counterpart in the eigenvalues of the curvature Hessian, evaluated in the vicinity of the optimum. These relate to appropriate linear

† *Data Reduction and Error Analysis for the Physical Sciences*, McGraw-Hill Book Company, New York, 1969.

combinations of correlated parameter variations whose effects are mutually orthogonal, i.e. numerically independent, in the curve-fitting. Positive eigenvalues locate, via their corresponding eigenvector set, the χ^2 minimizing principal directions; the largest eigenvalue corresponding to the axis of highest determinacy. In practical curve-fitting problems a finite number n_{max} of such positive eigenvalues will be found. This number corresponds to the information content or limit of the data for unique optimal parameter estimation: by increasing the number of parameters beyond this limit the correspondingly extended curvature Hessian is found to contain a negative eigenvalue. Its eigenvectors locate the direction of the local level region which has disrupted the determinacy of a uniquely optimizing parameter set.

A complete presentation of the results of a curve-fitting analysis should therefore include information on this optimal curvature Hessian, as well as the optimal parameter set itself. Of particular interest are the leading diagonal elements of the inverse of the curvature Hessian — the 'error matrix'. The square roots of these elements (multiplied by 2) measure directly the *correlated* error estimates of the parameter set. The Hessian's eigenvalue and eigenvector set indicate, by their orientations with respect to the parameter axes, which parameters, or parameter combinations, may be well determined from the data. The numerical operations on matrices required to perform such tasks can be readily obtained from computer algorithm distributors — e.g. the Numerical Algorithm Group (NAG).

APPLICATION TO VV ORI — The fitting function just described has been applied to a set of six ultra-violet light curves of the relatively bright, early type EB system VV Orionis, published by Eaton in 1975. The results are shown diagrammatically in Figure 8.3, corresponding to the parameter sets given in Table 8.2. The system was reviewed in some detail by Duerbeck in 1975, and the solution given in Table 8.2 subtantiates the generally accepted picture of a pair of relatively massive, unevolved B stars which are members of the Orion's Belt Young Star Association. A third, physically related, but lower mass A type star also contributes slightly to the light L_3 of the system. A constant estimate for the measurement standard deviation Δl was adopted for all except the 1550 Å light curve. The flux ratio J_1/J_2 can be used, in combination with standard model atmosphere predictions for the emergent flux of early type stars, to estimate the secondary effective temperature for an adopted primary value based on a good spectral type classification for this star. The absolute parameters which come from combining the solutions

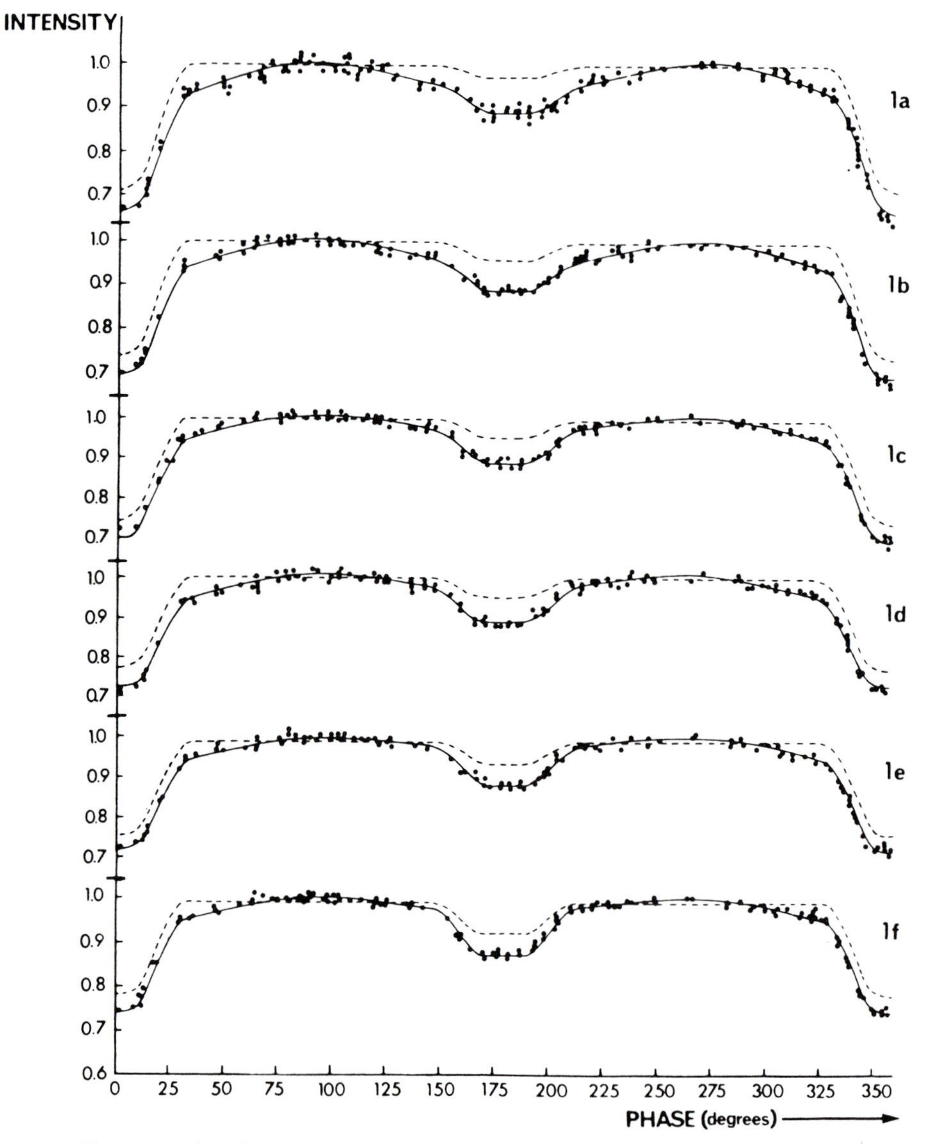

Fig. 8.3. The six ultra-violet light curves of Eaton (1975), and matching theoretical light curves corresponding to the parameters of Table 8.2. The wavelengths are as follows (Å): 1550, 1920, 2460, 2980, 3320 and 4250. The dashed curves indicate the underlying 'spherical model' onto which proximity effects are added.

of Table 8.2 with the spectroscopic data given by Duerbeck are presented in Table 8.3.

The points to be made here are that the solutions of Table 8.2 are the results of *five*-parameter simultaneous optimization runs. The six major

Table 8.2. *5 parameter curve-fits to the ultra-violet light curves of VV Ori.*

Parameter	Wavelength (Å)					
	1550	1920	2460	2980	3320	4250
L_1	0.975 ± 0.003	0.963 ± 0.002	0.954 ± 0.002	0.949 ± 0.002	0.943 ± 0.002	0.926 ± 0.002
L_2	0.025 0.003	0.036 0.002	0.045 0.002	0.049 0.002	0.055 0.002	0.069 0.002
L_3	0.000	0.000	0.001	0.002	0.002	0.005
r_1	0.367 0.004	0.369 0.003	0.370 0.003	0.373 0.003	0.375 0.003	0.371 0.003
r_2	0.178 0.005	0.174 0.004	0.173 0.004	0.176 0.004	0.177 0.004	0.168 0.005
i	88°.5 3°.9	89.1 2°.7	88°.4 1°.4	87°.3 1°.3	85°.7 1°.2	85°.8 1°.8
u_1	0.65	0.58	0.50	0.40	0.36	0.37
u_2	0.8	0.76	0.63	0.46	0.40	0.45
τ_1	0.95	0.79	0.64	0.56	0.52	0.46
τ_2	1.53	1.24	0.98	0.82	0.70	0.62
E_1	1.45	0.78	0.43	0.30	0.29	0.18
E_2	1.00	1.26	1.47	1.56	1.58	1.59
χ^2	287.5	201.0	153.1	231.6	211.6	170.0
N	274	207	182	204	179	180
Δl	0.01	0.007	0.007	0.007	0.007	0.007
J_1/J_2	8.52	5.85	4.52	4.22	3.73	2.91

Table 8.3. *Absolute parameters (mass, radius, temperature, luminosity) for the VV Ori binary components (solar units).*

	Primary	Secondary
M	7.6	3.4
R	4.51	2.13
T	25000	15200
L	7100	220

parameters of an eclipsing binary light curve were specified in Section 7.3. One of these, the correction to the nominal zero phase position to best fit the minima, is relatively independent of all the others, especially for circular orbits (which can be anticipated in close, strongly tidally interacting binaries, such as VV Ori) and uniform coverage of the light curve. This was fixed in early approximate trials, (at 0.25°) and dropped from later curve-fitting experiments, when its value remained more or less constant. The data-fitting could determine simultaneously a sixth parameter (primary gravity brightening — found to be not significantly different from its previously assigned theoretical value), but not a seventh (mass ratio). Hence, an unambiguous photometric confirmation of the spectroscopic mass ratio supplied by Duerbeck, together with a simultaneous determination of the main geometric parameters and the gravity brightening, is not possible for single light curves of this set. Increasing the parameter set from five to six caused significant increases in the error estimates. Quoted error estimates are thus, in a sense, always lower limits. They imply complete certainty about the values of those parameters which are adopted, rather than determined from the data.

8.4 Frequency domain analysis

A series of papers appeared after 1975 developing an idea, initiated by Kopal in the preceding decade, for expressing the information content of a light curve via frequency domain decomposition. In an idealized, proximity-effect-free, total eclipse minimum, strictly located with respect to axes of phase and intensity (i.e. $\Delta\theta_0 = 0$, $U = 1$), and with a given limb-darkening coefficient (u_1) of the eclipsed star, four integrals A_{2m}, given as

$$A_{2m} = \int_0^{\frac{\pi}{2}} (1 - l) d \sin^{2m} \theta, (m = 0, ...3),$$

are necessary and sufficient to determine the four remaining parameters of the spherical problem — r_1, r_2, i and L_1 (the other limb-darkening parameter u_2 is only involved in the other eclipse minimum). Indeed, the connection between the integrals and the parameters is explicit and fairly direct. This was the starting point.

Two other considerations promoted interest in this approach: (i) the idea of *filtering out* proximity effects in the general light curve problem, and (ii) that the integration process smoothes out the scatter of data points, so that the resulting parameter values should be stable against such noise (though they are not necessarily optimal, in the χ^2 minimizing sense). The proximity effects would have their more sizable components at low multiples of the orbital frequency, whereas noise is associated with the high-frequency end of the spectrum, i.e. at frequencies $\sim 1/\Delta\varphi$, where $\Delta\varphi$ is the mean spacing of the data points in phase.

In the general problem, where proximity effects are present, the approach has been to obtain 'empirical moments' A_{2m}, as given above, from numerical quadrature of the observational data. From these moments proximity effect integrals are subtracted to leave residues which can be related to the sought parameters along the lines of the spherical problem. Determination of the proximity integrals has usually involved representing the light curve in the regions outside eclipse minima as a series in powers of $\cos\theta$. The required integrals then become relatively simple expressions in the coefficients c_j of this series.

The plan is thus straightforward, with the forementioned points about separability and noise removal arguing in its favour. Certain complications have been encountered in practice, however, so a detailed exposition of a general method cannot be concisely spelled out here.

One particular source of awkwardness is in the predominating effect of the proximity integrals, because the $\sin^{2m}\theta$ weighting couples with the proximity effects carrying through all phases, unlike the eclipse effects which are confined to relatively low phases. Hence, the proximity effect representation has to be very precise, and though the $\cos^n\theta$ form has a nice algebraic tractability, the corresponding c_js are not well suited to accurate numerical derivation. Alternative strategies have been developed to deal with this, but, even so, the values of the eclipse residues, and therefore the sought parameters, remain sensitive to the proximity contributions, particularly with the higher $2m$ forms. We need not use integrals higher than A_4 to evaluate the basic geometric parameters if moments of both minima can be combined, and, in any case, the use of both minima provides an information advantage, allowing solution consistency to be checked.

Also, the direct connection between eclipse moments and parameters applies really only for total eclipses. For the annular case similar approximate expressions can be given, but the procedure definitely loses simplicity when the eclipses remain partial.

During the course of frequency domain investigations various interesting matters have come to light. A striking example is the uniform expression for the generalized eclipse function when expressed as a Hankel transform of the product of two Bessel functions. One of these latter functions arises from the two-dimensional Fourier transform of the light distribution over the 'aperture' of the eclipsed star, the other is the equivalent transform for the opacity of the eclipsing star. A great systematization of notation for a wide variety of eclipse types is thereby acheived. For further information on this, as well as detailed treatment of a variety of other procedures relating to the photometric effects of close binary systems in the frequency domain the reader is referred to Kopal's *Language of the Stars* (see Section 8.6).

8.5 Narrowband photometry of binaries

Eclipses offer significant opportunitites for sensitive narrowband probing of spectral line effects in stellar atmospheres (cf. Section 4.4). The eclipsing star, acting as a kind of screen, allows, for example, any difference in the centre to limb variation between a line and its surrounding continuum to be monitored. This is particularly effective in cases of Algol type systems showing total, or deep partial, eclipses, where one star, though of comparable size to the other, is of very different luminosity. Differential effects in the atmosphere of the brighter star then become detectable.

Algol systems represent a basic, and relatively well-understood, stage in *interactive* binary evolution (Section 7.2). There is good evidence that they are accompanied by sizable amounts of circumbinary matter — in the form of low density plasmas, sometimes associated with gaseous streams, with material being transferred from one star (the 'loser') to the other (the 'gainer'), or matter being expelled entirely from the system. Narrowband photometry may also discern particular effects in this context.

We shall explore some curve-fitting and relevant parameter estimation methods for narrowband data analysis, concentrating on phases in and near eclipses for Algol type binaries. For convenience, we continue with the notation of the frequently used β photometry.

For most Algols Doppler shifts associated with the spatial motion of the system as a whole — the 'γ-velocity' — are typically of order 1 Å or less at H_β. This is an order of magnitude less than the half-width of the narrowband H_β filter, so that the effects of such displacements, which are present even close to the eclipse phases when orbital radial velocities are minimal, can be neglected. Rotational Doppler-spread effects, which do not alter the equivalent width of line features in normal cases, can be similarly neglected in the major features of β-index light curves of these binaries.

Since the measured β-index corresponds to the difference of a pair of magnitudes in filters, of which the wider one is still only of intermediate bandwidth, we can deduce (Section 4.3) that different local systems can be related to a standard system by a linear formula such as

$$\beta_{standard} = a\beta_{local} + b, \tag{8.10}$$

where a and b are constants, determined from the measurements of standard stars in the local system.

The linear proportionality of the β-index to the equivalent width w_β (4.24) holds true for a spectrum made up of two or more components, where the net equivalent width still corresponds to (4.22). This $\bar{w}_\beta$ of the composite spectrum can be related to the equivalent widths $w_{\beta 1}$, $w_{\beta 2}$, $w_{\beta e}$ of individual spectral components by the formula

$$\bar{w}_\beta = \frac{L_{c1}w_{\beta 1}\{1 - [p\alpha_c - (p-1)\alpha_\beta]\} + L_{c2}w_{\beta 2} - w_{\beta e}(1-\alpha_e)}{L_{c1}(1-\alpha_c) + L_{c2} + L_{c3}}, \tag{8.11}$$

where L_{c1}, L_{c2}, L_{c3} represent fractional continuum luminosities of primary, secondary and circumstellar sources, respectively ($L_{c1} + L_{c2} + L_{c3} = 1$). The light loss functions α_c, α_β are of the same form as specified by (7.11), though α_β requires a quantity $\tilde{u}_\beta$ representing a linear limb-darkening coefficient for a spectral region centred on the line centre, whose effective extent is $pw_{\beta 1}$. p can be defined as $L_{c1}/(L_{c1} - L_{\beta 1})$, where $L_{\beta 1}/L_{c1}$ is the fractional depth of the line centre. In this way, we regard the line as equivalent to a separate little spectral region, having its own luminosity $L_{\beta 1}$ and limb darkening u_β. A typical value for p for early type primaries of Algols is 2.5. In any case, p and u_β are not independent. If p is large the line must be shallow, i.e. not so distinct from the surrounding continuum, so the limb-darkening coefficients u_c, u_β, must then tend to coincide in value.

The third component of the numerator represents an emission contribution of equivalent width $w_{\beta e}$, due to circumstellar plasma. Our methodology implies that we spell out appropriate forms for the light loss α_e for suitable models, using the data to constrain parameter values. The relative complexity of such models is measured by the number of parameters needed in the description, but the determinacy of this set depends on the quality of the available data. In practice, we should expect derivable parameter sets to

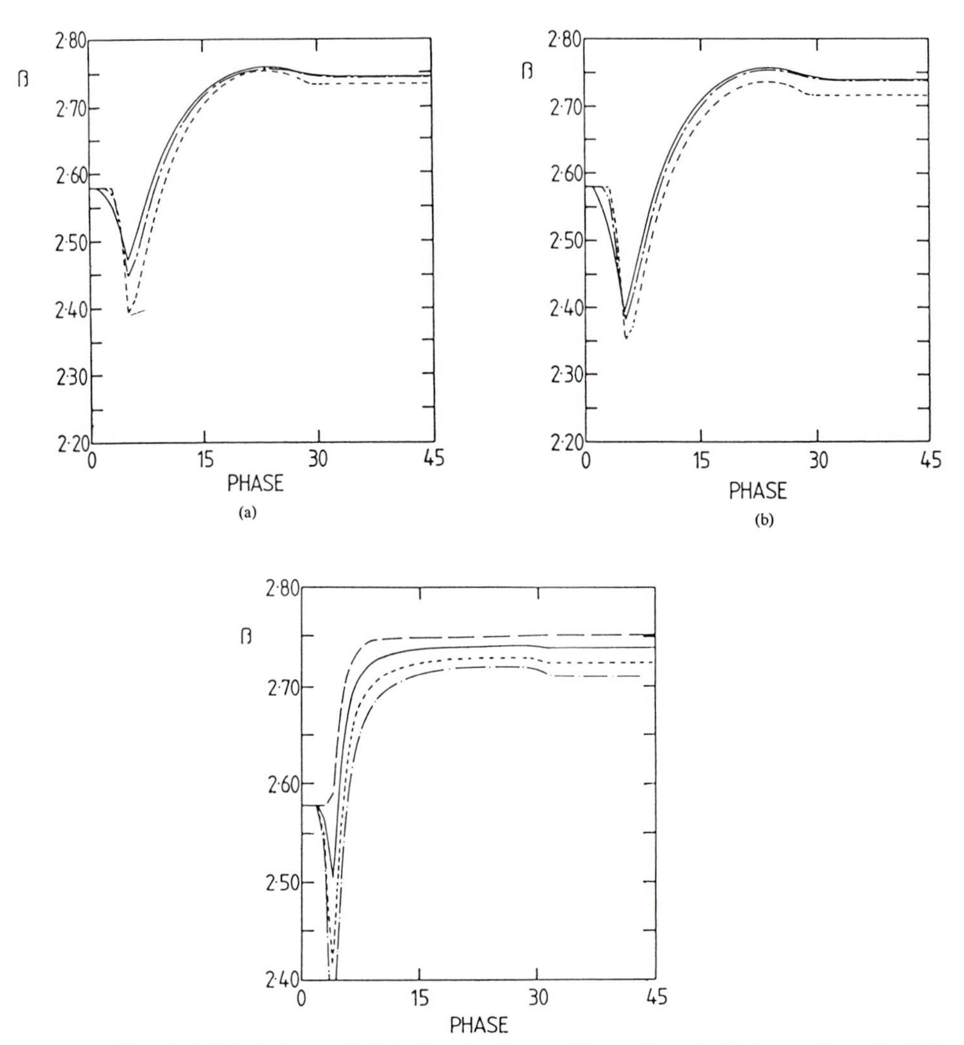

Fig. 8.4. Variation of β-index during eclipses for shell (upper) and disk (lower) type emitting sources, with various values of the four fitting parameters of Table 8.4. (See end of Section 8.6 for more details.)

correspond to fairly simple models. Effects around eclipses, observable with β-photometry, corresponding to uniform shell and disk models are indicated in Figure 8.4. The disk model is examined in what follows, and applied to β-photometry of the interactive Algol systems U Cep and U Sge.

DISK SOURCE — In this idealization the emission is associated with a uniform, optically thin disk surrounding the primary equator of extension h in units of the primary radius and fractional thickness t ($t \ll 1$); t can be taken to have a value of $\sim v_{sound}/\Omega$, where Ω is the orbital angular velocity. This

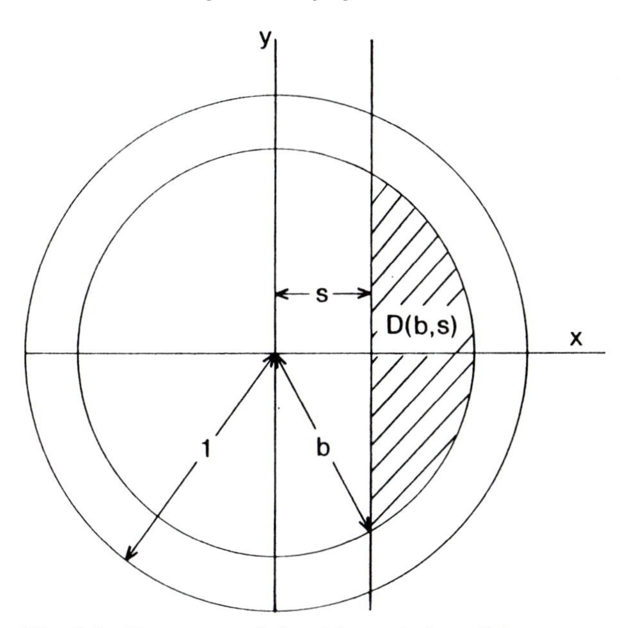

Fig. 8.5. Geometry of the thin emission disk.

kind of thin accretion structure has been theoretically anticipated in the interactive mass-transferring stage of Algol evolution. Its eclipse can be analysed with α-integrals. We will use the uniform intensity α_0^0, and express it via the intermediate variable s (7.15). In the present case s represents the projected distance between the shadow cylinder and the centre of the disk, in units of the disk radius (Figure 8.5). It is generally necessary to subtract the area $D(b, s)$ from the underlying star in this treatment, but in the range $1 > s > 1/(1 + h)$ only the unocculted edge of the disk is eclipsed.

When the disk is eclipsed by a larger object, of lateral extent k, we have α_e as,

$$\left.\begin{array}{lcll}
\alpha_e & = & 0 & (d \geq 1 + h + k) \\
\alpha_e & = & \alpha(s)/c & (1 + h + k > d \geq 1 + k) \\
\alpha_e & = & [\alpha(s) + \alpha(s_1) - \alpha(s')/(1 + h)^2]/2c & (1 + k > d > k - 1) \\
\alpha_e & = & 1 - (1 - \alpha(s))/c & (k - 1 \geq d > k - 1 - h) \\
\alpha_e & = & 1 & (k - 1 - h \geq d),
\end{array}\right\} \quad (8.12)$$

where the 'straight edge' α-function $\alpha(s)$ is given by

$$\alpha(s) = \frac{1}{\pi}\left(\arccos s - s\sqrt{1 - s^2}\right),$$

and the other quantities are as follows: $d = \sqrt{\sin^2\phi \sin^2 i + \cos^2 i}/r_1$, $k = r_2/r_1$, $s' = d - k$, $s_1 = 1/(1+h)$, so that $s = s's_1$, and $c = \alpha(s_1) + \frac{1}{2}(h^2 + 2h)s_1^2$, for primary and secondary relative radii r_1, r_2, phase ϕ and orbital inclination i. If i is significantly different from 90°, the foregoing formulae are still usable, but k is replaced by $k' = \sqrt{r_2^2 - \cos^2 i}/r_1$.

The same formulae also can be used for the 'transit' eclipse, when the eclipsing object is smaller than the lateral extent of the disk. In this case, to account for the reappearing part of the disk, we introduce $d' = d + 2k'$ and then

$$\alpha_{e,tr}(d) = \alpha_{e,oc}(d) - \alpha_{e,oc}(d'). \tag{8.13}$$

APPLICATION TO U CEP AND U SGE — The classical Algol binaries U Cep and U Sge have both been frequently studied. They are bright, totally eclipsing systems of about the same overall mass (7–8 times that of the Sun). The mass ratio (secondary/primary) is significantly greater in the case of U Cep, however. This implies that, though the original configurations of these two binaries may well have been quite similar, U Sge is older and somewhat further on (by ~6 million years) into the Algol stage of its evolution. Its secondary (the original primary) has lost a bigger fraction of its original mass. U Cep is much more 'active' as an Algol binary. Broadband photometry reveals small changes of shape of the light curve, particularly the primary eclipse minima, during spasmodic episodes which may last for weeks and are separated from each other by a few years. These episodes correlate with somewhat enhanced phases of mass-loss, the mean value for which is already high compared with most well-known Algols. This is evidenced by the relatively high rate of orbital period decrease for U Cep.

These ideas, many of the finer ramifications of which are still not properly established, are supported by information derivable from fitting the β-photometry. In Figure 8.6 we show some results of four-parameter curve-fits to such data. Corresponding parameter sets are listed in Table 8.4

Of the four parameters $w_{\beta e}$, h, u_β and p, involved in the curve-fitting, the main effect comes from $w_{\beta e}$. This quantity fixes the amplitude of the observed 'W' type variation. The phase at which this pattern starts and ends depends on h. The other two quantities u_β and p are interdependent, as already mentioned. Thus putting $p = 1$, which produces a better fit for the U Sge data, implies that the absorption component of the primary line is 'saturated' (zero central intensity). In this case, there is no effect of differential limb darkening for the absorption component — essentially no flux comes from it.

The U Sge data are, though, scarce in the central phases of the eclipse, and insufficient for a proper optimization exercise. The coverage is better with

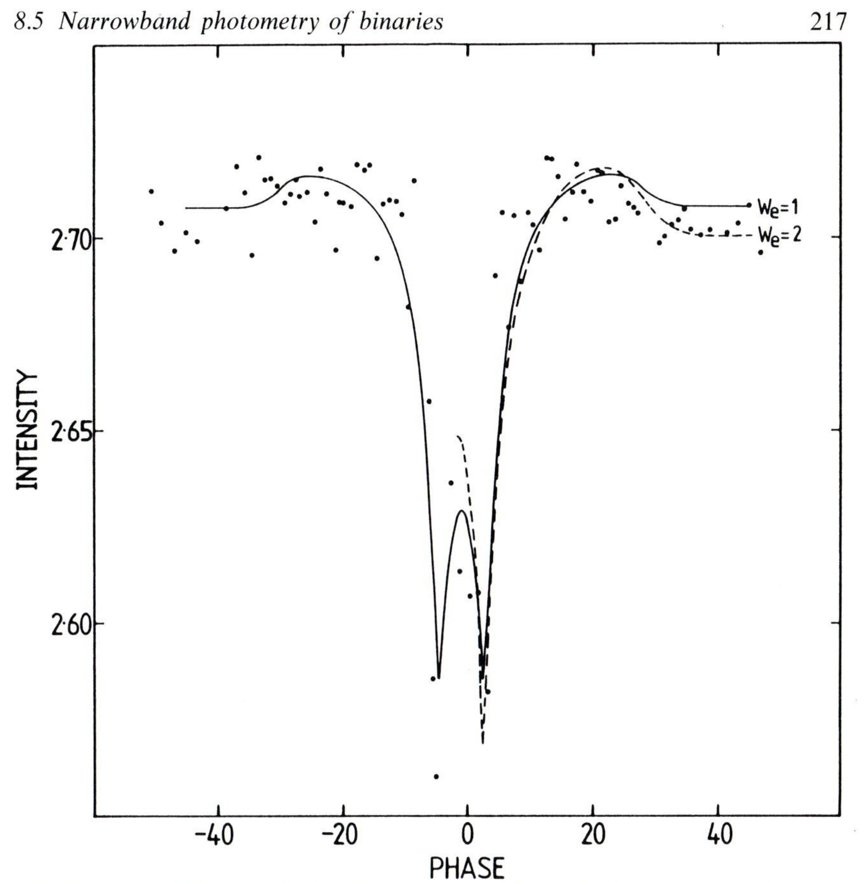

Fig. 8.6. The 'W'-shaped β-index variation through the primary eclipse of U Sge (upper) and U Cep (lower), as observed by Olson (1976).

Table 8.4. *Thin disk model parameters.*

Parameter	U Cep	U Sge
w_e	1.5	0.8
h	0.6	0.6
u_β	0.6	–
a	3	1

E.C. Olson's β photometry of U Cep, but in that case the clear asymmetry argues that a single uniform disk must be an oversimplification. In fact, the two halves of the W curves were matched separately in Figure 8.6, the values of $w_{\beta e}$ given in Table 8.4 being an average.

Though these models are relatively crude, they allow some quantifications

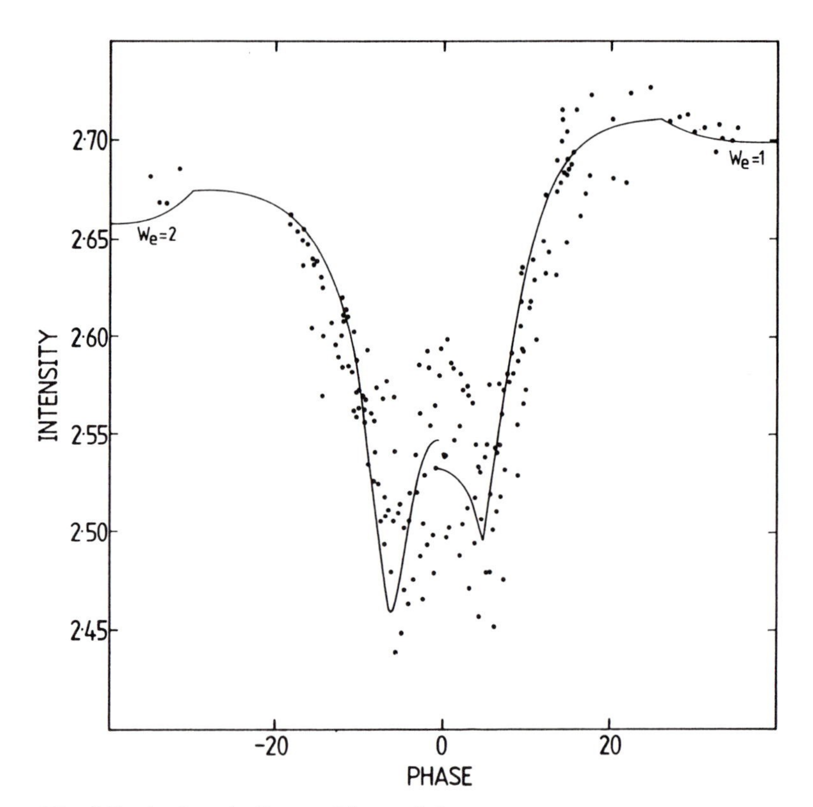

Fig. 8.7. A plot similar to Figure 8.6.

for the thin disk-like structures expected around the primaries of Algol binaries. Representative electron densities corresponding to the observed emission levels are around 10^{18} m^{-3}. The masses for the disks of both systems turn out to be about the same ($\sim 5 \times 10^{20}$ g), the physical size of the structure for U Sge compensating for its somewhat lower density. Both structures are weak by-products of the mass-transfer process: they are less massive than the amount of matter that the mass losers would shed even in one orbital period. They are probably rather unstable, and variable in their physical characteristics — particularly in the case of U Cep. Such findings indicate the potential of narrowband photometry for studying interactive binary evolution.

8.6 Bibliographical notes

Much of the literature on close binary systems cited in the preceding chapter's bibliography is still relevant.

Section 8.1 is more specifically based on E. Budding, *Astrophys. Space Sci.*, **26**, 371, 1974. The examples of eccentric eclipsing binaries analysed in that reference, parts of which reappear here, are from G. E. Kron and K. C. Gordon, *Astrophys. J.*, **118**, 55, 1953; Å. Wallenquist, *Ark. Astron.*, **1**, 59, 1949; D. Chisari and T. Saitta, *Publ. Oss. Astron. Catania*, NS No. 54, 1963; M. Rodonò, *Publ. Oss. Astron. Catania*, NS No. 98, 1967; D. J. K. O'Connell, *Vistas Astron.*, **12**, 271, 1970.

The following two sections draw similarly from E. Budding, *Astrophys. Space Sci.*, **29**, 17, 1974, and E. Budding, and N. N. Najim, *Astrophys. Space Sci.*, **72**, 369, 1980, where details and origins of the formulae presented are given. Reference should also be made to the careful paper by S. Söderhjelm, *Astron. Astrophys.*, **34**, 59, 1974. The example of VV Ori comes from J. A. Eaton *Astrophys. J.*, **197**, 379, 1975.

Section 8.4 is a very brief pointer to a subject which occupied a good deal of Z. Kopal's *Language of the Stars*, Reidel, Dordrecht, 1979. The Hankel-transform-based representation of α-functions was introduced by him in *Astrophys. Space Sci.*, **50**, 225, 1977. An earlier approach using Fourier-analysis techniques was that of M. Kitamura, whose *Tables of the Characteristic Functions of the Eclipse* was published by the University of Tokyo Press, Tokyo, 1967.

Alternative methods of light curve analysis are from A. M. Cherepashchuk, A. V. Goncharskii and A. G. Yagola, *Sov. Astron.*, **12**, 944, 1969 (integral equation solution technique); D. B. Wood, *Astron. J.*, **76**, 701, 1971 (WINK); R. E. Wilson and E. J. Devinney, *Astrophys. J.*, **166**, 605, 1971 ('W–D'); G. Hill, *Publ. Dom. Astrophys. Obs.*, **15**, 297, 1979 (LIGHT); and others. The triennial reports of Commission 42 (*Trans. Int. Astron. Union (Rep. Astron.)*) give statistics on the applications of such procedures.

Section 8.5 makes extensive use of material presented in E. Budding, *Close Binary Stars: Observations and Interpretation* (IAU Symp. 88, ed. M. J. Plavec, *et al.*), Reidel, Dordrecht, p299; E. Budding and N. Marngus, *Astrophys. Space Sci.*, **67**, 477, 1980; and M. A. Khan and E. Budding, *Astrophys. Space Sci.*, **125**, 219, 1986. The H_β data on U Cep come from E. Olson, *Astrophys. J.*, **204**, 141, 1976.

9

Spotted Stars

9.1 Introductory background

'Starspots' are not a new notion. At one time starspots were offered as an explanation for all stellar variability. From that extreme, in the last century, attention to the hypothesis had dwindled away almost completely, after the development of stellar spectroscopy, until, in a well-known pair of papers dealing with the cool binaries AR Lac and YY Gem in the 1950s, the photometrist Gerald Kron revived it. As events have turned out, there is now a good deal of evidence to support Kron's conjecture, for certain groups of cool variable.

The two stars that Kron referred to are good examples of somewhat different but related categories of 'active cool star' (cf. Figures 9.1–9.2) — stars of spectral type generally later than mid-F, usually having a relatively rapid rotational speed. In the cases of AR Lac and YY Gem, both close binary systems, this rapid rotation is a consequence of tidally induced synchronism between rotation and orbital revolution. The M type dwarf components of YY Gem are typical flare stars, characterized by Balmer line emissions which on occasions become very strong, reminiscent of solar flares, but with much greater relative intensity. AR Lac also shows 'chromospheric' emission lines, but its particular configuration, G5 and subgiant K0 stars in a $\sim$2 day period binary, places it as a standard RS CVn type binary.

Active cool stars display a wide range of interrelated phenomena at different wavelengths, which support the picture of physical processes basically similar to the electrodynamic activity of the Sun, but on a much enhanced scale, i.e. activity tracers are typically one or two orders of magnitude greater. From the accumulation of evidence it would now appear surprising if the broadband photometric variations attributed to large spots were not also present. Indeed, phase-linked distortions observed in selected spectral lines

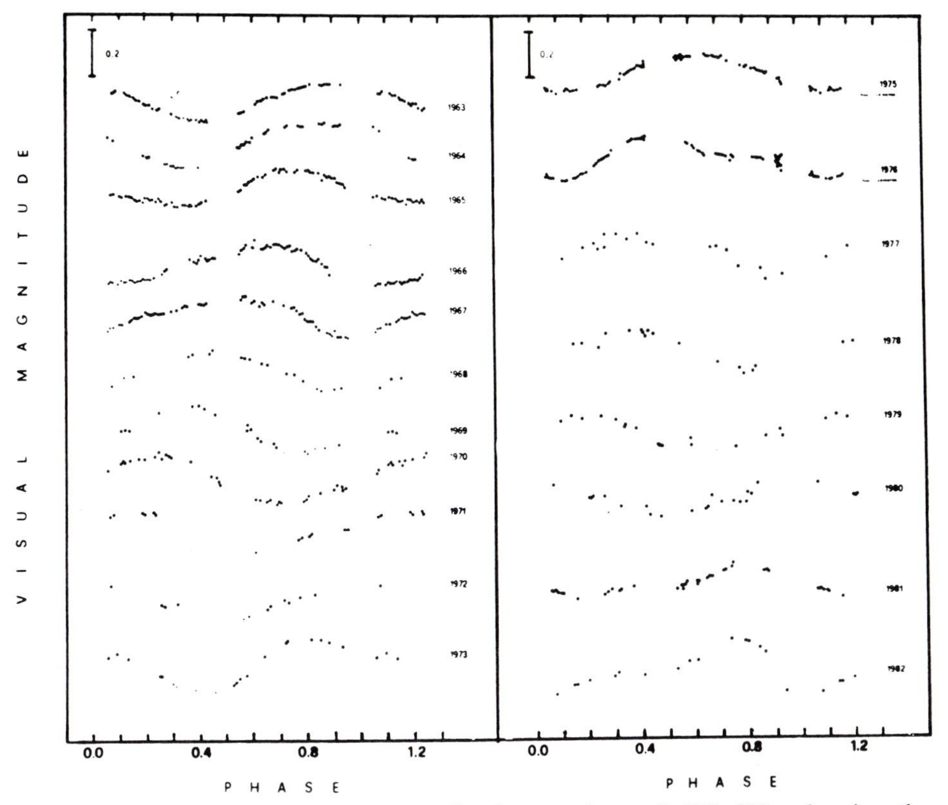

Fig. 9.1. Light curves of (out-of-eclipse regions of) RS CVn, showing the characteristic 'wave type distortions'.

of rapidly rotating stars, analysed, in recent years, with the technique of 'Doppler Imaging', provide an independent confirmation of the presence of large photospheric starspots. These 'maculation' effects, and the possibility to represent them quantitatively by starspots, offer a relatively direct route to monitor the activity of such stars, in a way comparable to the use of relative sunspot number for solar activity. Numerous studies have been made to relate such monitorings with other diagnostics of stellar activity, traced with a variety of multi-wavelength techniques.

The value of such statistical work is substantially enhanced with an extensive database of photometric coverage. There has thus been special interest in applying automatic photometric telescopes (APTs) to active cool stars, and, though such devices are increasingly associated with other kinds of astronomical monitoring, this work provided an early persuasive case for their support.

In what follows, a simple and direct method of systematically quantifying

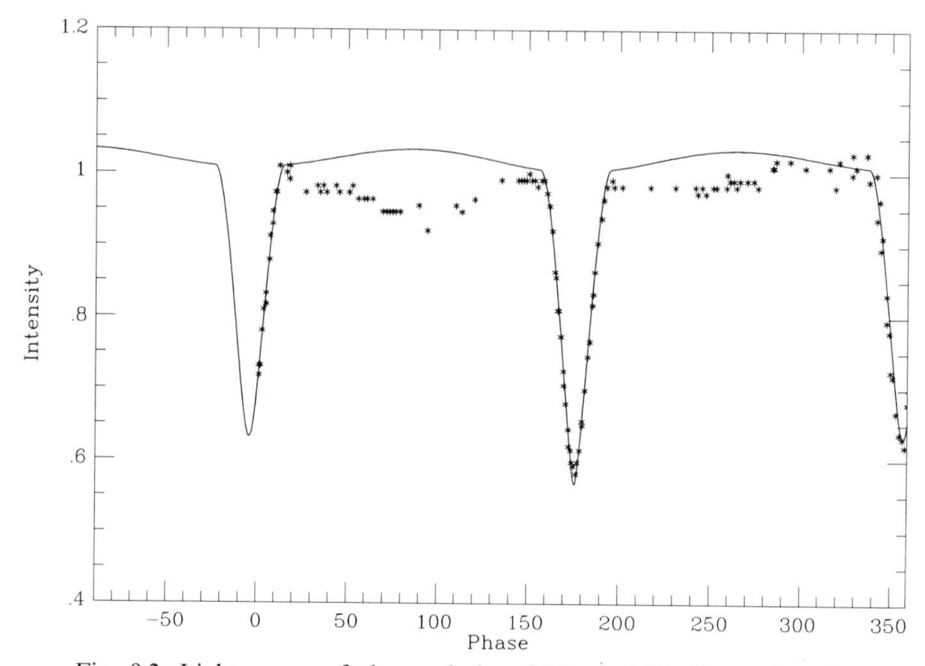

Fig. 9.2. Light curve of the cool dwarf binary YY Gem, also showing maculation effects. A standard eclipsing binary model curve-fitting is shown for comparison.

starspot effects will be described. The shortcomings of oversimplification of the real physical situation and lack of uniqueness in the modelling should be noted here, even though these limitations, unfortunately, are essentially a property of the data, if the representation adequately summarizes that data, without either loss of significant information, or overinterpretation. This can be achieved by careful inspection of the χ^2 corresponding to the model's fit. Model inadequacy is signalled by an improbably high value of χ^2. Overinterpretation implies non-positive definiteness of the χ^2 Hessian with respect to the model's adjustable parameters. A model which satisfactorily avoids either of these conditions is adequate, within the limitations of available accuracies. It should help guide ideas about underlying physical processes giving rise to the phenomena.

9.2 The photometric effects of starspots

The photometric effect produced by a dark spot on the photosphere of a star, moved around by the star's rotation, is a function of time t, or equivalent angular phase ϕ, and some set of n parameters a_j ($j = 1, ...n$), relating to the

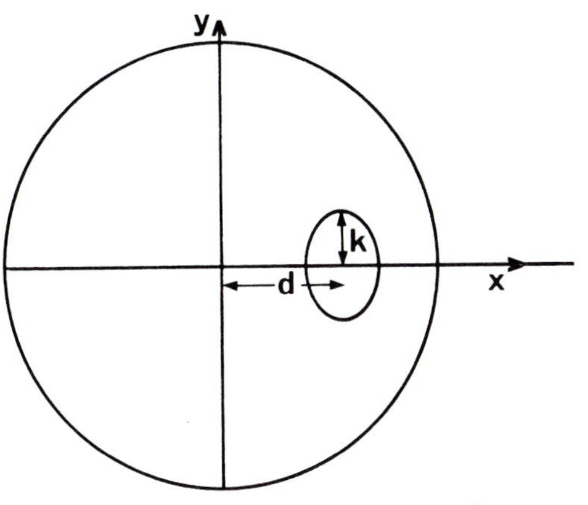

Fig. 9.3. Schematic representation of a starspot of radius k whose elliptical outline is centered about a point distant d from the centre of the star.

location and size of the spot, the relative intensity of its light in the line of sight, the inclination of this line to the axis of rotation, and so on.

A treatment can be devised which closely parallels that of the previous chapters dealing with eclipsing binaries. The same form of expression (7.1) can be used to measure the goodness of fit by χ^2, involving the differences between observed l_{oi} and calculated $l_c(a_j, t_i)$ light levels. The essential difference between the two cases, as far as the curve-fitting is concerned, lies in the use of a different *fitting function* for the maculation effect.

We aim here at simplicity in specifying a minimal number of parameters and assumptions to provide a satisfactory curve-fit, in the χ^2 sense. Hence, we take the spot to be of circular outline, i.e. formed by the intersection of a plane with a spherical star surface, with projected area A (Figure 9.3). This leads to integrals of the type

$$\pi\sigma_n^m = \iint_A x^m z^n dx dy,$$

where the z-axis coincides with the line of sight to the centre of the star, and the x-axis points in the direction from the centre of the stellar disk to that of the elliptical outline of the spot.

Listing a minimal set of parameters required for this specification we have: longitude of spot centre, λ; latitude of spot centre, β; inclination of the rotation axis to the line of sight, i; angular extent of the spot, γ. The apparent semimajor axis of the spot k is then given by $k = \sin\gamma$. The reference light level for the unspotted ('immaculate') state is U, which takes a nominal value of unity (Section 7.3). The luminosity of the spotted star as a fraction of U is L_1. The ratio of the mean flux over the starspot to the normal photospheric flux, over spectral window W, we denote κ_W. This is often a relatively small quantity ($\kappa_W \lesssim 0.1$) for the stars and spectral ranges encountered. u represents the linear limb-darkening coefficient (Section 7.3) for the spotted star's photosphere.

A coordinate transformation is required to relate the system ξ in which the coordinates λ, β are naturally expressed to the $\mathbf{x}(x, y, z)$ system of the σ-integrals. This is done using the rotation matrix notation of Chapter 4, thus:

$$\mathbf{x} = \mathbf{R}_z(\psi)\cdot\mathbf{R}_y(i)\cdot\mathbf{R}_z(-\phi)\cdot\xi. \tag{9.1}$$

In this way, we wind back through the elapsed phase angle $-\phi$, and then tilt the z-axis from the rotation axis down to the line of sight. The third rotation, about the line of sight, rotates the x-axis into alignment with the line passing from star to spot centre, in the observer's coordinate system.

Since the y coordinate of the spot centre is zero, we can write for the coordinates of this point in the $\mathbf{x}$ system (d', 0, z_0). It is convenient to have the radius of the spotted star as the unit of distance. It is then easily shown that the separation d of the *apparent* spot centre from the centre of the disk is related to the separation of the actual central point of the spot d' by $d' = d/\sqrt{(1-k^2)}$. We then also find $d = \sqrt{1-z_0^2}\cos\gamma$. The quantities k and d play a formally equivalent role in relation to the σ-integrals as do the corresponding quantities for the α-integrals in the eclipse problem (Section 7.4). It is more convenient in practice, however, to regard σ_n^m as dependent on the size parameter γ, and retain z_0 $(= \sqrt{1-d'^2})$ as the intermediate quantity which relates the spot's location to the parameters λ, β and i at a given phase ϕ. When multiplied out, the last row of (9.1) gives,

$$z_0 = \cos(\lambda - \phi)\cos\beta\sin i + \sin\beta\cos i \tag{9.2}$$

σ-INTEGRALS — The basic σ-integrals, σ_0^0 and σ_1^0, take the following forms:

(i) Annular case (i.e. entire outline visible), $d \le 1 - k^2$,

$$\sigma_0^0 = k^2 z_0, \tag{9.3}$$

$$\sigma_1^0 = \frac{2}{3}\left\{1 - \sqrt{1-k^2}\left[(1-k^2) + \frac{3d^2k^2}{2(1-k^2)}\right]\right\}. \tag{9.4}$$

(ii) Partial case (i.e. spot lies partly over the disk's perimeter), $d > 1 - k^2$,

$$\sigma_0^0 = \frac{1}{\pi}\left(\arccos s + k^2 z_0 \arccos v - d\sqrt{1-s^2}\right), \tag{9.5}$$

$$\sigma_1^0 = \frac{2}{3\pi}\left(\arccos\left(\frac{v}{s}\right) + \frac{\sqrt{1-k^2}}{2s}\left\{kz_0(3k^2-1)\sqrt{1-v^2} - \right.\right.$$

$$\left.\left. - [2s(1-k^2)+3dk^2]\arccos v\right\}\right). \tag{9.6}$$

Two intermediate quantities have been introduced here, namely $s = (1-k^2)/d$ and $v = (d-s)/kz_0$.

(iii) Total case (i.e. entire hemisphere covered), $d \leq 1-k^2$, $\gamma \geq \pi/2$, we have the trivial values:

$$\sigma_0^0 = 1, \tag{9.7}$$

$$\sigma_1^0 = \frac{2}{3}. \tag{9.8}$$

9.3 Application to observations

Using (7.1) for χ^2, we need the form of $l_c(a_j, \phi)$. This can be given as,

$$l_c(\phi) = U - L_1(1-\kappa_W)\sigma_c(u, \gamma, z_0(\lambda, \beta, i, \phi)), \tag{9.9}$$

where the various parameters a_j have been written out as λ, β, i, etc., as defined above. The light loss term σ_c accounts for the effect, with a given spot area, of varying light intensity over the photospheric disk. A spot near the limb, for example, would remove a greater proportion of light if the stellar disk was of constant brightness, relative to the normal case of some limb darkening. This entails that σ_c is a linear combination of σ-integrals with the appropriate limb-darkening coefficients, thus, in the first order form,

$$\sigma_c = \frac{3}{3-u}[(1-u)\sigma_0^0 + u\sigma_1^0]. \tag{9.10}$$

Formula (9.9), which has been usefully applied to data sets from cool stars apparently showing maculation effects, builds in an assumption that the limb darkening in the photosphere is the same as that in the cooler region of the spot. This will not be exactly true in practice, though in the optical region κ_W is usually so small that the difference is not serious. For a better rendering, however, we could write,

$$l_c(\phi) = U - L_1\sigma_c(u_p, \gamma, z_0), +\kappa_W L_1\sigma_c(u_s, \gamma, z_0), \tag{9.11}$$

where we separately subtract the photospheric light loss, darkened with coefficient u_p, and add in the spot's own contribution, with coefficient u_s, σ_c taking the same form as (9.10) in both cases.

The conversion of observed time data t_i to equivalent phase values ϕ_i requires provision of an ephemeris formula, i.e. $\phi = (t - t_0)/P$, where t_0 is an initial reference time of zero phase, and P is the period. In spotted stars which happen to be in synchronized binary systems the binary's orbital ephemeris can be used for this calculation, and particular data segments are assigned appropriate spot longitudes. An alternative procedure, where there is a sufficiently long and continuous data set covering an identifiable starspot wave, would be to regard the spot longitude as fixed and optimize the corresponding period. This raises the subject of differential rotation, which has often been discussed since D. S. Hall's pioneer review of the RS CVn stars in 1975. Investigators have commonly noted the presence of a 'migrating distortion wave', i.e. that a spot effect centred at a certain longitude would later have drifted, often to greater longitude. By analogy with similar effects observed for the Sun it is inferred that the underlying body has a rotation speed which depends on latitude. This question is potentially more troublesome when there is no available binary ephemeris. The reference period then becomes some average of various wave-cycle durations, derived from different data sets. In practice, it appears that inferred differential rotation rates in RS CVn stars are smaller, typically by an order of magnitude, than what is found for the Sun.

In Figure 9.4 we see a classic starspot light curve. It comes from Evans' observations of the dwarf flare star CC Eri, which were first published in 1959, and later analysed by the numerically integrated starspot representation of Bopp and Evans in 1973. The parameters characterizing this fit have been reproduced in Table 9.1, alongside another set obtained by χ^2 minimization techniques. The two corresponding theoretical curves are also shown, and we now make a few remarks from comparing the results.

Both analyses concur fairly well on the spot's longitude, and since this coincides (for a single spot) with the phase of maximum darkening, eye-inspection makes the result close to 150°. The analyses also agree about the size of the spot, which is closely linked to the depth of the feature, and, reading this on the intensity scale, we anticipate, using (9.3), a value of $\sim \arcsin 0.4$ by eye. That would only become really close for a centrally located spot on a zero limb-darkened photosphere, however.

The optimal model has the spot close to the centre of the disk when the loss of light is greatest, while the other scheme places it just less than 40° below centre. In the former case some part of the spot will always be on

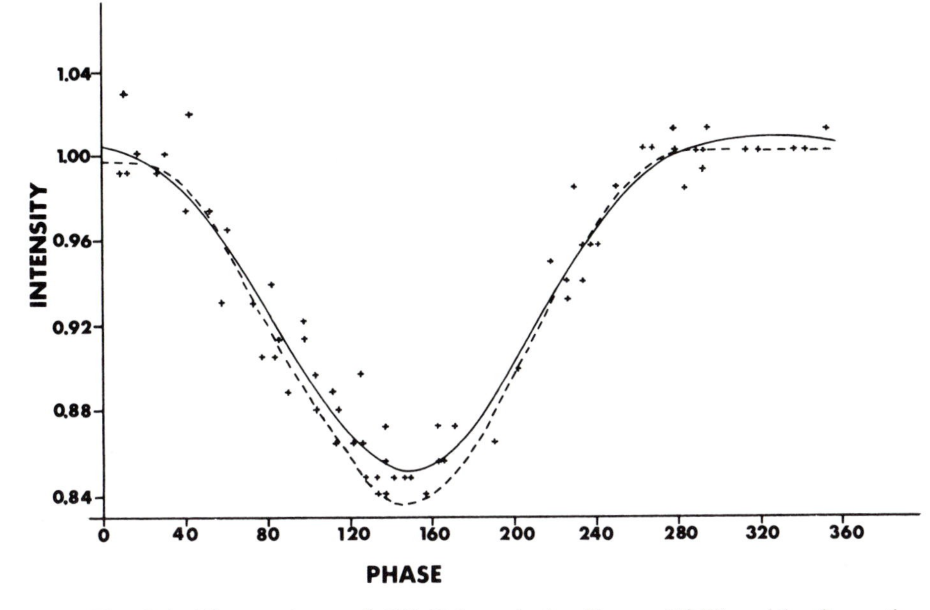

Fig. 9.4. Observations of CC Eri made by Evans (1959), with alternative theoretical curve-fits: dashed, that of Bopp and Evans (1973); continuous, that of Budding (1977).

Table 9.1. *Starspot parameter sets for CC Eri.*

Parameter	Optimal value	$\pm$	Bopp and Evans' (1973) value
λ	148.6	1.1	152.5
β	38.0	4.7	10
i	50.1	5.0	42
γ	21.1	0.2	21
m_0	8.755	0.005	8.76
κ_V	0.0		0.0
u	0.5		0.5
Δl	0.015		
χ^2/v	0.91		

the visible disk, even when the spot centre is at its furthest distance from the apparent central point, i.e. some 13.1° over the disk's edge. Hence, the light curve is one of continuous variation in the optimal model, though the variation is relatively slight when the spot centre is behind the visible disk. This will happen, from (9.2), when the phase angle, as measured from the maximum darkening, ψ, say, $(= \lambda - \phi)$ satifies

$$\cos \psi = -\tan \beta \cot i,$$

which acts as a loose constraint on the semiwidth of the depressed part of the light curve. From the data in Table 9.1 this yields $\psi \simeq 130°$, so the light curve should be fairly flat for ϕ greater than about 280°, or less than ~20°. More exactly, a spot would cease to have any effect if ψ can satisfy

$$\cos \psi = \frac{-(\sin \gamma + \sin \beta \cos i)}{\cos \beta \sin i},$$

which Bopp and Evans' model would do for $\psi \simeq \pm 140°$, i.e. $\phi \simeq 290$, or 10°, and this can be observed on the dashed curve in Figure 9.4.

We may also note from Table 9.1 that the reference level of luminosity, U, when no spot is present, has been reconverted to an apparent magnitude scale. However, since some part of the spot is always visible in the optimal model, this light level is not actually observed, and the amplitude of light variation is consequently somewhat reduced from what is possible with a spot of this size. The same point also goes toward explaining why the Bopp and Evans curve has a lower minimum than the optimal one, even though its spot, of essentially the same area when normally projected, is not as centrally placed at minimum light, and would therefore have a somewhat foreshortened area. The different representations of the spot's shape may also have a slight effect on the form of apparent variation.

The values of flux ratio κ_V and limb-darkening coefficient u were taken from the Bopp and Evans model, so that the two models could be more closely compared in the quantities derived. The reduced χ^2 implies an acceptable underlying hypothesis, at the adopted accuracy of observations, for the optimal model — however, such inferences have to be regarded cautiously, since we already know of oversimplifications with regard to the flux ratio and limb darkening, let alone the shape of the spot.

A better fit can be expected when more parameters can be adjusted. In the optimization procedure there was one more of these than for the Bopp and Evans treatment, where the inclination was derived from separate spectroscopic data.

Further refinements of the model can be made by similar σ-integral-related expressions in the fitting function. Perhaps easiest is the extension to two or more spots, where we write, in place of (9.9),

$$l_c(\phi) = U - L_1(1 - \kappa_W) \sum_{j=1}^{m} \sigma_c(u, \gamma_j, z_0(\lambda_j, \beta_j, i, \phi)), \tag{9.12}$$

for the m spots required. Care is needed in ensuring that the spots do not overlap if κ is to preserve its distinct meaning in (9.12). This is dealt with by the condition $\arccos(\xi_j \cdot \xi_k) > \gamma_j + \gamma_k \quad (j \neq k)$, where ξ_j, ξ_k represent the position vectors to the centres of the jth and kth spots ($1 \le j, k \le m$), respectively.

Such a constraint can be built into the fitting function algorithm, so that a pair of close spots are forced not to overlap. This has a physical interpretation, in that spot pairs are often found on the solar photosphere. An alternative constraint, also with known solar counterpart, is to fix two spots to the same centre coordinates, such that $\gamma_2 > \gamma_1$, say, and that $\kappa_2 + \kappa_1 \simeq 1$, with $\kappa_2 \sim 0.7$ — i.e. the spot has a 'penumbral' perimeter region. These possibilities point up the difficulty, with available data accuracies, of resolving any detail in the shape of a feature; such experiments as have been carried out with spot pairs have not found crucial differences in consequent light curve forms from those of a single spot of the same total area.

In effect, most available data sets for spotted stars seem to require a maximum of two well-separated spots to match determinability to data accuracy. This entails a total of eleven parameters, i.e. adding the positional coordinates λ, β and radius γ of the second spot to the preceding eight of Section 9.2.

The flux ratio parameter κ is much better determined when there are light curves at more than one wavelength — particularly, if one of these is at red or infra-red wavelengths. In this case we can expect to see maculation effects of the same general shape in the two wavelength regions, but the amplitude should be smaller at the longer wavelength, where the intensity contrast is less (Figure 9.5). This effect forms good supporting evidence that relatively large cool surface regions are indeed responsible for the observed effects, and leads on to enable the temperature decrement of the spot region to be determined. The fact that a light curve at a single wavelength can be explained by bright regions in antiphase to the dark regions of the adopted hypothesis harks back to the non-uniqueness issue of the introductory section. Coordinated multi-wavelength data are therefore desirable to resolve this, and help provide a meaningful interpretation in terms of starspots.

One procedure for determining spot temperatures starts by analysing the light curve with the largest amplitude of variation (due to maculation) using an assigned value of κ, i.e. adopted photospheric and spot temperatures, seeking essentially only the positions and sizes of spots. Fixing the derived geometrical parameters into the fitting function, one then uses the longer wavelength light curve to determine only the best fitting value of κ. This enables an improved estimate for the spot temperature — the photospheric temperature being usually fixed from separate evidence, involving spectral information, or the mean colour at light maximum. One substitutes the derived temperature back to get an improved value of κ to fix into the fitting

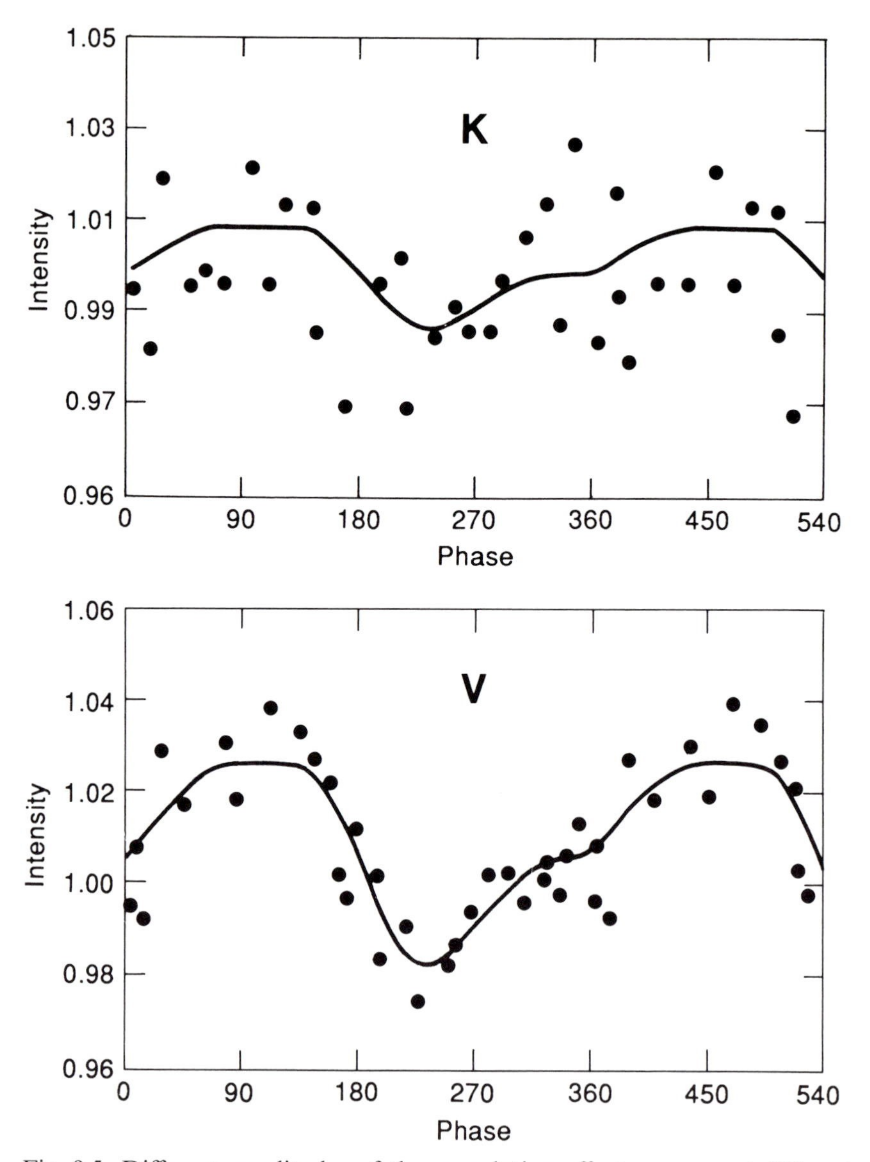

Fig. 9.5. Different amplitudes of the maculation effect are seen at different wavelength data sets for UV Psc, consistent with a temperature deficit of ~1500 K

function, which can then be reapplied at the shorter wavelength to rederive the geometric parameters. This completes the first iterative cycle, which may be repeated. One or two iterations usually suffice to bring the variation of successive spot temperature values well within their probable errors.

9.4 Starspots in binary systems

Many of the active cool stars on which maculation effects are observed are in binary systems. This is partly because close (eclipsing) binary systems tend to be singled out for more intensive photometric investigation. However, binarity may also drive activity in certain stars, because of tidally enforced synchronism speeding up the rotation, and therefore the underlying electromagnetic 'dynamo'.

Since eclipsing binaries are informative on basic astrophysical parameters, it is convenient when active stars also eclipse. For example, the inclination and sometimes limb-darkening values can then be separately found, or confirmed. Binarity may release further circumstantial evidence. The eclipses themselves may allow surface features to be scanned and localized. Binaries also often have appreciable differences in component luminosities, allowing detectable spot effects to be predominantly associated with the brighter component.

The procedures developed separately hitherto, as eclipse and maculation effects, can be followed through into combination. The treatment becomes somewhat complicated, but a formal approach to the fitting function allows a more economic exploration of the parameter manifold. We also have a chance to see if deductions from the general uneclipsed regions of the light curve can be confirmed or extended by the scanning of the eclipsed star's surface as the eclipse proceeds.

Now σL_1 expresses the loss of light of a star of luminosity L_1 due to a circular spot, though a proportion $\kappa_W \sigma L_1$ is added back. In a similar way, αL_1 expresses the loss of light due to eclipse. When both of these effects are present at the same time the respective light losses are additive, and we write for the instantaneous light level $l(\phi)$,

$$l(\phi) = L_2 + L_1[1 - (1 - \kappa_W)\sigma_u - \alpha]. \tag{9.13}$$

The σ-function in (9.13) is suffixed with u to denote that the loss comes from the *uneclipsed* part of the spot (which may, of course, be all of it). We can generalize this by writing $\sigma_u = \sigma(1 - \alpha_s)$, where σ is the normal σ-function as in (9.10), and α_s denotes the eclipsed fraction of the spot.

The evaluation of α_s is complicated by two issues: firstly, determining the general geometrical circumstances under which the starspot becomes eclipsed; and secondly, expressing the general form of the light loss integral when there is a limb-darkened distribution of light across the eclipsed star's photosphere. The former problem boils down to finding the appropriate equation for the common chord. A cubic has to be solved, and the appropriate root extracted. In devising a computer algorithm for this, care is required to distinguish between various alternative sign possibilities, particularly where square roots and inverse trigonometric expressions are involved. The second matter is more difficult to deal with formally. However, since observed starspots are usually appreciably smaller than component radii, a convenient approximation may be made by computing only the undarkened expression for the light loss, and weighting that with an appropriate z coordinate for the spot to obtain a corresponding limb-darkened form.

The undarkened form for α_s is made up of two 'D-shaped' areas, formed by the common chord and the perimeter of the eclipsed region of the spot and the boundary of the eclipsing star. These areas are both basically of the form:

$$D = k^2 \left[\arccos\left(\frac{c}{k}\right) - \frac{c}{k}\sqrt{\left(1 - \frac{c^2}{k^2}\right)} \right], \tag{9.14}$$

though the detailed expressions for the integral limit c and, for the elliptical boundary, the equivalent to the arc radius k are awkward to spell out.

In Figure 9.6 a sequence showing the effects of a large spot positioned at different longitudes on an eclipse minimum is presented.

9.5 Analysis of the light curves of RS CVn stars

Many active cool stars are in close binary systems. A group including numerous well-known eclipsing binaries showing distorted light curves and spectral emission features, particularly in the H and K lines of ionized calcium, were designated RS CVn stars by Douglas Hall. We will briefly follow through a method for the photometric analysis of such stars.

We first note the important concept of *separation* of additive contributions to the net light level. In the previous chapter the photometric effects due to mutual proximity of components were treated like this. Maculation effects, in general, are also separable. This is firstly because the fitting function for a standard (non-eccentric) eclipsing binary model (SEBM) (Chapter 8) is even about phase zero, whereas spots can be centred at any phase. Secondly, the proximity effects are determined by stellar properties which are also reflected in the eclipse fitting. The SEBM is thus characterized by five basic photometric determinables (U, L_1, r_1, k, i), and perhaps also the mass ratio q, though this is usually supplied from separate, i.e. spectroscopic, evidence, as are the temperatures, needed to fix the scale of proximity effects through (8.8) and (8.9). The basic parameters are all determined by eclipses only in the spherical model (Section 7.3), i.e. the proximity effects, having components

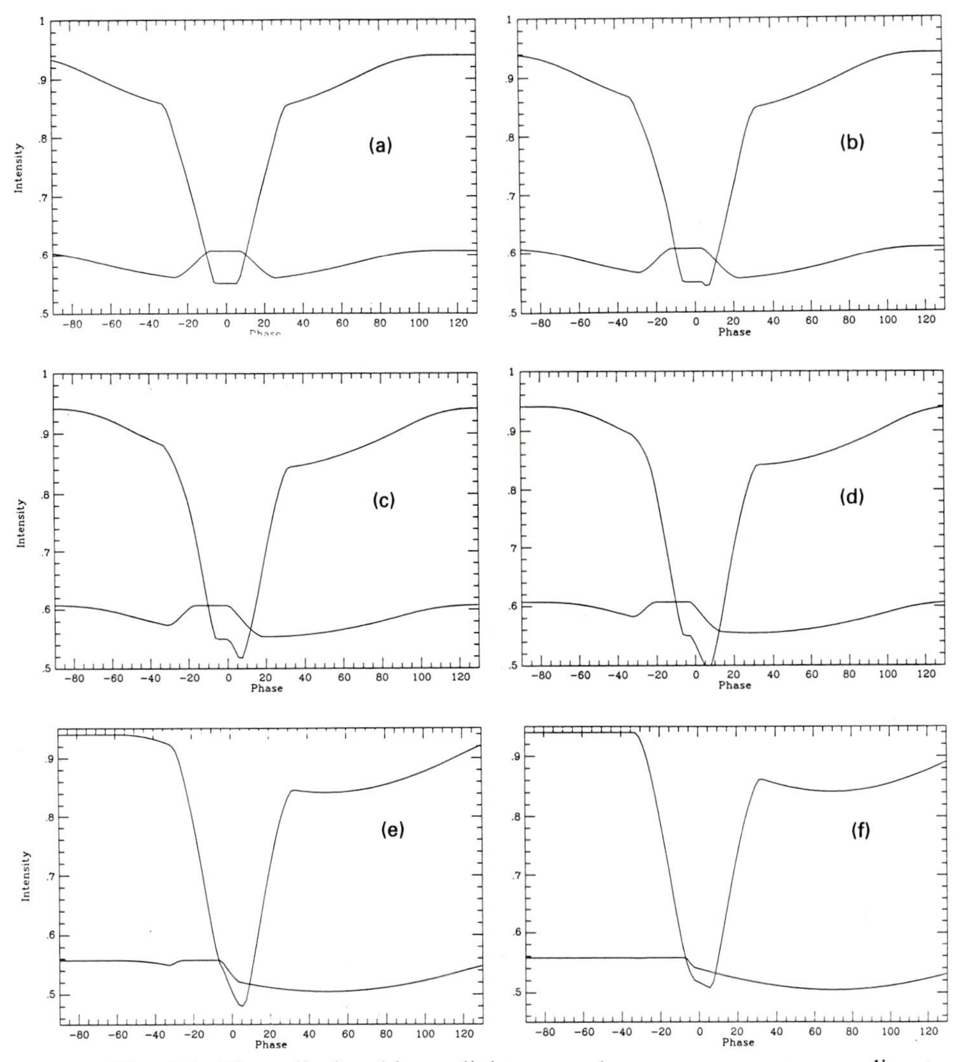

Fig. 9.6. The eclipsing binary light curve (upper curve, corresponding to ordinate scale) has been modelled on the parameters of SV Cam (Table 9.2). A spot of radius 20° and at latitude 6° is located at longitude: (a) 1, (b) 10, (c) 20, (d) 30, (e) 50 and (f) 70°. The maculation effect is shown below at half scale, and arbitrary height on the intensity axis. The eclipses start at around orbital phase –32°, and the spot starts to be eclipsed about 5° later in (a). It becomes completely eclipsed by about –8°, so that the maculation effect is completely lost and the primary minimum looks shallower than it otherwise would. With the spot at 20° (c), it starts to emerge about half way through the annular minimum, wherupon the bottom of the primary eclipse drops down. By 50° (e), the main maculation minimum is appreciably outside the eclipse, and the effect of the foreshortened spot on the eclipse becomes much less obvious, though the bottom of the minimum still has a steep slope. By 70° (f), though it must still be eclipsed during the ingress, the net effects of *spot eclipse* are practically insignificant.

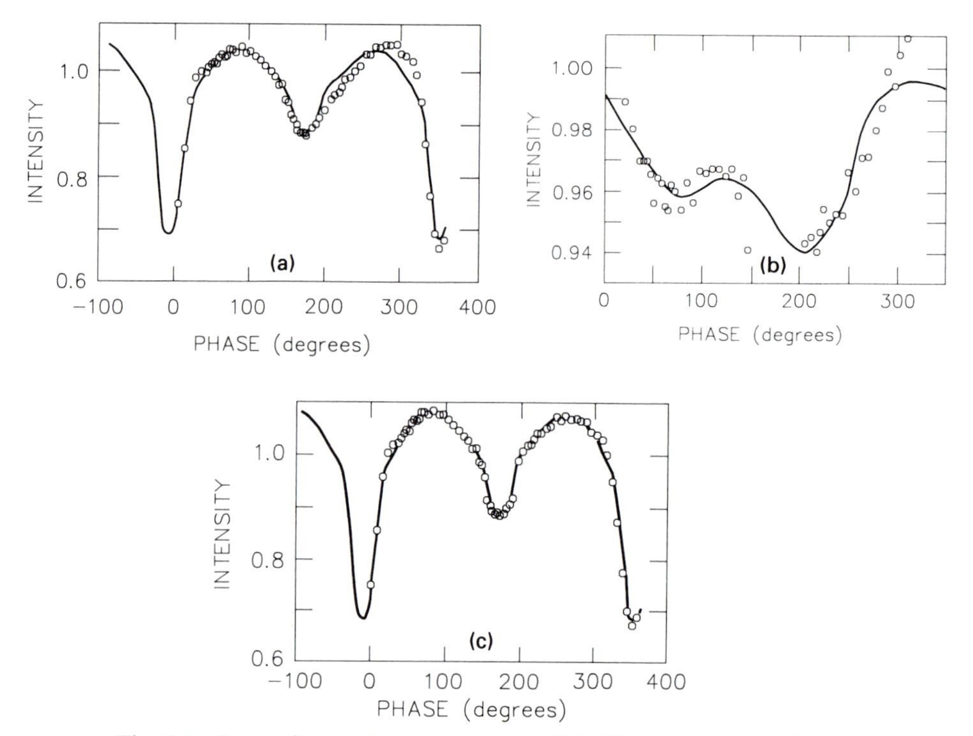

Fig. 9.7. Curve-fits to the short period RS CVn type binary XY UMa: (a) original data fitted with SEBM; (b) difference curve fitted with a pair of circular spots; (c) 'cleaned' curve fitted with SEBM.

at low multiples of the fundamental period of variation, are interdependent with the eclipses, which affect only a relatively narrow range of phases. On the other hand, a spot usually affects ~90° of phase on either side of its longitude.

Another point affecting the procedure is that eclipses generally provide the largest *amplitude* of variation. This is the primary effect, matched in the first attempt to fit the light curve. A practical method is then to proceed by successive approximations, separating out first the SEBM, and then fitting the difference curve with a spot model. The theoretical form of this is then added back to the original data, to produce a 'cleaned' eclipsing binary light curve, for which a second fitting can be made. In principle, this second model is again subtracted from the original data, to leave an improved approximation to the difference curve, which can again be fitted by a spot model, and the whole process repeated. In practice, fitting the second approximation to the difference curve does not yield spot-model parameter sets significantly different from those of the first difference curve — at least, with currently available photometric accuracies (cf. Figure 9.7). The results of such analysis directed to eight (short period) RS CVn systems are presented in Table 9.2.

Note that it is only the *differential* effect of spots which are modelled here. In practice model starspot effects are superposed on an adopted unit of light (with binary effects removed), but there is likely to be an additional constant contribution to the maculation. This has been interpreted either as a 'polar spot', or a more or less uniform longitudinal distribution of spots. This contribution emerges from the analysis, because the final modelling of the clean eclipsing binary light curve includes the quantity U, which determines the value of the spot-free luminosity in magnitude units, i.e. as m_0 in Table 9.1.

This reference magnitude m_0 is derived using that of the comparison star, since the magnitude differences Δm_i of comparison from the variable at the observed phases, make up the original light curve. The conversion of this variation to one of linear intensity, $10^{0.4(\Delta m_0 - \Delta m_i)}$, involves an assigned reference magnitude difference, Δm_0, which, if exactly correct, would result in $U = 1$; conversely, the difference between U and 1 can be fed back to correct Δm_0. By careful monitoring of this reference magnitude from one light curve to another, knowledge of any secular photometric variation can be gained. It would be clearly advantageous, when comparing light curves in this way, if the different data sets all originate from the same photometer and filter system, because slight differences in the calibrations from one observatory to another complicate the issue. A notable example of a spotted star (not an eclipsing binary), where such secular effects have been monitored, is the southern RS CVn-like star AB Dor (Figure 9.8).

The interpretation to be put on any such trends is not clear-cut, because apart from any variation of an apparently constant contribution to the spottedness, the level of mean background photospheric brightness may itself show a trend in dependence on the relative level of activity. Such effects have recently been quantified for the Sun, where the mean photospheric brightness tends to increase at the maximum of the solar cycle, when spots are most frequent.

This extension of the 'solar laboratory' to other comparable stars in order to form a fuller picture of stellar magnetodynamic activity is still relatively underdeveloped. Coming decades should allow increased application of dedicated automatic photometric telescopes to present much more complete photometric data sets on active RS CVn stars, from which a better general awareness should grow.

Table 9.2. *Parameter sets characterizing binary and maculation effects in some short period RS CVn stars (from Budding and Zeilik (1987)).*

Curve Fitting Parameters—λ-Dependent and T-Dependent Quantities

Parameter	XY UMa	SV Cam	RT And	CG Cyg	ER Vul	BH Vir (Sadik)	BH Vir (Scaltriti)	WY Cnc (Chambliss)	UV Psc (Sadik)	UV Psc (Capilla)
					Distorted Light Curves					
L_1	0.955	0.893	0.857	0.581	0.354	0.595	0.625	0.975	0.803	0.866
	±0.004	±0.01	±0.01	±0.01	±0.01	±0.002	±0.003	±0.002	±0.008	±0.02
L_2	0.045	0.107	0.143	0.419	0.646	0.405	0.375	0.025	0.197	0.134
					"CLEANed" Light Curves					
L_1	0.930	0.901	0.866	0.745	0.652	0.573	0.634	0.976	0.811	0.878
	±0.003	±0.008	±0.008	±0.01	±0.01	±0.003	±0.01	±0.004	±0.008	±0.015
L_2	0.070	0.099	0.134	0.253	0.348	0.427	0.366	0.024	0.189	0.122
					Adopted Auxiliary Parameters					
u_1	0.70	0.64	0.65	0.65	0.7	0.83	0.83	0.65	0.65	0.70
u_2	0.70	0.88	0.75	0.70	0.7	0.80	0.80	0.88	0.80	0.80
τ_1	1.20	1.27	1.07	1.31	1.14	1.19	1.12	1.20	1.15	1.14
τ_2	1.27	1.70	1.27	1.40	1.20	1.14	1.14	1.77	1.38	1.32
E_1	0.96	2.23	0.53	1.43	0.92	1.02	1.02	2.50	0.54	0.63
E_2	1.58	0.97	2.54	1.27	1.48	1.26	1.26	0.85	2.93	2.37

TABLE 3C

Curve Fitting Statistics

Parameter	XY UMa	SV Cam	RT And	CG Cyg	ER Vul	BH Vir (Sadik)	BH Vir (Scaltriti)	WY Cnc (Chambliss)	UV Psc (Sadik)	UV Psc (Capilla)
					Distorted Light Curves					
Δl	0.01	0.007	0.01	0.01	0.02	0.01	0.015	0.01	0.02	0.03
$\Delta l'$	0.014	0.021	0.02	0.03	0.01	0.007	0.015	0.01	0.02	0.02
N	67	98	109	120	133	265	153	56	115	65
χ^2	71.7	96.1	86.6	99.6	118.8	193.0	177.9	60.5	162.1	56.2
ν	61	93	103	114	127	260	147	50	110	60
χ^2/ν	1.18	1.04	0.84	0.87	0.94	0.74	1.21	1.21	1.47	0.94
					"CLEANed" Light Curves					
$\Delta l'$	0.01	0.014	0.014	0.02	0.007	0.007	0.01	0.01	0.02	0.02
N	67	98	106	120	133	265	153	56	115	65
χ^2	51.2	88.35	115.3	108.0	139.2	203.9	173.8	35.8	69.4	47.0
ν	61	93	110	115	126	260	147	50	110	60
χ^2/ν	0.84	0.95	1.15	0.94	1.10	0.78	1.18	0.72	0.63	0.78

TABLE 4

Starspot Parameters

Parameter	XY UMa	SV Cam	RT And	CG Cyg	ER Vul	BH Vir (Sadik)	BH Vir (Scaltriti)	WY Cnc (Chambliss)	UV Psc (Sadik)	UV Psc (Capilla)
L_1	0.95	0.89	0.86	0.59	0.5	0.58	0.58	0.96	0.82	0.90
$i(°)$	88.2	90.0	88.9	82.8	71.6	90.0	86.8	...	90.0	86.0
u_1	0.70	0.64	0.70	0.70	0.7	0.83	0.83	0.65	0.65	0.70
κ_λ	0.0	0.0	0.0	0.0	0.0	0.0	0.0	0.0	0.0	0.0
$\lambda_1°$	81.4	212.0	127.8	120.8	100.8	94.0	97.4	256.1	224.6	187.9
	±3.2	±1.6	±8.0	±2.7	±5.1	±2.9	±5.7	...	±4.0	±2.0
$\beta_1°$	45	60	45	45	45	45	45	45	45	...
$\gamma_1°$	13.0	29.9	8.3	20.8	9.2	16.8	10.7	9.7	18.3	15.9
	±0.3	±0.3	±0.1	±0.5	±0.3	±0.4	±0.8	...	±0.4	±0.3
$\lambda_2°$	202.8	...	...	219.4	234.9	259.5	234.7	...	...	288.8
	±2.5	...	...	±6.8	±3.1	±2.9	±2.2	...	...	±1.6
$\beta_2°$	45	...	...	45	45	45	45	...	...	45
$\gamma_2°$	15.9	...	...	12.5	10.9	20.7	16.7	...	...	17.4
	±0.4	...	...	±0.6	±0.3	±0.5	±0.4	...	...	...
$\Delta l'$	0.007	0.01	0.01	0.014	0.0	0.015	0.02	0.005	0.01	0.007
N	44	66	88	84	105	174	111	42	72	52
χ^2	32.0	50.8	86.4	74.4	88.5	212.2	103.6	44.0	61.6	57.0
ν	39	63	85	79	100	166	106	39	70	47
χ^2/ν	0.82	0.81	1.02	0.94	0.89	1.28	0.98	1.13	0.88	1.21

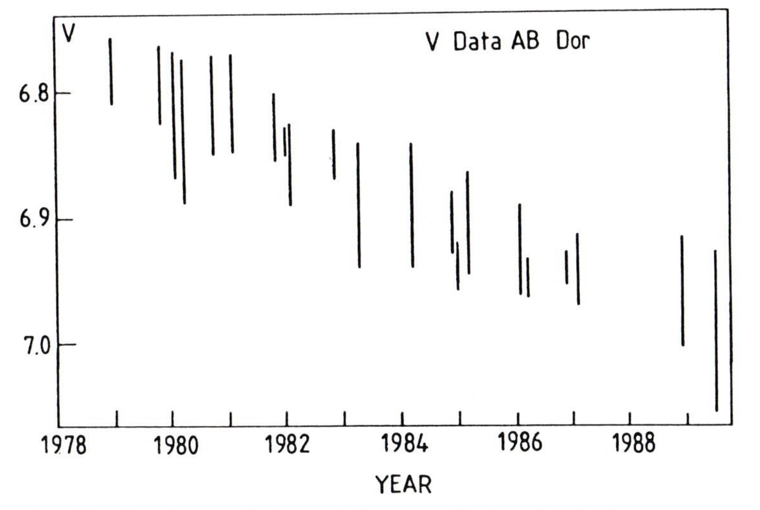

Fig. 9.8. Secular variation of the magnitude of AB Dor.

9.6 Bibliographical notes

A very extensive literature on spotted stars has developed since D. S. Hall's review in the *Proc. IAU Colloq. 29: Multiple Periodic Variable Stars* (ed. W. Fitch) Reidel, Dordrecht, p287, 1975. The *Cambridge Workshops on Cool Stars, Stellar Systems and the Sun*, held annually since the early eighties, have discussed a great deal of related material, and reference still more. A number of the proceedings of these conferences have been published in the Lecture Notes in Physics series of Springer Verlag (Berlin, Heidelberg) e.g. Vol. 291, 1988 (ed. J. L. Linsky and R. E. Stencel), or, more recently, *Surface Inhomogeneities on Late Type Stars* (ed. P. B. Byrne and D. J. Mullan) Vol. 397, 1992. An earlier review was that of S. Baliunas and A. M. Vaughan *Annu. Rev. Astron. Astrophys.*, **23**, 379, 1985. Through such sources the development of the spotted star idea, and related phenomena can be traced over recent years.

The light curves of RS CVn come from Blanco *et al.*, in *Activity in Red-Dwarf Stars* (ed. P. B. Byrne and M. Rodonò), Reidel, Dordrecht, 1983, p387; that of YY Gem is due to a recent compilation of C. J. Butler (private communication).

Sections 9.2 and 9.3 are based on E. Budding, *Astrophys. Space Sci.*, **48**, 287, 1977. The cited data of B. W. Bopp and D. S. Evans is in *Mon. Not. R. Astron. Soc.*, **164**, 343, 1973, (see also D. S. Evans, *Mon. Not. R. Astron. Soc.*, **154**, 329, 1971). Section 9.4 essentially comes from E. Budding, *Astrophys. Space Sci.*, **143**, 1, 1988, though Figure 9.6 is new.

The studies referred to in Section 9.5 develop from E. Budding, and M. Zeilik, *Astrophys. J.*, **319**, 827, 1987. A recent similar study is M. Zeilik *et al.*, *Astrophys. J.*, **354**, 352, 1990. Figure 9.8 is from G. Anders, *IBVS* 3437, 1990.

10
Pulsating Stars

10.1 Introductory background

The eclipsing binaries and spotted stars discussed in the previous three chapters still represent only about a quarter of all variables. The largest class of variable stars are those having some inherent physical variation in luminosity, as distinct from a variation of geometrical aspect. They are referred to collectively as pulsating stars. They contain examples whose light pattern repeats in measurably the same form for many cycles, with a periodicity of comparable constancy to that of many eclipsing variables. There are others whose light level aimlessly wanders up and down, with no apparent pattern or predictability. Many are in between these extremities, with quasi-periodic variations of a 'semiregular' nature. Some examples of the different light curves are shown in Figure 10.1.

The application of spectroscopy to such variables demonstrated that the variations in apparent magnitude are linked with changes of net radius. By studying the Doppler shifts of absorption lines in the more regular cases it was deduced that the entire body of the star pulsates inward and outward in the same period as the brightness cycle, the star being faintest close to, but somewhat before the time when it is smallest. The rise to maximum brightness tends to take place over a relatively short phase range, and by the time the star reaches its maximum extension it is already well past its optically brightest point, which, by convention, occurs at phase zero. In fact, the velocity of expansion tends to be greatest at this phase, the radial velocity curve being usually a reflection of the shape of the brightness curve.

From such evidence, the important but separate role of the temperature cycle could be deduced. A given relative variation of surface temperature is easily seen to have a more important overall effect than the same relative variation of radius, since the net luminosity is proportional to T^4, compared

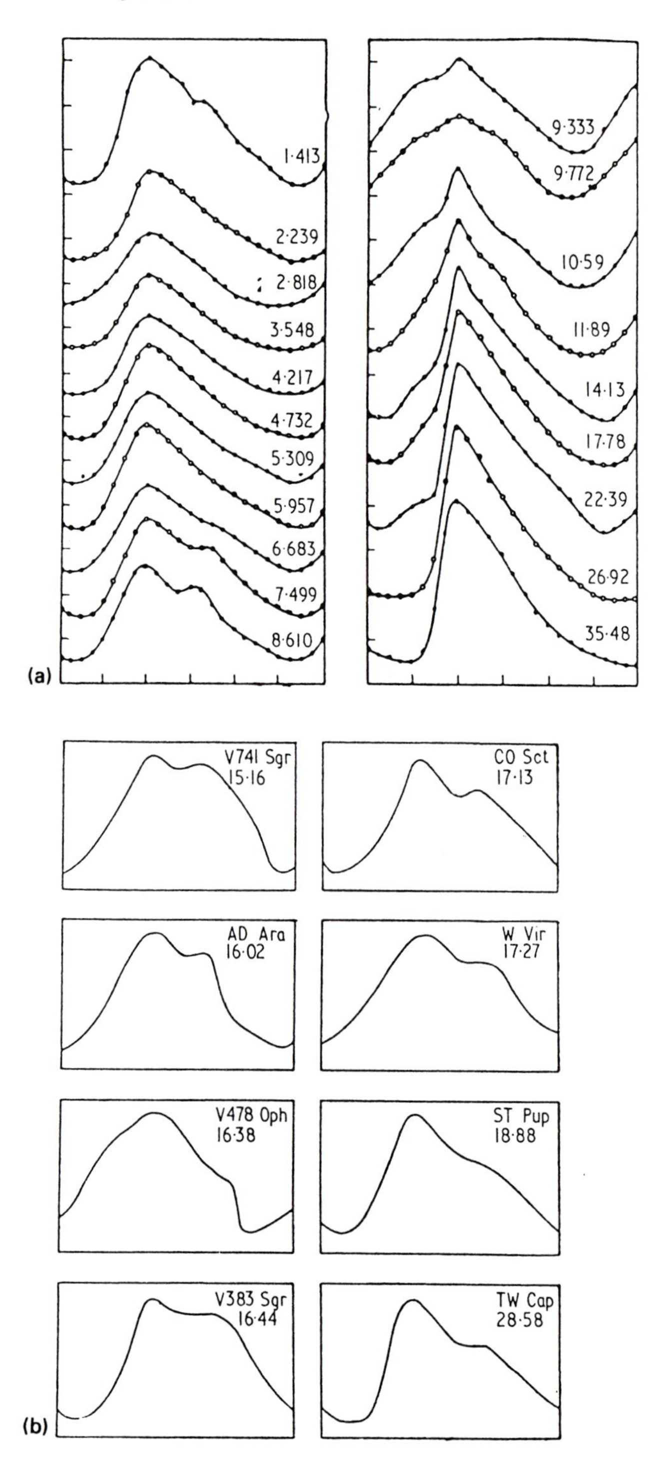
1·413
2·239
2·818
3·548
4·217
4·732
5·309
5·957
6·683
7·499
8·610
9·333
9·772
10·59
11·89
14·13
17·78
22·39
26·92
35·48
(a)
V741 Sgr
15·16
CO Sct
17·13
AD Ara
16·02
W Vir
17·27
V478 Oph
16·38
ST Pup
18·88
V383 Sgr
16·44
TW Cap
28·58
(b)

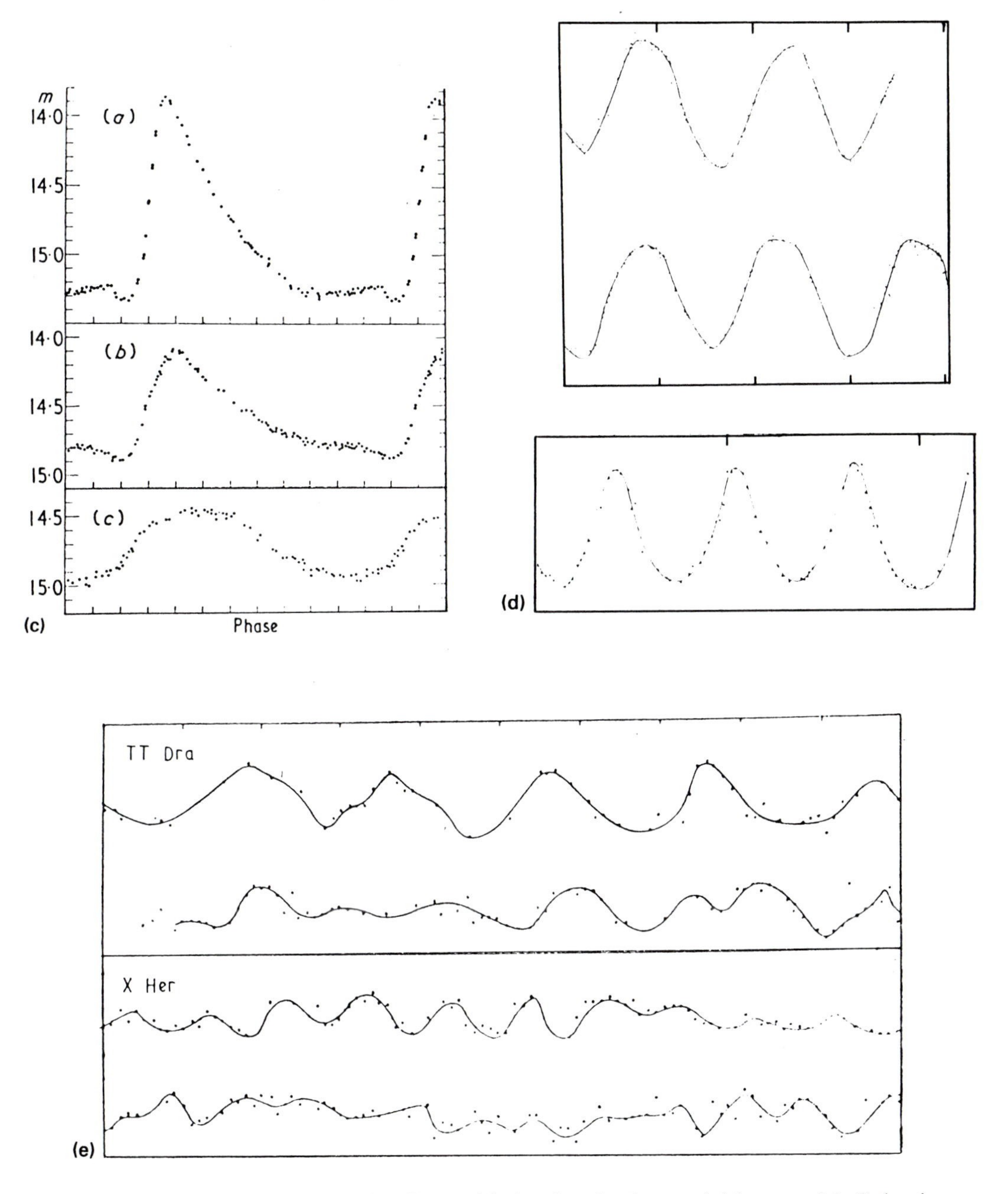

Fig. 10.1. Some examples of different kinds of pulsating variable star: (a) Galactic cepheids, (b) W Virginis type, (c) RR Lyrae types, (d) Mira stars, (e) Semiregular variables.

with the R^2 dependence on radius. The separation in phase of the two cycles suppresses a simplistic interpretation that maximum temperature should occur with maximum compression, i.e. minimum radius, though accounting for the observed value of the 'phase lag' has provided something of a challenge to theory.

To separate the contributions of radial and thermal variation, and thereby acquire a more incisive view of the physical processes underlying them, is the aim of that combination of spectroscopic and photometric (magnitude and colour) techniques associated with the names Baade and Wesselink. In practice, the procedure is not so clear cut, and a number of different versions of it have been pursued. The essence of the Baade–Wesselink methodology, however, retains the notion of integrating the observed radial velocity and combining this with temperature, determined from colour photometry, and, by comparing the result with the observed luminosity variation, to elicit the values of unknown parameters in the linking equations. A procedure for doing this will be outlined in more detail in the next section.

Apart from Baade–Wesselink analysis, which requires simultaneous (or near simultaneous) spectroscopy, the photometric periodicity itself provides a central point of interest in the study of pulsating variables. Collected observational data are searched for frequency structure, which can be compared with an extensive range of theoretical models. A fairly basic example of the significance of period determination, discovered many years ago by Henrietta Leavitt, is in the period–luminosity law for those variables whose pattern of variation resembles the prototype δ Cep — the cepheids. Such supergiant stars have highly repetitive light curves characterized by periods in the range 1–$\sim$70 days. The cepheids with longer periods have a greater mean luminosity.

There are, in fact, two major groupings in this period range (Figure 10.2). Apart from the majority of 'classical cepheids' a sizable proportion ($\sim$25% of the whole) obey a separate period–luminosity law. These stars, called after their prototype W Virginis, are associated with the older population of stars in the Galaxy.

It is inferred that the longer period stars are of greater mean radius, taking into account the mean surface temperatures across the range of observed periods. It is not difficult to square this finding with the theoretical result that periods are generally proportional to the inverse of the square root of the mean density of the star. It is deduced that the slow variables are bigger objects, containing a total amount of material not that much different from the faster pulsators, but much more diffuse in structure.

The tendency toward intrinsic variation is strongly related to the position

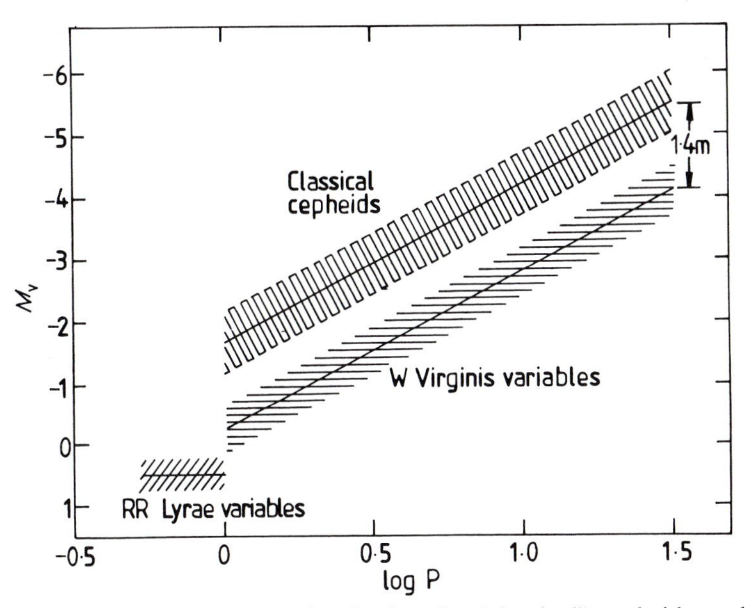

Fig. 10.2. The period–luminosity law for ('classical') cepheids, and the comparable, but somewhat rarer, W Virginis type stars.

which the star occupies on a colour–magnitude diagram (Figure 10.3). There is a certain relatively narrow but linearly extended region of this plane, which intersects the Main Sequence almost normally in the range of spectral types around A2–A8, called the instability strip. Many of the more well-known types of pulsating variable are located in, or close by, this strip. The point should be noticed that these variables are not unstable in the full sense. They remain intact, though vibrating: 'overstable' is a special term used for their condition.

The existence of the instability strip is interpreted in terms of the relative locations of ionization zones of the principal constituents of stellar matter, i.e. hydrogen and helium. For parallel strips to the left of the instability strip the ions would begin to recombine nearer to the surface; for parallel strips to the right the zones where these atoms become largely ionized are located at deeper levels in the star. The specific locations of ionization concentrations, relative to the internal energy density distribution, is of basic significance to the mechanism which drives the oscillation on to the relatively dramatic scale of variation observed in some instances.

Of course, all stars can, and probably do, oscillate over a wide range of different frequencies, and in non-radial as well as purely radial modes. One set of these more general oscillations is characterized by rapid longitudinal variations of pressure — the *p*-modes; another group is associated with the

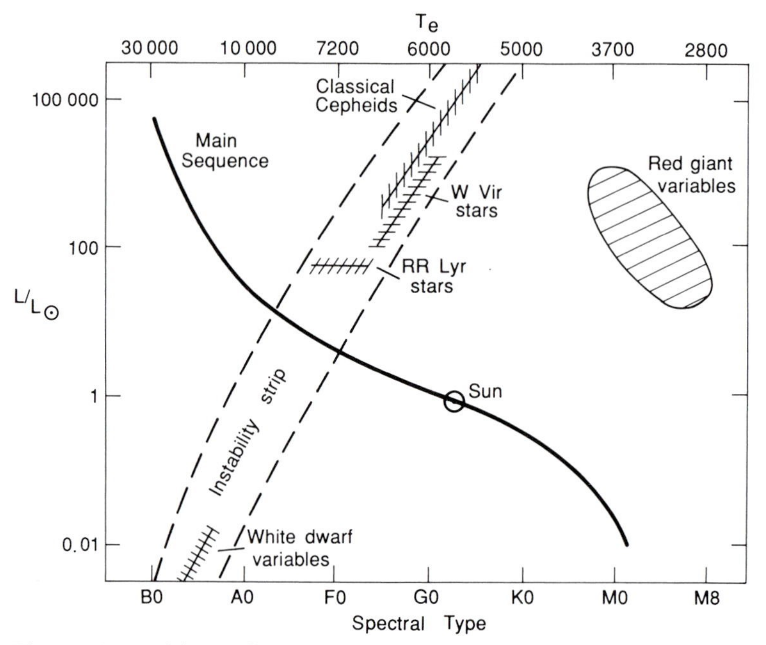

Fig. 10.3. Position of various variable stars in the Hertzsprung–Russell diagram

action of gravity in restoring density differences over spheres concentric with the star — the *g*-modes. In the majority of cases, however, such oscillations would tend to be damped out by non-conservative actions. Under certain circumstances they are reinforced, when a suitable driving mechanism is present. For this some 'negative dissipation' (Eddington's term) has to operate. This has been associated with appropriately phased withdrawals from the reservoir of ionization energy, particularly that of doubly ionized helium in the case of classical radial pulsation, when the star is in the right evolutionary stage and the ionizing layer is in the right position.

A key task lies in decomposing the more general observed variability patterns in terms of component oscillations of ascribed mode and frequency, and relating the results to stellar models of given net mass, composition and evolution, confirming or otherwise, their calculated structures. A number of different approaches to this were neatly reviewed in W. A. Cooper and E. N. Walker's book,† and this topic will not be pursued further here.

† *Getting the Measure of the Stars*, Adam Hilger, Bristol, 1989.

10.2 The Baade–Wesselink procedure

We spell out one version of this method of analysis, influenced by the setting of photometric and radial velocity data on classical cepheids in which it originated. The procedure is shown as another optimal curve-fitting exercise, comparable to those discussed in previous chapters.

In order to specify the radius at a given phase, we approximate the observed radial velocity curve, consisting of n data points, by a Fourier series

$$\dot{r}_a = a_0 + \sum_{i=1}^{h} a_i \cos 2\pi i\phi + \sum_{i=1}^{h} b_i \sin 2\pi i\phi, \qquad (10.1)$$

where the component of the star's radial motion in the line of sight $\dot{r}_a$, as measured from line shifts, is expressed in terms of the phase ϕ, and a_0, a_i, b_i are constants to be determined. This formula should match the observations to within their quoted accuracies. The number of terms to be retained in the summations is chosen by minimizing the reduced χ^2 for the corresponding number of degrees of freedom of the data $n - (2h + 1)$.

The radial motion $\dot{r}_a$ in (10.1) is normally given in kilometres per second. It is convenient to multiply this by a scaling factor s, which yields the radial motion in units of the solar radius $R_\odot$ per unit phase. First removing a_0, which simply represents the line of sight spatial motion of the star as a whole, and then integrating, we obtain for the apparent position of the line forming layer

$$R_a(\phi) = R_{a0} + s\left(\sum_{i=1}^{h} \frac{a_i}{2\pi i} \sin 2\pi i\phi - \sum_{i=1}^{h} \frac{b_i}{2\pi i} \cos 2\pi i\phi\right), \qquad (10.2)$$

where $R_a(\phi)$ measures the radial position of the line forming layer averaged over the Earthward hemisphere at phase ϕ, and R_{a0} is the mean position of this layer. The scale factor s has the value $86400P/R_\odot$, where P is the period of oscillation in days and $R_\odot$ is in kilometres.

We now relate (10.2) with the radius of the photosphere R_p. A suitable formula is

$$R_p(\phi) = R_{p0} + ps\left(\sum_{i=1}^{h} \frac{a_i}{2\pi i} \sin 2\pi i(\phi + \epsilon) - \sum_{i=1}^{h} \frac{b_i}{2\pi i} \cos 2\pi i(\phi + \epsilon)\right), \qquad (10.3)$$

where p represents another scale factor, resulting from the averaging out of $R_a(\phi)$ over the Earthward hemisphere, i.e. the observed radial velocity shifts are less than the true radial motion. ϵ allows for a difference in phase between photometrically detected radial movements of the photosphere and the spectroscopically derived motion of the line forming layer. A positive value of ϵ implies that the absorption line forming layer lags behind the photosphere.

When the difference in radii of the photosphere and line forming layers may be neglected, and the Doppler shifted spectrum of the radially expanding layer fades from the disk centre according to a linear limb-darkening law, a simple-minded approach yields $p = (6-2u)/(4-u)$, where u is the coefficient of linear limb darkening. With a standard limb-darkening coefficient of 0.6 we then find the result $p = \frac{24}{17}$, frequently cited in older texts. More thorough analysis shows this to be rather an overestimate. The profile is rendered asymmetric by radial motion, the sense of the asymmetry tending to put the measured shift somewhat greater than the foregoing would predict, to an extent which depends on inherent line broadening, instrumental resolution and measuring technique. Detailed calculations have led to the common adoption of $p = 1.31$ as the best estimate for this factor for classical cepheids.

The photosphere has a certain apparent magnitude m at effective wavelength λ

$$m(\lambda) = M_\odot(\lambda) - 2.5\left\{2\left[1 + \log\left(\frac{R_p}{\rho'}\right)\right] + \log\left[\frac{F_p(\lambda, T_p)}{F_\odot(\lambda, T_\odot)}\right]\right\}, \quad (10.4)$$

where M denotes absolute magnitude, and suffixes $\odot$ and p are applied to solar and stellar photospheric quantities. The quantity ρ' denotes what the distance to the star would be in the absence of interstellar extinction. If we take this extinction into account, we have, for the actual distance ρ,

$$\log\rho = \log\rho' - \frac{A(\lambda)}{5}, \quad (10.5)$$

where $A(\lambda)$ is the interstellar extinction at effective wavelength λ. The flux ratio $F_p/F_\odot$ involves, in addition to the wavelength λ, the temperatures $T_\odot$ and T_p. If these are brightness temperatures, then, by definition, the flux ratio is simply expressed as

$$\left(\frac{F_p}{F_\odot}\right)_\lambda = \frac{\exp(c_2/\lambda T_\odot) - 1}{\exp(c_2/\lambda T_p) - 1}, \quad (10.6)$$

where c_2 has the value 0.014388 mK (3.9).

We need the value of T_p in order to use (10.4) as a fitting function. Continuing with the Baade–Wesselink methodology, we interpret this in terms of the observed colour of the star, though this step raises difficulties about whether temperature can be uniquely related to colour for cepheid atmospheres.

Many investigators introduce here the effective temperature, T_e, since it forms the natural independent variable in the study of model stellar atmospheres from which flux values at particular wavelengths, and therefore colours, are determined. Conversely, appropriately chosen observed colours C, defined at wavelengths λ_1 and λ_2, can determine effective temperatures on the basis of given model stellar atmospheres. In practice, one of the wavelengths, λ_1 say, will be that of the fitting equation (10.4).

Empirical relationships between colour and effective temperature of the form

$$\theta_e = aC + \theta_{e1}, \tag{10.7}$$

where we have introduced the conventional reciprocal effective temperature $\theta_e = 5040/T_e$, and a and θ_{e1} are constants, have been found to be reasonably valid over restricted temperature ranges with selected colour filters for certain equilibrium stellar atmospheres. The form should hold good over most phase ranges for cepheid type variable stars, when thermal response rates for the photosphere are significantly faster than the temperature variation rates associated with the pulsations, though calculation of the steep rise of temperature just before brightness maximum may be flawed on this count. From the practical point of view, departures from the form are likely to be small compared with the probable errors of colour determinations. Moreover, the small range of phases associated with steep temperature (and gravity) variation could be given low weight, or omitted from the fitting.

Since the Planck version of the colour–temperature relationship is of the same form as equation (10.7) to a close approximation, (cf. equation (3.24)), it suggests relating brightness and effective temperatures by writing e.g. $\theta_b = k\theta_e$, where k is an empirical constant. The sought relationship between colour and brightness temperature is then also of the form

$$\theta_b = \kappa C + \theta_{b1}, \tag{10.8}$$

where

$$\kappa = \frac{5040\lambda_1\lambda_2}{2.5c_2(\lambda_2 - \lambda_1)\log_{10} e}. \tag{10.9}$$

If a single brightness temperature value does not hold over the wavelengths defining the colour index for a more realistic model, a small change of θ_b at λ_1 can still be approximated to a corresponding change of colour by a relationship of the type (10.8) if we write,

$$\kappa = \frac{5040\lambda_1\lambda_2}{2.5c_2\left[\lambda_2 - \partial\theta_b(\lambda_2)/\partial\theta_b(\lambda_1)\,\lambda_1\right]\log_{10} e}, \tag{10.10}$$

where $\partial\theta_b(\lambda_2)/\partial\theta_b(\lambda_1)$ is a quantity that can be derived from relevant model atmosphere calculations. For appropriately chosen broadband filter combinations, giving a smooth response in colour to temperature variation, we should reasonably expect that $\partial\theta_b(\lambda_2)/\partial\theta_b(\lambda_1) \approx 1$, and is slowly varying. Over the 1000 K or so surface temperature variation typical of cepheid pulsation it can be approximated by a constant.

Suppose we have a set of N photometric observations of a cepheid at two wavelengths λ_1, λ_2; i.e. m_i and C_i, at assigned phases ϕ_i. The C_is are substituted into (10.8), to yield brightness temperatures θ_{b_i}, which, when converted into T_{p_i}, go into the flux term (10.6). Equation (10.3) is used to represent photospheric radii R_p at the given ϕ_i. In this way, the observed m_i values are matched by a corresponding set of calculated values m_{ci} using (10.4) and the six quantities p, R_{p0}, ϵ, κ, θ_{b1} and ρ', which thus play the role of parameters in an optimization problem.

These six parameters are not independently determinable solely from (10.4), because p, R_{p0} and ρ' appear only in the combination p/ρ' and R_{p0}/ρ'. However, since the values of R_{p0} and ρ' are of particular interest, it is convenient to keep the foregoing notation and take it that p can be assigned a sufficently accurate value from theory.

The value of κ should not be wildly different from that given by (10.9). The reference temperature θ_{b1} is, by (10.8), constrained from a suitable comparison source of known colour and brightness temperature, e.g. the Sun. We have, in this case,

$$\theta_{b1} = \theta_{b\odot} - \kappa C_\odot. \tag{10.11}$$

Concerning the value of ϵ — real differences in phase between the pulsations of the continuum and absorption line forming layers for a cepheid should be barely detectable ($\lesssim 0.01$) according to detailed analysis. General calculations by M. Schwarzschild *et al.* for the atmosphere of the cepheid η Aql indicated that a phase lag of up to 3° would be encountered in moving from the continuum forming layer up a distance of 1% of the mean radius, i.e. some 100 × the photospheric pressure scale height. This implies no detectable effect on ϵ. Some models of non-classical cepheids (e.g. W Vir), however, indicate that the atmospheric pulsation lag may become quite appreciable ($\sim 0.1\times$period).

A non-zero value of ϵ could be appropriate if the radial velocity and photometric curves have not been properly reduced to the same epoch and period for some reason; but it is also possible that relaxation of the condition $\epsilon = 0$ results in improved curve-fits but with unphysical phase shifts. It has been known for a long time, for instance, that a reasonable match to the observed luminosity variations can be produced, even on the basis of a black-body hypothesis for the observed colour changes, provided the phase of minimum radius is displaced to coincide with that of maximum luminosity. Moving the minimum photospheric radius onto this brightest phase, however, implies a shift in the wrong direction, i.e. that the photospheric minimum occurs *after* the radial velocity zero (about its mean).

In this optimization problem, then, we have five independent parameters, of which two only — R_{p0} and ρ' — are treated as relatively free unknowns. The fitting naturally arranges itself into a two-stage procedure. At first R_{p0} and ρ' are determined on the basis of assigned p, κ, θ_{b1} and ϵ. In the second stage we would relax the values of the latter three parameters; though, since ϵ should be effectively zero for a classical cepheid, and θ_{b1} is constrained to κ by (10.11), the thrust of the procedure lies in finding the optimal value of κ. This is of particular interest in testing model atmospheres, e.g. by formulae such as (10.10).

The role of the temperature variation can be found from the variation δm of $m(\lambda)$ about its mean value during the course of the pulsation, i.e.,

$$\delta m = \frac{\partial m}{\partial C}\delta C + \frac{\partial m}{\partial R_p}\delta R_p. \tag{10.12}$$

$\partial m/\partial R_p$ is simply given by

$$\frac{\partial m}{\partial R_p} = \frac{\partial}{\partial R_p}(-5 \log R_p) = -\frac{2.172}{R_p}. \tag{10.13}$$

As for $\partial m/\partial C$, from the reasoning which preceded (10.8) (cf. Section 3.5) we have

$$\frac{\partial m(\lambda_1)}{\partial C(\lambda_1, \lambda_2)} \approx \frac{1/\lambda_1}{[(1/\lambda_1) - (1/\lambda_2)]}. \tag{10.14}$$

10.3 Six-colour data on classical cepheids

We will follow through the foregoing method with some classic observations of cepheids made in the six-colour system of Stebbins and Whitford. In particular, it is the G and I filters of this system which are used to derive temperatures. This may be compared with the $V - R$ (Johnson) temperature measurer used more frequently recently (Section 3.7). Note that the wavelengths usually quoted for these filters are mean wavelengths. The effective wavelengths are obtained by folding the spectral response of the cell plus filter with the energy distribution of the source. This has little effect with a star of around solar temperature on the mean wavelength of the G filter

(5700 Å), but the effective wavelength of the I filter is significantly displaced to shorter wavelengths, dropping from $\lambda_0 \simeq$ 10300 Å to $\lambda_{eff} \simeq$ 9700 Å. Using these values, we find from (10.9) $\kappa = 0.446$, and from (10.14) we have

$$\frac{\partial m_G}{\partial(G-I)} \approx 2.43. \tag{10.15}$$

In place of equation (10.12) we therefore write,

$$\delta m_G \simeq 5.5\kappa\delta C - \frac{2.2}{R_p}\delta R_p. \tag{10.16}$$

The major features of cepheid optical light variation may be assessed from this equation, recalling the phase difference between minimum R_p and minimum C (maximum temperature), the latter occurring typically about 0.1 period after the former. In the case of δ Cep, for example, an apparent brightness variation of 0.84 mag is observed in the G-filter observations of Stebbins. This is matched by a colour variation $(G - I)$ of 0.44 mag. The temperature variation thus has over four times the effect of the radial variation at visual wavelengths.

From (10.16) it follows that the *amplitude* of the colour variation, not the absolute value of the colour, is significant as far as the shape of the light curve is concerned. Hence, interstellar reddening, which would subtract a certain, nearly constant, amount from the apparent magnitude observed with each filter (cf. Section 4.1.3), should not have any great part to play in the curve-fitting other than fixing the mean light level. On the other hand, interstellar extinction will, as indicated by (10.5), affect the meaning of the obtained value of ρ'. Where the extent of reddening is available for a particular cepheid it can be initially subtracted from the observed colours C_i. The apparent magnitudes m_i must then be made correspondingly brighter to obtain a consistent representation. This correction can be determined from (10.14), i.e. for the six-colour system,

$$\Delta m_{Gred} = -\frac{\partial m_G}{\partial(G-I)}\Delta(G-I)_{red} \approx -2.4\Delta(G-I)_{red}.$$

When such corrections are initially made to the data, the resulting value of ρ' is equivalent to the true distance to the star, and the separate calculation (10.5) is not required.

A correction for binarity, when known to be present, could be similarly made *a priori*. The proportional contibution of the companion at the wavelength of each filter requires the addition of corresponding magnitude corrections, rendering data appropriate for the cepheid alone to have magnitudes m_c, somewhat greater than those observed m_o. This may entail a successive approximations procedure, depending on the relative scale of the companion's light, but if this is small (as would be normally the case), we can write

$$m_c \simeq m_o + 1.086 \times 10^{0.4\Delta m_o} \left(\frac{L_B}{L_{Amax}} \right), \tag{10.17}$$

where L_B/L_{Amax} is the ratio of the luminosity of the companion to that of the cepheid at maximum. The magnitude differences Δm_o are measured from that at phase zero, i.e. the lowest m_o.

L_B/L_{Amax} will, in general, be different at different wavelengths, so that anomalous colour variations occur for a cepheid of a given amplitude if this ratio becomes significant. This point has given rise to some methods of determining binarity, as a result of which it appears that an appreciable proportion — at least $\sim$30% — of cepheids are in binary systems. The complications thereby introduced for photometric analysis tend to lose significance at longer periods, where the cepheid's luminosity increasingly dominates, in agreement with the period–luminosity law.

At the initial stage the parameters κ, θ_{b1}, p and ϵ are fixed at 0.446, 0.951, 1.31 and 0, respectively. Equation (10.16) shows that the G light curve is well conditioned to determine a value of R_{p0}, and the mean apparent magnitude constrained then to yield a value of ρ'. As already noted, however, these quantities directly depend on the adopted value of the scaling parameter p; the real accuracy of their determination thus cannot be greater than that with which p can be specified, which is probably to within 10%.

10.3.1 Application to δ Cep and η Aql

We are now ready to turn to two famous examples of cepheid variables — δ Cep and η Aql — whose variability was discovered in the eighteenth century by the young amateur astronomer friends John Goodricke and Edward Piggott.

RADIAL VELOCITIES — Since both stars are quite bright, it is not surprising that relatively good radial velocity curves were obtained many years ago by T.S. Jacobsen. For δ Cep, this was published almost twenty years before the six colour photometry of Stebbins, and so some tailoring of the phases of the two data sets was required. This is complicated by a drift in period of δ Cep (5.366282 days at the time of Stebbins' photometry), but the shortening has been found to be essentially linear, so that phases for the radial velocity on the same ephemeris as the photometric variation are derived with reasonable confidence. The result is shown in Figure 10.4 (c),

Table 10.1. *Fourier fit (equation (10.1)) to Jacobsen's (1926) radial velocity curve for δ Cep.*

coefficients in km s^{-1}			
a_1	0.42	b_1	−15.5
a_2	0.61	b_2	−6.76
a_3	−0.21	b_3	−3.65
a_4	0.13	b_4	−2.27

$a_0 = -15.74$, $\Delta\dot{r} = 1.5$,
$\chi^2 = 10.69$, $v = 10$.

Table 10.2. *Fourier fit to the combined Schwarzschild* et al.*'s (1948) and Jacobsen's (1926) radial velocity curves for η Aql (with Schwarzschild* et al.*'s phases).*

coefficients in km s^{-1}			
a_1	−8.59	b_1	−13.22
a_2	−8.81	b_2	−0.25
a_3	−1.32	b_3	2.78
a_4	0.14	b_4	1.18
a_5	0.69	b_5	0.52

$a_0 = -74.82$, $\Delta\dot{r} = 1.0$,
$\chi^2 = 29.91$, $v = 36$.

which shows the smoothly changing radius with phase. This comes from integrating the Fourier represention (10.1) with the coefficients (in Jacobsen's ephemeris) listed in Table 10.1.

In the case of η Aql Jacobsen's data were supplemented by observations by Schwarzschild *et al.*, made close to the time of Stebbins *et al.*'s photometric observations. An eleven-term Fourier fit gives a very good representation of the radial velocity, the details of which are given in Table 10.2. The radius of η Aql rises to its larger maximum in a more drawn out way than that of δ Cep, but the location in phase and shape of the minima are very similar.

INTERSTELLAR REDDENING — The period–luminosity law for cepheids (Figure 10.2) has allowed them to play an important 'yardstick' role in gauging distances, which, since they are intrinsically very luminous, can be effective over large spatial extents. It is necessary, though, to have a clear awareness of the role of interstellar reddening, particularly as classical cepheids are, in general, concentrated towards the galactic plane. A number of studies have been made of this, and the averages of measured colour excesses, including some in the six-colour system itself, and others interpolated to it, turn out as $\Delta(G - I)_{red} = 0.09$ for δ Cep and $\Delta(G - I)_{red} = 0.19$ for η Aql. The corresponding absorptions are $A(G) = 0.22$ for δ Cep and $A(G) = 0.48$ for η Aql.

BINARITY — There is a visual companion to δ Cep, about 3–4 mag fainter, of spectral type A0, and some 40 arcsec south of the variable. Possible light contamination from this star is seldom mentioned. It can be assumed that experienced photometrists have excluded its light, and has therefore been neglected in the present analysis.

In 1980 some astronomers using the IUE satellite were able to show, from a conspicuous excess at far ultra-violet wavelengths, that η Aql has an unresolved binary companion. It was determined to be an A0 type main sequence star. This component must be included in all photometry of the cepheid; however, there is a difference of 4–5 magnitudes in absolute visual magnitude between the two stars, so that its effects can be at most a few per cent in the G and I ranges. Indeed, its presence was not detected by earlier anomalous colour effect methods, so that neglect of it cannot be too serious when judged against available photometric accuracy. The derived colour temperatures are presumably slightly high, and the required radii would be incremented as a result, but any such variation is within the error arising from other sources of uncertainty.

OPTIMAL PHOTOMETRIC PARAMETERS — The unreduced photometric data of Stebbins and his colleagues were analysed. The standard deviation of a single observation is about 0.007 mag. This means that the colour determinations correspond to a standard deviation of 0.010, and from (10.16) we therefore expect colour-based luminosity predictions to be accurate to within $\Delta m = 0.024$ mag, neglecting the effect of observational errors on the derived radius variations. This quantity, used in the calculation of $\chi^2 = \sum_{i=1}^{N} [(m_i - m_{ci})/\Delta m]^2$, permits a χ^2 test on the feasibility of the fit.

The comparison star used for δ Cep was ϵ Cep, while β Aql was similarly used for η Aql. In order to convert known V apparent magnitudes to the G scale, the formula $G = V - 0.0099(B - V)$ was used. Taking V and $B - V$ values from the Arizona-Tonantzintla Catalog, we therefore find: $G(\epsilon\,\mathrm{Cep}) = 4.17$ and $G(\beta\,\mathrm{Aql}) = 3.64$. The absolute magnitude of the Sun in the G filter was similarly found to be 4.81.

With these preliminaries adopted, the results of optimal fittings to the observed magnitudes are presented in Tables 10.3 and 10.4. Three sets of results are given for each star. These correspond to (i) the single brightness temperature hypothesis; (ii) κ being given by (10.10) and $\partial\theta_b(I)/\partial\theta_b(G)$

Table 10.3. *Optimal parameter sets — δ Cep.*

Period = 5.366282 d					
Element	(i)	(ii)	$\pm$	(iii)	$\pm$
R_{p0} ($R_\odot$)	42.9	49.4	1.8	53.2	2.7
ρ' (pc)	256	298	9	311	13
M_V	–3.06	–3.41	0.06	–3.50	0.08
κ	0.446	0.4135		0.3876	0.003
θ_{b1}	0.951	0.9447		0.9395	
χ^2	111.4	38.14		18.93	
ν	23	23		22	

$p = 1.31$, $\epsilon = 0.0$, $\Delta m = 0.0235$

Table 10.4. *Optimal parameter sets — η Aql.*

Period = 7.176678 d					
Element	(i)	(ii)	$\pm$	(iii)	$\pm$
R_{p0} ($R_\odot$)	47.3	54.5	1.1	56.5	1.6
ρ' (pc)	266	310	5	323	7
M_V	–3.29	–3.63	0.03	–3.72	0.05
κ	0.446	0.4135		0.4084	0.003
θ_{b1}	0.951	0.9447		0.9437	
χ^2	69.6	43.19		41.29	
ν	41	41		40	

$p = 1.31$, $\epsilon = 0.0$, $\Delta m = 0.0235$

evaluated numerically from model stellar atmosphere data ($\simeq 0.945$); and (iii) κ allowed to be a free parameter. Observed and calculated data are shown in Figures 10.4 and 10.5.

Probable errors have been assigned in Table 10.3 and 10.4 for the parameters corresponding to cases (ii) and (iii). The large χ^2 value for case (i) demonstrates that this is an unrealistic model, so that formal errors would here be clearly inappropriate. These errors were determined from the diagonal elements of the error matrix. The extent of correlation between κ and R_{p0} can be judged from the decreasing precision with which R_{p0} is estimated when we allow κ to be a free parameter. In the case of η Aql the optimal solution with free κ is only slightly different from the case (ii) solution. The same is not true for δ Cep, where the model stellar atmosphere prediction of κ is significantly different from the optimal solution.

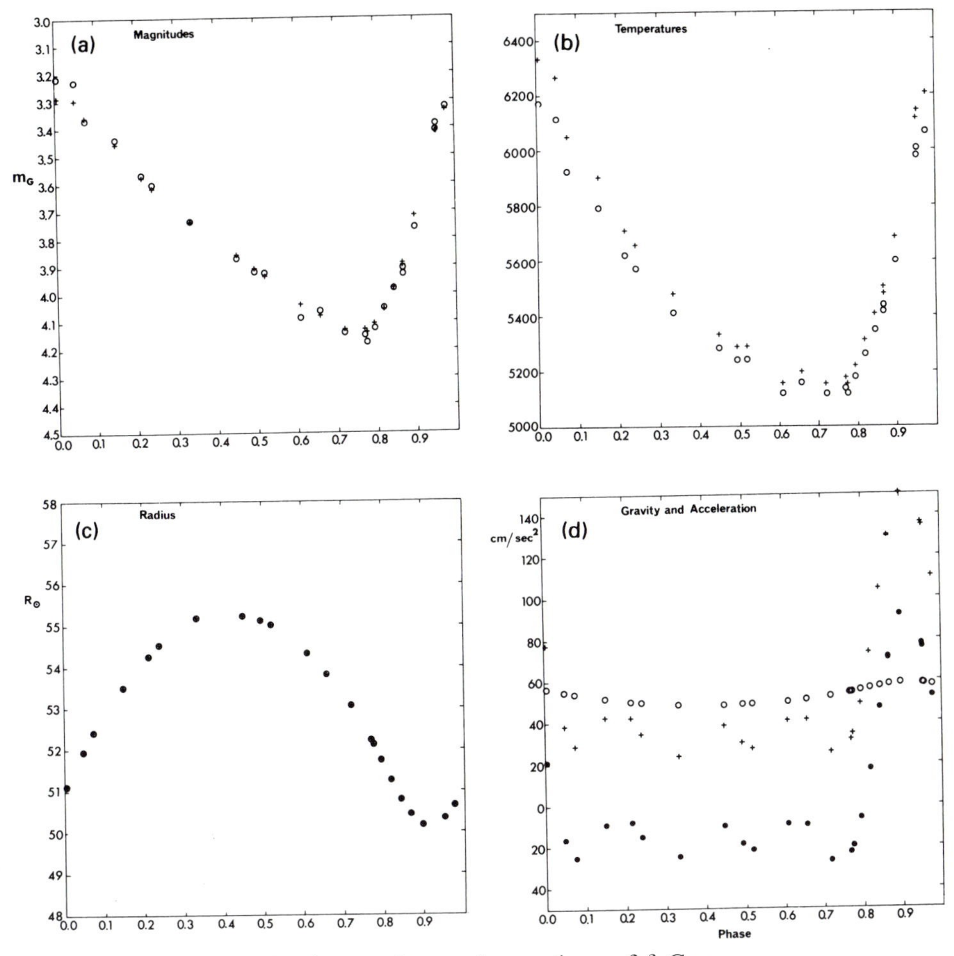

Fig. 10.4. Optimal curve-fits to observations of δ Cep.

It should again be emphasized that the quoted errors refer to the mathematical properties of the fit, and do not properly reflect more general physical uncertainties. Thus the values of R_{p0} and ρ' directly depend on the value of p, while ρ' depends on certain other assumed quantities which have been entered into the fitting equation.

Effective temperatures — The temperature variable in equation (10.4) is the brightness temperature at the wavelength of the light curve. Effective temperatures are, however, of more general interest. Where information is available to connect brightness and effective temperatures, as in case (ii) of the last section, we can obtain an appropriate empirical relationship. κ was calculated for this case by comparing the variation of brightness temperature

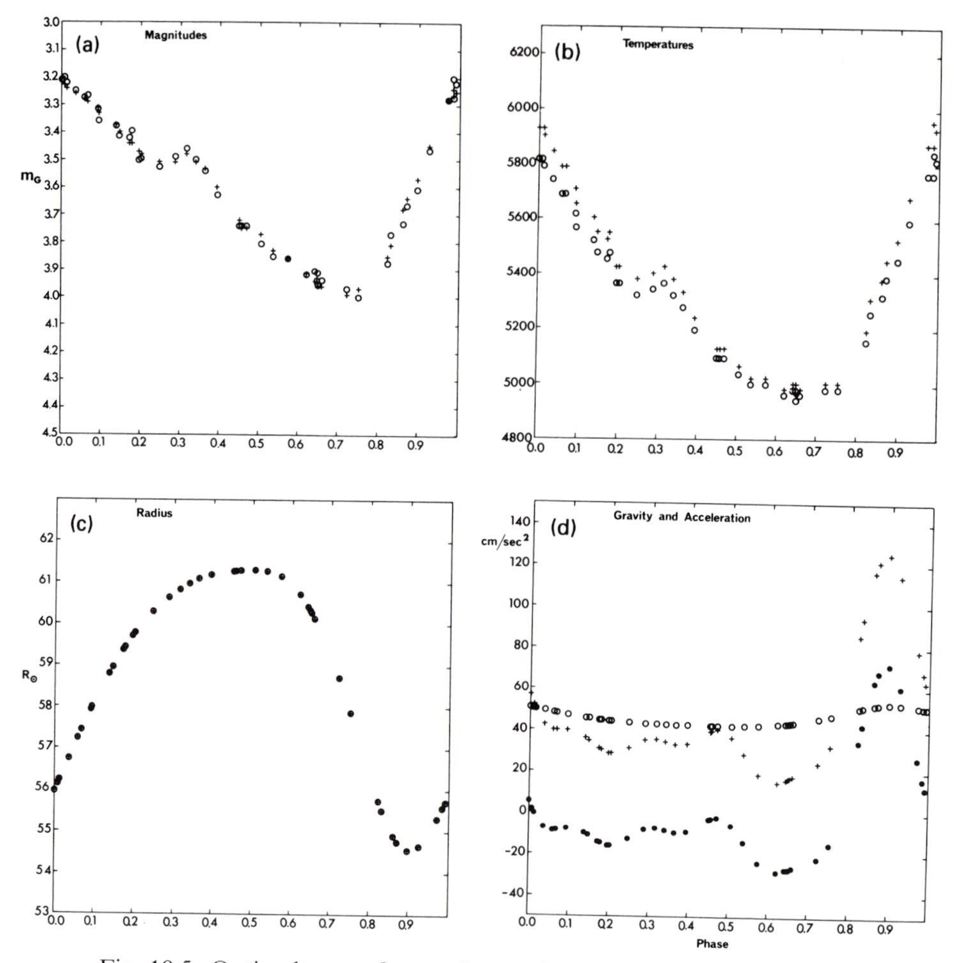

Fig. 10.5. Optimal curve-fits to observations of η Aql.

at the I wavelength with that in the G region as the effective temperatures varied from 5000 to 6000 K, using the model atmosphere data of D. F. Carbon and O. Gingerich† (with $\log g_{eff} = 2$). In this temperature range a linear relation between fluxes and *effective* temperatures (neglecting any effect of gravity variation) yields

$$\frac{\partial(G-I)}{\partial\theta_e} = 0.448.$$

Again using solar values to calibrate, we find

$$\theta_e = 0.448(G-I) + 0.951.$$

† See *Theory and Observation of Normal Stellar Atmospheres*, MIT Press, Boston, 1969.

Of course, colour–effective temperature relationships of this type, if valid at all, are likely to be so only under restricted ranges of the variables involved.

CONCLUSIONS — The primary objective of the method has usually been the determination of parallax and mean radius, but a secondary one arises in comparison of theoretical and optimal values of the colour–temperature constant κ. The simple hypothesis of one and the same brightness temperature valid at both wavelengths defining the colour has been ruled out in accounting for the light curves of δ Cep and η Aql. The χ^2 test discriminates against this hypothesis with over 99% confidence.

The static model stellar atmospheres used, while able to provide a plausible fit for η Aql, are less reliable in the case of δ Cep, though the data coverage for η Aql, both spectroscopic and photometric, is greater than for δ Cep. There is some uncertainty attached to the phase consistency of radial velocity and luminosity variations with δ Cep, due to a large difference in observation epochs. The behaviour of the optimal solution has been found to be sensitive to such phase differences, e.g. a variation of ϵ by 0.01 could produce a variation in the optimal value of R_{p0} by 2–3%.

The values of κ obtained by optimizing the fit of calculated to observed apparent magnitudes appear less than theoretically expected. This could be explained by a greater role of blocking in the G region at the low gravities encountered in cepheid atmospheres. The mean effective gravities $\log(GM/R_p^2 + \ddot{R}_p)$ of δ Cep and η Aql are about 1.5 and 1.6 respectively. Line blanketing, at the effective temperatures involved, increases in the visible region as gravity decreases. Some colour choices appear better for giving a constant value of κ than others. Barnes and Evans found $V - R$ to be significantly less sensitive to gravity than $B - V$, for example (Section 3.7). Other authors have made different claims for the most generally reliable colour to use for results which are consistent with theory, or alternative empirical tests. More extensive theoretical data could allow this to be checked more thoroughly, and, though χ^2 for the η Aql fit indicates that existing static models are already acceptably close, matters may be still improved by relaxing (10.8) or (10.10) to include gravity dependence.

The Baade–Wesselink method can estimate mean photospheric radii of classical cepheids in a self-consistent way to within 10%, perhaps 5%. The position is less certain with regard to parallaxes. The uncertainties of interstellar reddening data are not negligible — often of the order of 0.1 mag. The consequent effects on $\log \rho'$ are about half this figure. This is in addition to the error correlated with that of R_{p0} and those associated with the constants in the colour–temperature relation. Nevertheless, cepheids provide

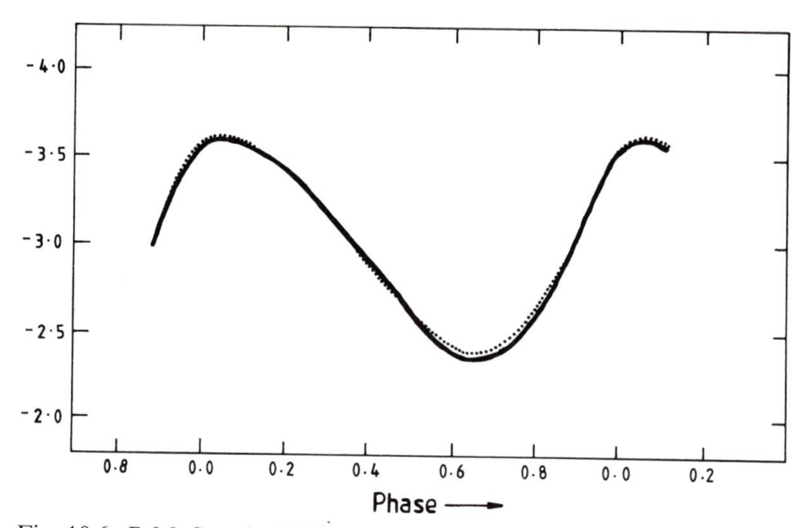

Fig. 10.6. R.M. Scott's (1942) comparison of the mean bolometric luminosity variation of Mira (full curve) with one calculated by using a Baade–Wesselink technique (dotted curve).

one of the best-known means of probing great distances in space, and it is important to obtain absolute magnitudes of at least a few reference stars. Optimal curve-fits to high quality photometric observations, where reliable interstellar reddening values are provided, offer a promising means of doing this.

10.4 Pulsational radii

Apart from its application to classical cepheids, the Baade–Wesselink analysis procedure has also been used to study the sizes and distances of W Virginis stars, RR Lyrae variables, Miras (Figure 10.6) and variables of the δ Scuti type. It has been generalized to take account of non-radial modes of oscillation, which this latter class of variable sometimes indicates. In principle, it could be applied to any star undergoing a measurable radial expansion or contraction, whose brightness in different wavelength ranges can be monitored, and thereby corresponding surface temperature changes inferred. It has even been extended to the photospheric expansion stage of a supernova. However, the domain of classical cepheids has been its main testing ground. Cepheids are among the best known and understood of the pulsating variables, so there is a good development of theory at hand with which to compare results.

Efforts to improve and extend the technique continue to be made, despite built-in difficulties connected chiefly with (a) the colour–temperature conversion formula, and (b) uncertainties in the connection between the observed radial velocity variation and the derived photospheric changes of radius, e.g. via the parameter p. Precise calculations for moving (spherical) atmospheres imply that p should be in turn dependent on other model parameters. Indeed, putting p as a constant may be a physical oversimplification for exact matching of the entire cycle, though some mean value must exist which will provide the proper mean radius.

In a critical survey of almost 400 determinations of cepheid radii for 55 stars by different people using different versions of the Baade–Wesselink technique on different data sets, J. D. Fernie found a fair measure of agreement ($\Delta R_{p0}/R_{p0} \sim 5-10\%$) over the entire range of periods. As Fernie pointed out, however, it is possible to check these results independently, by comparing the slope of the period–radius relation using Baade–Wesselink radii with that obtained from a group of cepheids taken to be all at effectively the same distance, such as in clusters, or the Magellanic Clouds. Even if there are doubts about the constant term in these latter formulae, no good reason has been advanced as to why the best-fitting slope should not be a reliable test. In fact, the slope is significantly different for the alternative radii, at least at longer periods. This is in the sense that the Baade–Wesselink radii are too small — the error appears to reach up to 50% at periods $\sim$ 50 d.

Though this discrepancy is still not properly resolved, it has been found that a better accord is obtained when a version of the Baade–Wesselink method is used which extends beyond the single dependence of θ_b on C, as indicated by (10.8) with κ constant, to include also a gravity dependence. In an early study of RR Lyrae stars using the Baade–Wesselink approach Richard Woolley and Ann Savage pointed out the desirability of including gravity effects. The RR Lyrae stars show similar or greater radial velocity amplitudes than the classical cepheids, but their periods are typically an order of magnitude or more shorter: they must experience very marked surface accelerations as a result, and an enhanced role for the effective gravity is understandable.

Some versions of the Baade–Wesselink method focus their attention on matching radii, rather than the light curve, as done in the preceding sections. In this case, instead of (10.3) being substituted into (10.4), the latter equation is rearranged to provide an expression for R_p, so that (10.3) becomes the fitting equation. The light curve may also be Fourier decomposed so that different phase arguments may be used than those of the raw data. Another version, developed by some European authors, treats the problem as an integral equation. The magnitude colour and log(flux) colour products are integrated around the variability cycle. The integral of the latter product is small (zero, in the classical Baade–Wesselink formulation, in which flux is single-valued with colour), but non-negligible in general. The sum of these two integrals is then matched with a corresponding log(radius) colour product integrated around the light cycle. A certain mean radius will balance the sum. This has been called the CORS method, and its results appear to give the required higher slope to the log(period) versus log(radius) relation.

No doubt, continued experimentation will yield advances both in results and methodology, seen against the required purposes. This chapter has shown that the light curves of pulsating stars can be treated along essentially similar lines to the optimal curve-fitting of those of other types of variable. Apart from the Fourier coefficients of the radial velocity representation, whose physical significance was not explored, the derived parameter set was small — only R_{p0}, ρ' and κ in the examples considered. It remains to be seen whether more extensive modern observational data sets will allow extension of the number of derivable parameters to include gravity, microturbulence, or perhaps other parameters in the fitting equation. Alternatively, more extensive calculations of model atmospheres, covering the relevant temperature, gravity and microturbulence ranges, would allow a more accurate flux representation to be implanted, thereby concentrating the optimization to the mean radius and distance. It is challenging that independent means also exist whereby results thus derived can be checked.

10.5 Bibliographical notes

The book on *Variable Stars* by C. Hoffmeister, G. Richter and W. Wenzel, Springer Verlag, Berlin (1984) provides a comprehensive background to the physical variables considered in this chapter. Other books which could be referred to are S. Rosseland's *The Pulsation Theory of Variable Stars*, Clarendon Press, Oxford, 1949, which has a nice chapter on the Baade–Wesselink technique; S. A. Zhevakin's *Theory of Stellar Pulsation*, Leningrad, 1957; W. Strohmeier's *Variable Stars*, Pergamon Press, Oxford, 1972; *Pulsating Stars* (ed. B. V. Kukarkin) Keter Publishing House, Jerusalem, 1975; M. Petit's *Variable Stars*, John Wiley, Chichester, 1987 and *IAU Colloq. 82, Cepheids: Theory and Observation* (ed. J. Madore) Reidel, Dordrecht, 1985. At a more intensive level, J. P. Cox in *Rep. Prog. Phys.*, **37**, 563, 1974, gives a full review of the physics of cepheid-like pulsating stars – at least, in the linear theory. The diagrams used in Figures 10.1–10.3 come from that paper.

Among the large background of papers on the Baade–Wesselink idea are: W. Baade, *Astron. Nachr.*, **228**, 359, 1926; K. F. Bottlinger, *Astron. Nachr.*, **232**, 3, 1928; W. Becker, *Z. Astrophys.*, **19**, 219, 1940; A. van Hoof, *Koninklijke Vlaamsche Acad. Weten.*, **5**, No. 12, 1943; A. J. Wesselink, *Bull., Astron., Inst., Netherland*, **10**, 91, 1946; J. Grandjean and P. Ledoux, *Ann. Astrophys.*, **17**, 161, 1954; R. Cavanaggia and J. C. Pecker, *Ann. Astrophys.*, **18**, 151, 1955; C. Whitney, *Astrophys. J.*, **121**, 682, 1955; J. B. Oke, *Astrophys. J.*, **133**, 90 and **134**, 214, 1961; C. F. Keller and J. P. Mutschlecner, *Astrophys. J.*, **167**, 127, 1971.

R. M. Scott, *Astrophys. J.*, **95**, 58, 1942, compared radiometrically derived radial variations of *o* Ceti by Pettit and Nicholson (cf. Chapter 3) with those from the Baade–Wesselink technique, though there are still problems in obtaining a suitable infra-red wavelength base to obtain reliable colour temperatures for such cool stars. The technique was extended to RR Lyrae stars in the cited study by R. Woolley and A. Savage, *R. Obs. Bull.*, No 170, 365, 1971, and new work on this class of star was introduced by T. Liu and K. A. Janes in *Astrophys. J. Suppl.*, **69**, 593, 1989. The study of W Virginis stars referred to is that of E. Böhm-Vitense, *Astrophys. J.*, **188**, 571, 1974.

An updated review was given by A. Gautschy, *Vistas Astron.*, **30**, 197, 1988. The particular treatment presented in Sections 10.2 and 10.3 derives from E. Budding, *Astrophys. Space Sci.*, **48**, 249, 1977. The data studied are from rather old sources (photometry — J. Stebbins, *Astrophys. J.*, **101**, 47, 1945, J. Stebbins *et al.*, *Astrophys. J.*, **115**, 292, 1952; radial velocities — T. S. Jacobsen, *Lick Obs. Bull.*, **12**, 138, M. Swarzschild *et al.*, *Astrophys. J.*, **108**, 207); but the stars (δ Cep and η Aql) are sufficiently bright that a relatively high accuracy was obtainable even in the early days of photometry and radial velocity curve derivation: sufficiently accurate to show some inadequacy in the simpler forms of the Baade–Wesselink procedure, as seen in the results.

A full review of pulsational radii obtained by a number of, to some extent, independent techniques was provided by J. D. Fernie in *Astrophys. J.*, **282**, 641, 1984. Another review is that of E. Böhm-Vitense *et al.*, *Astrophys. J.*, **343**, 343, 1989.

Appendix

Frequently encountered constants

For convenience very commonly met with numbers are here listed in one place. Information on mathematical constants can be found in e.g. M. Abramowitz and I. A. Stegun's *Handbook of Mathematical Functions*, Dover Publications, New York, 1964. Much more of the relevant physical and astronomical data can be located in e.g. C. W. Allen's *Astrophysical Quantities*, Athlone Press, London, 1973, or (more recently) in G. W. C. Kaye and T. H. Laby's *Tables of Physical and Chemical Constants*, Longman, 1986.

Mathematical

π	3.14159265358979
$\pi/2$	1.57079632679490
$\sqrt{\pi}$	1.77245385090552
1 radian	$57\overset{\circ}{.}2957795130823$
e	2.71828182845905
$\log_{10} e$	0.43429448190325

Physical

Velocity of light c	2.997924580×10^8 m s^{-1}
Planck's constant h	6.626176×10^{-34} J s
Boltzmann's constant k	1.380662×10^{-23} J K^{-1}
Gravitational constant G	6.6720×10^{-11} N m^2 kg^{-2}
Electron charge e^-	1.602189×10^{-19} C
Electron mass m_e	9.109534×10^{-31} kg
Proton mass m_p	1.672648×10^{-27} kg
Radius of first Bohr orbit a_0	5.291771×10^{-11} m
Energy of 1 eV E_0	1.602189×10^{-19} J
Corresponding frequency ν_0	2.417969×10^{14} Hz
Corresponding wavelength λ_0	1.239852×10^{-6} m
Corresponding temperature E_0/k	11604.5 K
Photon energy of wavelength λ	$1.986478 \times 10^{-25}/\lambda$ J
Astronomical unit (AU)	$1.49597870 \times 10^{11}$ m
Tropical year (yr)	3.155692548×10^7 s
Light year	9.460528×10^{15} m
Parsec (pc)	3.085678×10^{16} m
Mass of Sun $M_\odot$	1.990×10^{30} kg
Radius of Sun $R_\odot$	6.9598×10^8 m
Luminosity of Sun $L_\odot$	3.827×10^{26} W
Sun's visual magnitude $m_{v\odot}$	$-26.75 \equiv 1.25 \times 10^5$ lx
Sun's bolometric magnitude $m_{bol\odot}$	$-27.82 \equiv 1360$ W m^{-2}

Subject index

a-parameter, 79
absolute magnitude – see magnitude
absolute (all sky) photometry, 142
abundances, 65, 69, 80, 95
  [Fe/H] ratio, 77
accuracy
  Baade–Wesselink method, 257, 259–60
  magnitude – see magnitude accuracy
  parameter, 70, 78, 100, 151 – see also curve fitting
  timing, 145
  tracking, 111
  – see also resolution
achromatic beamsplitter, 93
actinic (in photography), 64
æther, 13–14, 26, 29
airmass, 57–61
albedo, 87, 89, 92
Algol paradox, 178
Almagest, 13, 15
α function, 185, 215, 231
α integral, 90, 187–9, 215, 231–2
α table of values, 185
alt-azimuth coordinates, 59
amateur astronomers, 2, 9, 251
amplification, 38, 117
  CCD, 130
  DC, 132–4
  pulse-counting, 138–9
analyser – see photopolarimetry
anomaly – mean (M), true (v) 199–200
aperture
  input, 111
  iris type, 111
  stop, 140
areal detection (imaging), 114, 116, 120, 126–32
astrometry, 1
atmospheric seeing, 111, 116
atmospheric transparency, 15, 56–8, 164–5
automatic photometric telescope (APT), 3, 6, 221
averted vision, 37

Baade–Wesselink method, 5, 245–60
backwarming, 69
Balmer absorption edge (limit), 64, 76
Balmer continuum, 65
Balmer decrement (jump), 24, 68, 70
Balmer lines (series), 64, 80–4, 93, 94, 220 – see also β photometry
band
  conduction, 117–19
  gap, 118–19
  valence, 117–19
bandwidth
  frequency (f_b), 138, 139
  limitation, 138
  selection, 115
  – see also filters
Barnes–Evans relation, 52–3, 54, 55, 257
beam collimation, 115
beam focusing, 139
beam multiple, 111
beam readout, 128
beam splitting, 115
β Lyrae paradox, 178
β photometry, 79, 80–4, 212–18
binary stars, 15–16
  Algol (EA) systems, 101–4, 175–7, 212–16
  β Lyr (EB) systems, 175–7, 207
  circumbinary matter, 101, 212
  close, 5–6, 104, 174, 177, 179, 199–219
  common envelope, 179
  detached, semidetached, contact classification scheme, 179
  disk sources in, 214–16
  dwarf novae, 179
  gaseous streams, 103, 212
  geometry, 101–5, 174–5, 201
  interactive evolution, mass transfer 5, 175, 212, 216–18
  proximity effects, 201–5
  RS CVn type, 5, 179, 220–2
  synchronous rotation in, 220
  W UMa systems, 175–8
  X-ray, 204

– see also stars – binarity, eclipsing binary systems
Black Birch outstation, 145, 157
black body radiation, 30, 33–4, 35, 39, 40, 45, 54, 68, 204–5
blanketing, 69
BL Lac type objects, 12
blocking, 69
bolometer, 117
bolometric correction (*BC*), 23, 25, 38–40, 42, 45, 53
Bouguer's law, 15, 57–8
brightness
  descriptive introduction, 3, 9–12
  of stars and magnitude units, 16, 22
  surface, 15
  temperature, 43
broadband filter, 45–51
broadband photometric systems, 64–73

c_1, $[c_1]$ (Balmer decrement) index, 24, 77–80
calibration coefficients (ϵ, μ, etc.), 143, 148
calibration curves, 17, 19
calibration of photometric system, 10, 142–57, 213
Cassegrain focus, 111
cataclysmic variables, 2, 12
cepheids, 12, 240, 242–4, 249–58, 259, 260–1
  binarity in, 250–1, 253
  distances from, 251, 257
  period–luminosity law, 242–3
  period–radius law, 259
  phase lag, 248, 250
  position in Galaxy, 252
  position in H–R diagram, 243–4
charge coupled devices (CCDs), 120, 121, 129–32, 141
  channel stops, 129
  chip architecture, 120, 121
  depletion regions, 129
  frame transfer, 130–1
  signal carriers, 129
  wavelength sensitivity, 131
check star, 159
χ^2 – see curve fitting
chromospheric emission, 221
Clairaut stability criterion, 204
clock, 115
coherence length, 26
colour, general introduction, 9–10, 16, 27, 28
colour diagram, 63, 67-9
colour – effective temperature relation – see temperature – colour, effective
colour excess, 24, 78, 84
colour index, 23, 24
colour–magnitude diagram, 12, 62, 64, 66–9
coloured glass, 10, 47
column-density, 62
comparison star, 159
component
  cooling, 128, 131, 134
  stability, 125, 135, 137, 139
computer
  capabilities, 3, 98, 140, 141, 161, 170, 219
  control, 110, 115, 138, 161, 164
  data handling, 130–2, 135, 138, 148, 161, 167
  light curve analysis, 186–96, 203 – see also eclipsing binary systems, spotted stars, pulsating stars
  model atmospheres, 39, 44, 247, 248, 249, 254, 256, 257, 259, 260
  personal (micro), 2, 161, 205
convective heat transport, 204
corpuscular theory, 26
CORS method, 260
cosmic ray incidence, 124
cross-section (absorptive), 61
current
  dark, 124–5, 130
  to frequency converter (CFC), 135, 136, 161
  measurement (DC method), 132–5
curve fitting, 172–4, 205–7
  χ^2 minimization 189–93, 206–7, 226, 253
  error evaluation, 192–3, 206–7, 210, 257
  fitting function, 172–3, 223, 225, 228, 229, 231, 246
  parameter optimization, 172, 190–2, 206–7, 216, 222, 226, 229, 251, 253–5, 257
  separability of effects, 232–4, 248
  – see also least squares method, eclipsing binary systems, spotted stars, pulsating stars

data – see photometric data processing
data file, 161–3, 166–8
data logger, 162
data space, 172
data spacing, 211
DC measurement, 132–5
DC photometry, examples, 144–5, 161–2
dead time (τ_d), 138–40
depolarizer – see photopolarimetry
detector, 116–32, 140–1
  dynamic range, 121
  efficiency (*Q*), 115, 117, 120, 121, 124, 127, 129, 132, 152
  event capacity, 121, 132
  linearity, 109, 115, 116, 119, 121, 122, 127, 135, 138, 139
  noise (detection), 123–5 – see also noise
  saturation effect, 121, 129
  size, 121
  sky limited, 116, 124, 126
  temperature, 118, 120, 123, 124, 130, 134
differential photometry, 142, 159–68
diffusion approximation, 204
discrimination (input signal), 135, 136
Doppler imaging, 221
Doppler shift, 212, 239
double stars 15 – see also binary stars
dwarf – see stars, types
dynodes, 121–4

eclipsing binary systems, 5, 12, 15, 44, 172
  8 parameters (spherical) model, 184, 193, 202
  11 parameters model, 203
  16 parameters (standard eclipsing binary – SEBM) model, 203–7, 232
  basic parameter derivation, 180–5
  computer analysis, 186–96
  eccentricity, 175, 199–201
  flux ratio, 183, 207
  frequency domain analysis, 210–12
  geometry of eclipses, 174, 181, 199
  gravity brightening (τ), 204–5, 210
  internal, external tangencies, 183
  light curve modelling, 199–219
  light curve morphology, 175–7
  mass ratio, 203, 210
  narrowband photometry, 212–18
  numerical quadrature, 203, 211
  orbital phase determination, 182, 210
  period determination, 182
  period variation, 216
  proximity effects, 179, 201–5, 211
  reflection coefficients (E), 204–5
  terminology (primary, secondary, annular, partial, total, transit, occultation, radii, inclination, eccentricity) 174–5
effective wavelength, 18, 25, 49–50
eigenvalues, eigenvectors, 205, 206–7
electromagnetic spectrum, 9, 11
electrometry, 122, 131, 132
electron affinity, 118
electron concentration, 120, 136, 138
electron emission, 117, 121, 127
electron energization, 117–19, 120, 133
electron (hole) migration, 119, 122–3, 129
electron pressure, 65, 68
elliptic integrals, 188–9, 205
elliptic motion, 200–1, 202
ellipticity effect, 202
enclosure (black), 30–1, 33
equations
  normal, 146
  of condition, 146, 151
  well determined, 146
equipotential surfaces, 179–80
equivalent width, 213
errors, effects of, 145–59 – see also curve fitting, least squares method
exotic objects, 12
exposure duration, 114, 132
extended object photometry, 1, 12, 84–98
extinction
  and airmass, 164–5
  atmospheric, 15, 56–8, 85, 105, 147, 155, 164–5
  coefficients ($k(\lambda)$, A_λ), 25, 51, 56–8, 61–3
  colour dependent terms, 58, 63
  definition, 25
  effect on distance determination, 22, 61
  primary, 57–8, 150
  second order terms, 57–8, 63, 152
  interstellar, 22, 61–4, 65, 68, 70, 72–3 86, 92, 93, 105–6
eye, 2, 9–12, 15, 17, 18, 27, 57, 64, 85
eye sensitivity
  photopic (foveal vision), 34–7, 45
  scotopic (rod vision), 34, 37
eyepieces, 111–14

Fabry lens, 116
fast photometry, 114–15, 125, 159
Fermi level (E_F), 118, 120
filters
  bandwidth, 23, 57, 63
  coloured glass, 47, 63, 74
  design, 115
  interference type, 74, 80
  neutral, 140
  specification, 152, 155
  transmission curves, 46–7, 70, 72, 73, 74–5, 81–2, 94, 96, 106
finding chart, 160
flow chart, 149, 190–1
flux
  and apparent brightness, 85
  black body, 33–4, 54
  calibration of magnitude scales, 37–9
  collector, 38, 110–11, 123
  correlation with colour, 51–3, 55
  definition, 25
  extended surface, 85, 95
  general introduction, 15–17
  gradient, 24, 42–3, 49, 50, 51, 76
  redistribution, 69
  relationship to other terms, 31–2
  solar, 35–7
  temperature relation, 40–5
focal ratio, 111
focusing, 116, 127, 128, 139
Fourier series, 206, 245, 252, 259, 260
Fourier transform, 212

G band, 84
galactic arms, 61–3, 96, 97
galactic nebulae, 92–5
galaxies
  brightness, 87
  colour, 97
  general photometry of, 95–8
gating, 114
giant – see stars, types
goodness of fit criterion (χ^2), 172
  – see also curve fitting
gradient – see flux – gradient

H and K lines (CaII), 232
Heliocentric Julian Date, 165–6
Hertzsprung Russell (H–R) diagram, 67, 83, 244
Hessian matrix, 205, 206–7, 222
hole, 119 – see also electron

$I^{\gamma}_{m,n}$–integrals, 205

illuminance, 25 – see also flux
illumination geometry, 31, 85, 87–8, 101
image
  dancing , 116
  display, 114, 131
  dissection technique, 1
  formation, 111, 116, 127
  handling, 132
  intensification (enhancement), 115, 126–9, 141
  transfer, 130, 131
Image Photon Counting System (IPCS), 128
impurities – see semiconductors
information
  advantage, 110, 120, 136
  content, limit, 173, 206
  transfer, 120, 128, 129–31, 132–4, 140
infra-red, 9, 11, 37, 72, 110, 117, 120, 124, 129
instrumentation, 109–41
intensity
  definition, 31
  moments of, 31–2
  surface mean, 40–1
  surface measurement (μ) 85–7, 95–7
interferometry, 26
intermediate bandwidth filters, 23, 73–80
International Amateur–Professional Photoelectric Photometry (IAPPP) association, 3, 8
interstellar extinction, 61–4, 65, 68, 70, 72, 73, 86, 92, 246, 250
interstellar medium, 61–4, 65, 68, 70
interstellar reddening, 25, 62, 63, 67–9, 70, 72, 78, 84, 250, 252–3, 257
inverse square law, 15, 36, 87
irradiance, 25 – see also flux
isochrone, 84
isophotal wavelength, 50
IUE satellite, 253

Kepler's equation, 201

Lambert's law, 88, 91
least squares method, 145–7
  error evaluation, 146–7, 152, 158–9
light curve
  analysis, 3, 5, 172 – see also curve fitting
  distortions, 7, 177, 202–3, 205, 220, 226
  model adequacy, 5, 192, 213–14, 222, 261
  numerical integration, 203, 205
  of a galaxy, 96
  of a supernova, 258
  of variable stars, 168–9 – see also eclipsing binary, spotted and pulsating stars
  production, 159
light emitting diodes (LEDs), 139
limb darkening, 15, 88, 91, 183, 204, 212 225
Lindblad criterion, 84
line index, 24 – see also β photometry
local photometric system, 142–3
longitude of periastron, 199
luminosity
  class, 64
  definition, 25
  distribution, 87, 96
  effect, 68
lunar occultations, 44, 115
Lyman series (Ly$_\alpha$), 93, 94

m_1 [m_1] (metallicity) index, 24, 77–80
maculation, 221
  – see also spotted stars
Magellanic Clouds, 19, 87
magnetodynamic activity, 179
magnitude
  accuracy, 3, 11, 13, 15, 17, 20, 21, 37 42, 57, 83, 116, 139, 145, 147, 156, 160 165, 168, 171, 184, 193, 235, 257
  aperture curve, 97
  bandwidth, 23, 24
  definitions (absolute, apparent, bolometric, monochromatic, standard) 22–3
  introduction by Hipparchos, 1, 12
Main Sequence – see sequence
Maxwell's equations, 26, 98
mean free path, 117
meridian photometer, 17, 18
microchannel plate, 127
middle ages, 13
Milky Way, 12, 87
model atmospheres – see computer
moments (cosine) of intensity, 31–2
Moon
  brightness, 11, 12, 86
  New Moon visibility, 86
Mueller matrix, 100, 101
multi-channel photometry, 111
multi-wavelength techniques, 221, 229 251, 253, 261
Munich Image Data Analysis System (MIDAS), 132

narrowband photometry
  analysis, 80–4
  application, 5–6, 80–4
  binary stars, of, 212–18
  definitions, 23, 24
  sampling problem, 74
nebulae
  and extinction, 63
  galactic, 92–5
noise – see also S/N ratio
  amplification, 21, 38
  limitation, 114, 120, 123, 134, 136
  low noise detection, 72
  multiplication, 124, 135, 136
  poissonian, 120, 123
  read-out, 128, 129, 131
  scintillation, 126
  shot noise, 123, 124, 133, 164
  sky, 116, 126
  sources, 121–6, 133, 134
  thermal (Johnson), 131, 133
non-blackness, 39
normal points, 186
North Polar Sequence – see sequence

offset guiding, 114
Olbers' paradox, 85
optical activity, 100
optical box, 111
optical depth, 32
optical filter properties, 47–51 – see also filters
optical range of spectrum, 1, 9–12
optically thin medium, 101–3
optimization, 172 – see also curve fitting

p_g, p_v scales, 18
parameter improvement – see curve fitting
parameter space, 172
passband – see also filters
  effects of finiteness, 57–8, 74
pen (chart) recorder, 132, 134–5, 145, 161
phonons, 118
photocathodes, 109, 117–18, 120, 121 123, 125, 127, 129, 132
photoconductive effect, 117, 141
photodiode, 114, 119, 124, 134
photoelectric cell, 19, 20
photoelectric methods, 11, 19–20, 21, 22, 27, 29, 65, 109, 116–41, 184
photoemission, 117, 118, 120
photographic photometry, 17–20, 28, 69, 121, 126–7
photometer design, 109–16
photometric data processing, 143–71
photometric filter systems
  four colour (*uvby*), 74–80
  RGU, 69
  H_β (or β), 80–84
  six colour (*UVBGRI*), 70–2
  $UB_1B_2V_1G$ (Golay), 72
  UBV, *UBVRIJKLMN* (Johnson), 21, 22, 45–7, 63, 64–9, 72–3
photometric nights, 142
photometric units
  apostilb, candela, jansky, lambert, lumen, lux, nit, phot, stilb, 31–2, 54
  – see also brightness, magnitude
photomultipliers, 21, 109, 115, 121–5, 134, 137, 141
photon count rates, 51, 94
photon registration, 117, 121, 123, 129
photopolarimetry, 98–105, 108, 115
  analyser, depolarizer, quarter-wave plate retarder, Stokes-meter, 99–100
photovoltaic effect, 118
pixel (picture cell), 85, 121, 131–2
Planck formula, 33, 247
plane parallel approximation, 58
poissonian distribution (statistics) 115, 120, 123, 164
polarization parameters (p, q), 99
  – see also Stokes parameters
positional coordinates, 11, 145
postdetection stage, 138 – see also DC measurement, pulse counting
preamplifier, 111
predetection stage, 110–16
pressure broadening, 82
pressure of radiation, 32
pulsating stars, 239–61
  6 parameters model, 248–9
  flux ratio (stellar/solar), 246
  Fourier analysis, 245, 252, 259, 260
  gravity variation, 255, 256, 257, 259
  infra-red photometry, 261
  ionization zones, role of, 243
  light curve modelling, 246–51
  light curve morphology, 239–41
  non-radial modes, 258
  numerical quadrature, 260
  parameter determination, 253–5
  period variation, 239, 251
  periodicity study, 242
  phase lag, 242, 248, 250
  terminology (instability strip, negative dissipation, overstable, non-radial (p and g) modes) 243–4
pulsational radii, 258–60
pulse-counting, 114, 125, 135–40
  after pulses, 137
  giant pulses, 124
  pulse pile-up, 138
  pulse spreading, 124

Q parameter (*UBV* system), 69
Q, U curve for Algol binaries, 104–5
quality factor, 164
quantization to h, 33
quantum efficiency (Q)
  – see detector efficiency
quantum physics, 27, 33

radial velocity curve, 239, 245, 251–2
radiative transport, 56, 204
radio-astronomy, 1, 32
Rayleigh–Jeans law, 34, 54
Rayleigh scattering, – see scattering
recombination, 69, 93
red leak, 115, 120, 155
red sensitivity, 115, 120, 132
reddening, – see also interstellar
  lines 62–3, 68–70
reflection
  effect, 102, 202
  laws, 87–92
relative gradient, 43, 49
resolution
  spatial, 110, 121, 126, 132
  time, 114, 115, 134, 138
response time, 119, 134
retarder, – see photopolarimetry
Roche lobes, 5, 179–80
Roche model, 202
rotation matrices
  coordinate system transformations, 59, 224
  electron scattering geometry, 102
  orientation of planetary illumination, 87–9

scattering
by Moon's surface, 86
electron, 101–2
geometry, 88, 89, 101–3
inverse power-law, 58, 61
models, 92
pure, 204
Rayleigh, 37, 58
scintillation, 126
screening, 139
secondary emission, 120, 121, 123 125, 127
selenium cell, 19–20
semiconductors, 117–20, 129
diode, 114, 119, 124, 125, 134
doping (impurities), 118, 120, 127
germanium, 117
intrinsic (undoped), 119
n-type, p-type, 119
reverse biasing, 119
silicon, 117, 129
semidetached, 5
sequence
dwarf (Main), 68
giant, 64
Main, 53, 64, 68, 78, 80, 243
North Polar, 18–19, 21, 28
secondary, 19
South Polar, 19
standard 159
σ function, 225, 228, 231
σ integrals, 90–2, 223–5
signal to noise S/N ratio, 2, 20, 115, 120, 123, 125, 133, 136
single-channel photometry, 109, 111, 120, 123–6
sky brightness, 12, 85, 86, 97, 107, 126, 159
sky noise, 111, 116, 160
sky subtraction, 114, 132
sky transparency, 159
solar laboratory, extensions to, 5, 221, 235
solar system
general photometry of planets, 87–92
phase law, 91–2
solid angle, 31
solid state detectors, 115
– see also CCDs
spectral illuminance, 25
spectral line
effect on colours, 69, 74
profile asymmetry, 246
– see also Balmer, H and K (CaII), nebulae, radial velocity curve, star
emission
spectral type, 23
spectrophotometry, 1, 10
spherical astronomy, 61, 161
spotted stars, 6, 12, 220–38
11 parameters model, 229
binary systems containing, 231–6
computer analysis, 225–30
differential rotation, 226
flux ratio (spot to photosphere), 224, 229–30
geometry of spots, 223, 228–9
period and phase determination, 226
square degree, 86
standard stars, 10, 22, 23, 46–8, 55, 76, 82, 106, 114, 115, 125, 139, 142–57, 159, 167, 171, 213
stars, characteristics of
ages, 79, 84
axial rotation, 80, 82, 220
binarity, 5, 12, 20, 44, 52, 53, 54, 67, 80, 100, 104
clusterings, 10, 62–3, 67, 68, 77, 80, 86–7
composition, 24, 65, 67, 69, 77, 79–80, 82
disks around, 178
emission lines, 82–3, 213, 221
evolution, 6, 12, 65, 78, 83–4
formation, 97
luminosity, 25, 38, 64, 65, 68, 72, 76, 78, 82, 84, 87
magnitude, – see magnitude
metallicity, 24, 65, 69, 72, 78
microturbulence, 80, 260
population, 65, 69, 70
size, – see luminosity, dwarf and giant
surface gravity, 24, 44, 65, 69, 72, 82, 255, 256, 257
surface temperature, 24, 35, 44, 58, 64, 68, 69, 70, 175, 205, 239, 242, 258
stars, types
active (cool), 220–1
barium, 179
binary – see binary stars
cool, 39, 49
cepheid, 240, 242–4, 249–58
δ Scuti, 258
dwarf, 39, 64–5, 68, 77
flare, 220
giant, 64–5, 68, 70, 77
hot, 5, 49
Mira, 2, 39, 52, 241, 258
nova, 179
pulsating, – see pulsating stars
RR Lyrae, 241, 243–4, 259
spotted, – see spotted stars
subdwarf, 204
subgiant, 177
symbiotic, 179
W Virginis, 240, 242–4, 258
X-ray, 179
starspots, – see spotted stars
Stefan's law, 33
Stokes parameters (I, Q, U and V), 99
Stokes-meter – see photopolarimetry
stray pick-up, 136
Sun
basic parameters, 263
brightness (intensity), 85
effective temperature, 34–5
general photometry, 34–8
magnitude, 37, 45
solar constant, 35, 37–8
spectrum, 36
sunspots, 221, 235

supernova, 3–4, 14, 258
surface photometry
  – see extended object photometry
synchrotron emission, 93

telescope
  large, 41, 70, 96
  medium, 51
  small, 3, 8, see also flux collector
television, 114, 120, 128
temperature
  ambient, 118, 123, 129, 130, 134
  brightness, 43, 74, 246–8, 253, 255, 257, 261
  colour, 16, 43, 54, 74, 77, 203, 232, 242, 247–8, 249, 253, 257, 259
  effective, 34–5, 39, 43–5, 51–4, 72, 74, 177, 207, 247, 255–7
  gradient, 204, 205
  working, 120, 124, 130
  starspot, 229–30
  Zanstra, 94
Thomson (electron) scattering, 102
thermionic emission, 118, 122, 124, 136
three-beam photometer, 113
transformation of photometric systems
  – see calibration
transmission, – see also filters
  of system, 148, 150, 152–7
two-beam photometry, 115
two-dimensional photometry, 93, 96 – see also extended object photometry

UBV filter system, 45–51
UBV photometry, 110, 115, 142, 171
  – see also photometric filter systems
ultra-violet, 9, 11, 24, 27, 37, 64, 72, 110, 111, 132, 155
uniqueness problem, 193, 221
unreddened indices, 68–9, 77, 78

variable stars, 2, 3, 7, 9, 12, 13, 14, 15, 16, 20, 21, 23, 39, 52, 126, 132, 160, 165, 166, 168–9, 171, 197, 220, 239, 240–1, 244, 258, 260
visual photometry, 2–3, 9, 17, 18, 21, 22, 27, 32, 34, 36–8, 39, 45, 74, 85–7
Vogt–Russell theorem, 178
voltage
  divider, 122
  to frequency converter (VFC), 135
  working, 125, 128, 137
von Zeipel law, paradox, 204

W type variation, 216–18
W Ursa Majoris paradox, 178
work function, 117, 118, 124, 127
wave theory of light, 26–7, 29
wave type distortion, 220–2, 226
Wien approximation, 34
Wien's displacement law, 33, 49, 50

X-ray astronomy, 1, 204
X-rays, 11, 37

Zeeman Imaging, technique, 5
zenith angle (distance), 57–60, 145, 150

Object index

Extended Objects

Galaxy – The Milky Way, 12, 86–7

Magellanic Clouds – LMC, SMC, 19, 87

nebulæ, clusters and galaxies
NGC 224 – Great Andromeda Galaxy (Nebula) (M31), 13, 87
NGC 869 – h Per, 86
NGC 884 – χ Per, 86
NGC 1952 – Crab (M1), 93
NGC 4565, 97
NGC 7293 – Helix, 95

sky, 1, 3, 9, 11, 12, 15, 19, 22, 86, 87, 97, 101, 102, 103, 104, 107, 111, 114, 116, 124, 126, 132, 142, 143, 144, 145, 147, 150, 159–63, 166–8, 182, 183

Solar System
Jupiter, 19
Moon, 12, 14, 19, 44, 86, 107
Sun, 9, 12, 15, 22, 34–8, 39, 42, 43, 45, 49, 63, 64, 67, 85–6, 87, 89, 93, 165, 166, 177, 179, 216, 220, 226, 235, 248, 253, 263
Venus, 19

Stars[†]

Algol, see Per β
And RT, 236
Aql β, 253
Aql η, 251–7, 261
Ara R, 161, 162, 166–8
Ara AO, 240
Ari α (Hamal), 48

† The name ordering within alphabetically listed constellations is conventional. HR identifications are equivalent to the BS designations used elsewhere.

Aur β (Menkalinan), 45
Boo α (Arcturus), 45
Cam SV, 233, 236
Cnc β, 48
Cnc WY, 236
CVn RS, 6, 179, 221, 237
CMa α (Sirius), 45
CMi α (Procyon), 45
Cap TW, 240
Cas YZ, 184–5, 186, 192, 193–6, 198, 205
Cen α, 43
Cep δ, 242, 251–7, 261
Cep ϵ, 253
Cep U, 5, 177, 214, 216–18
Cep CQ, 5, 6, 8
Cet o (Mira), 2, 14, 39, 52, 241, 258, 261
CrB α (Gemma), 174
CrB ϵ, 48
CrB R, 2
Cyg P, 14
Cyg CG, 236
Cyg V477, 202
Dra TT, 241
Dor AB, 235, 237
Eri CC, 226–7
Gem γ (Alhena), 45
Gem μ, 45
Gem YY (Castor C), 45, 220, 237
Her α (Ras Algethi), 45
Her τ, 48
Her u, 179
Her X, 240
Hya η, 48
Lac 10, 48
Lac AR, 220
Lac BL, 12
Lib β (Zuben E'schamali), 48
Lyr α (Vega), 45
Lyr β (Sheliak), 175, 176, 177, 178
Mira – see Cet o
V748 Oph, 240
Ori α (Betelgeuse), 45
Ori VV 207–10, 219

Peg β (Scheat), 45
Per β (Algol), 15, 20, 21, 28, 45, 104, 105, 175, 176, 177, 178
Psc UV, 230, 236
Pup ST, 240
Sco α (Antares), 45
Sco μ_1, 45
Sge U, 5, 177, 214, 216–18
Sgr α, 15
Sgr β, 15
Sgr V383, 240
Sgr V741, 240
Sct CO, 240
Ser α (Unuk Al Hay), 48
Tau α (Aldebaran), 45
Tri R, 52
Tuc γ – see HR 8848
UMa W, 175, 176, 178
UMa XY, 234, 236
Vir W, 240, 242, 248
Vir BH, 236
Vul ER, 236
Boss 4342, 41
HR 100 (κ Phe), 144
HR 373 (39 Cet), 144, 148
HR 531 (χ Cet), 144
HR 811 (π Cet), 144
HR 875, 48
HR 1030 (o Tau), 144, 153, 154
HR 6114, 166, 167
HR 8181 (γ Pav), 144
HR 8353 (γ Gru), 144
HR 8431 (μ PsA), 144
HR 8551 (35 Peg), 144
HR 8630 (β Oct), 144
HR 8675 (ϵ Gru), 144
HR 8832, 48
HR 8848 (γ Tuc), 144, 148, 153, 154
HR 8969 (ι Psc), 144
HR 9076 (ϵ Tuc), 144
HR 9091 (ζ Scl), 144
SN 1987a, 3, 4, 8